Lecture Notes in Economics and Mathematical Systems

677

More information about this series at
http://www.springer.com/series/300

Martin Gavalec • Jaroslav Ramík •
Karel Zimmermann

Decision Making and Optimization

Special Matrices and Their Applications in Economics and Management

Springer

Martin Gavalec
University of Hradec Kralove
Hradec Kralove
Czech Republic

Jaroslav Ramík
Silesian University in Opava
Karvina
Czech Republic

Karel Zimmermann
Charles University in Prague
Prague
Czech Republic

This work has been supported by the project of GACR No. 402090405

ISSN 0075-8442 ISSN 2196-9957 (electronic)
ISBN 978-3-319-08322-3 ISBN 978-3-319-08323-0 (eBook)
DOI 10.1007/978-3-319-08323-0
Springer Cham Heidelberg New York Dordrecht London

Library of Congress Control Number: 2014951713

Printed on acid-free paper

Springer is part of Springer Science+Business Media (www.springer.com)

Preface

This book deals with properties of some special classes of matrices that are useful in economics and decision making problems.

In various fields of evaluation, selection, and prioritization processes decision maker(s) try to find the best alternative(s) from a feasible set of alternatives. In many cases, comparison of different alternatives according to their desirability in decision problems cannot be done using only a single criterion or one person. In many decision making problems, procedures have been established to combine opinions about alternatives related to different points of view. These procedures are often based on pairwise comparisons, in the sense that processes are linked to some degree of preference of one alternative over another. According to the nature of the information expressed by the decision maker, for every pair of alternatives different representation formats can be used to express preferences, e.g., multiplicative preference relations, additive preference relations, fuzzy preference relations, interval-valued preference relations, and also linguistic preference relations.

Many important optimization problems use objective functions and constraints that are characterized by extremal operators such as maximum, minimum, or various fuzzy triangular norms (t-norms). These problems can be solved using special classes of matrices, in which matrix operations are derived from the binary scalar operations of maximum and minimum instead of classical addition and multiplication. We shortly say that the computation is performed in max-min algebra. In a more general approach, the minimum operation, which itself is a specific t-norm, is substituted by any other t-norm T, and the computations are made in max-T algebra.

Matrices in max-min algebra, or in general max-T algebra, are useful in applications such as automata theory, design of switching circuits, logic of binary relations, medical diagnosis, Markov chains, social choice, models of organizations, information systems, political systems, and clustering. In these applications, the steady states of systems working in discrete time correspond to eigenvectors of matrices in max-min algebra (max-T algebra). The input data in real problems are usually not exact and can be rather characterized by interval values. For systems described by interval coefficients the investigation of steady states leads to computing various types of interval eigenvectors.

In Chap. 1 basic preliminary concepts and results are presented which will be used in the following chapters: t-norms and t-conorms, fuzzy sets, fuzzy relations, fuzzy numbers, triangular fuzzy numbers, fuzzy matrices, alo-groups, and others.

Chapter 2 deals with pairwise comparison matrices. In a multicriteria decision making context, a pairwise comparison matrix is a helpful tool to determine the weighted ranking on a set of alternatives or criteria. The entry of the matrix can assume different meanings: it can be a preference ratio (multiplicative case) or a preference difference (additive case), or, it belongs to the unit interval and measures the distance from the indifference that is expressed by 0.5 (fuzzy case). When comparing two elements, the decision maker assigns the value from a scale to any pair of alternatives representing the element of the pairwise preference matrix. Here, we investigate transitivity and consistency of preference matrices being understood differently with respect to the type of preference matrix. By various methods and from various types of preference matrices we obtain corresponding priority vectors for final ranking of alternatives. The obtained results are also applied to situations where some elements of the fuzzy preference matrix are missing. Finally, a unified framework for pairwise comparison matrices based on abelian linearly ordered groups is presented. Illustrative numerical examples are supplemented.

Chapter 3 is aimed on pairwise comparison matrices with fuzzy elements. Fuzzy elements of the pairwise comparison matrix are applied whenever the decision maker is not sure about the value of his/her evaluation of the relative importance of elements in question. We particularly deal with pairwise comparison matrices with fuzzy number components and investigate some properties of such matrices. In comparison with pairwise comparison matrices with crisp components investigated in the previous chapter, here we investigate pairwise comparison matrices with elements from alo-group over a real interval. Such an approach allows for generalization of additive, multiplicative, and fuzzy pairwise comparison matrices with fuzzy elements. Moreover, we deal with the problem of measuring the inconsistency of fuzzy pairwise comparison matrices by defining corresponding inconsistency indexes. Numerical examples are presented to illustrate the concepts and derived properties.

Chapter 4 considers the properties of equation/inequality optimization systems with max-min separable functions on one or both sides of the relations, as well as optimization problems under max-min separable equation/inequality constraints. For the optimization problems with one-sided max-min separable constraints an explicit solution formula is derived, a duality theory is developed, and some optimization problems on the set of points attainable by the functions occurring in the constraints are solved. Solution methods for some classes of optimization problems with two-sided equation and inequality constraints are proposed in the last part of the chapter.

In Chap. 5 the steady states of systems with imprecise input data in max-min algebra are investigated. Six possible types of an interval eigenvector of an interval matrix are introduced, using various combination of quantifiers in the definition. The previously known characterizations of the interval eigenvectors that were restricted to the increasing eigenvectors are extended here to the non-decreasing eigenvectors, and further to all possible interval eigenvectors of a given max-min matrix. Classification types of general interval eigenvectors are studied and characterization of all possible six types is presented. The relations between various types are shown by several examples.

Chapter 6 describes the structure of the eigenspace of a given fuzzy matrix in two specific max-T algebras: the so-called max-drast algebra, in which the least t-norm T (often called the drastic norm) is used, and max-Łukasiewicz algebra with Łukasiewicz t-norm L. For both of these max-T algebras the necessary and sufficient conditions are presented under which the monotone eigenspace (the set of all non-decreasing eigenvectors) of a given matrix is nonempty and, in the positive case, the structure of the monotone eigenspace is described. Using permutations of matrix rows and columns, the results are extended to the whole eigenspace.

Hradec Kralove, Czech Republic — Martin Gavalec
Karvina, Czech Republic — Jaroslav Ramík
Prague, Czech Republic — Karel Zimmermann
December 2013

Contents

Part I
Special Matrices in Decision Making

Chapter 1
Preliminaries

Abstract In this chapter basic preliminary concepts and results are presented which will be used in the following chapters: t-norms and t-conorms, fuzzy sets, fuzzy relations, fuzzy numbers, triangular fuzzy numbers, fuzzy matrices, abelian linearly ordered groups and others.

1.1 Triangular Norms and Conorms

The notion of a triangular norm was introduced by Schweizer and Sklar in their development of a probabilistic generalization of the theory of metric spaces. This development was initiated by K. Menger [9], who proposed to replace the distance $d(x, y)$ between points x and y of a metric space by a real-valued function F_{xy} of a real variable whose value $F_{xy}(\alpha)$ is interpreted as the probability that the distance between x and y is less than α. This interpretation leads to straightforward generalizations of all requirements of the standard definition of a metric except for that of the triangular inequality. Elaborating Menger's idea, Schweizer and Sklar [14] proposed to replace the triangular inequality by the inequality

$$F_{xy}(\alpha + \beta) \geq T(F_{xy}(\alpha), F_{yz}(\beta)) \tag{1.1}$$

where T is a function from $[0,1]^2$ into $[0,1]$ satisfying the following conditions (1.2)–(1.5).

Definition 1.1. Let $T : [0,1]^2 \to [0,1]$ be a function satisfying the following properties:

$$T(a,b) = T(b,a) \quad \text{for all } a, b \in [0,1], \tag{1.2}$$

$$T(T(a,b),c) = T(a, T(b,c)) \quad \text{for all } a, b, c \in [0,1], \tag{1.3}$$

$$T(a,b) \leq T(c,d) \quad \text{for } a, b, c, d \in [0,1] \text{ with } a \leq c,\ b \leq d, \tag{1.4}$$

$$T(a,1) = a \quad \text{for all } a \in [0,1]. \tag{1.5}$$

The chapter was written by "Jaroslav Ramík"

M. Gavalec et al., *Decision Making and Optimization*, Lecture Notes in Economics and Mathematical Systems 677, DOI 10.1007/978-3-319-08323-0_1

A function $T : [0,1]^2 \to [0,1]$ that satisfies all these properties is called the *triangular norm* or *t-norm*. A t-norm T is called *strictly monotone* if T is a strictly increasing function in the sense that

$$T(a,b) < T(a',b) \quad \text{whenever } a, a', b \in]0,1[\text{ and } a < a'.$$

Triangular norms play an important role also in many-valued logics and in the theory of fuzzy sets. In many-valued logics, they serve as truth degree functions of conjunction connectives. In the fuzzy set theory, they provide a tool for defining various types of the intersection of fuzzy subsets of a given set. For a detailed treatment we refer to [7].

The axioms (1.2), (1.3), (1.4) and (1.5) are called *commutativity*, *associativity*, *monotonicity* and *boundary condition*, respectively. From the algebraic point of view, a triangular norm is a commutative ordered semigroup with unit element 1 on the unit interval [0, 1] of real numbers. Therefore the class of all triangular norms is quite large. Let us consider some important examples.

In connection with a problem on functional equations, Frank [6] introduced the following family $\{T_s \mid s \in]0, +\infty[, \ s \neq 1\}$ of t-norms:

$$T_s(a,b) = \log_s \left(1 + \frac{(s^a - 1)(s^b - 1)}{s - 1}\right). \tag{1.6}$$

The limit cases T_M, T_P and T_L defined by

$$\begin{aligned} T_M(a,b) &= \lim_{s \to 0} T_s(a,b) = \min\{a,b\}, \\ T_P(a,b) &= \lim_{s \to 1} T_s(a,b) = a \cdot b, \\ T_L(a,b) &= \lim_{s \to \infty} T_s(a,b) = \max\{0, a + b - 1\} \end{aligned}$$

are also t-norms. In the literature on many-valued logics, the t-norms T_M, T_P and T_L are often called the *minimum* (or *Gödel*), *product* and *Łukasiewicz* t-norm, respectively. Two other interesting examples are

$$T_F(a,b) = \begin{cases} \min\{a,b\} & \text{if } a + b > 1 \\ 0 & \text{otherwise,} \end{cases}$$

introduced by Fodor [32], and the so-called *drastic product*

$$T_D(a,b) = \begin{cases} \min\{a,b\} & \text{if } \max\{a,b\} = 1 \\ 0 & \text{otherwise.} \end{cases} \tag{1.7}$$

It can easily be seen that the minimum t-norm T_M is the maximal t-norm and the drastic product T_D is the minimal t-norm in the pointwise ordering, that is, for every t-norm T,

$$T_D(a,b) \le T(a,b) \le T_M(a,b) \quad \text{whenever } a,b \in [0,1]. \tag{1.8}$$

However, the class of t-norms is not linearly ordered by this pointwise relation. For example, the product t-norm T_P and the Fodor t-norm T_F are not comparable.

A class of functions closely related to the class of t-norms are functions $S : [0,1]^2 \to [0,1]$ such that

$$\begin{aligned} S(a,b) &= S(b,a) \quad \text{for all } a,b \in [0,1], \\ S(S(a,b),c) &= S(a,S(b,c)) \quad \text{for all } a,b,c \in [0,1], \\ S(a,b) &\le S(c,d) \quad \text{for } a,b,c,d \in [0,1] \text{ with } a \le c, b \le d, \\ S(a,0) &= a \quad \text{for all } a \in [0,1]. \end{aligned}$$

The functions that satisfy all these properties are called the *triangular conorms* or *t-conorms*.

It can easily be verified, see for example [7], that for each t-norm T, the function $T^* : [0,1]^2 \to [0,1]$ defined for all $a,b \in [0,1]$ by

$$T^*(a,b) = 1 - T(1-a, 1-b) \tag{1.9}$$

is a t-conorm. The converse statement is also true. Namely, if S is a t-conorm, then the function $S^* : [0,1]^2 \to [0,1]$ defined for all $a,b \in [0,1]$ by

$$S^*(a,b) = 1 - S(1-a, 1-b) \tag{1.10}$$

is a t-norm. The t-conorm T^* and t-norm S^*, are called *dual* to the t-norm T and t-conorm S, respectively. For example, the functions S_M, S_P, S_L and S_D defined for $a,b \in [0,1]$ by

$$\begin{aligned} S_M(a,b) &= \max\{a,b\}, \\ S_P(a,b) &= a + b - a.b, \\ S_L(a,b) &= \min\{1, a+b\}, \\ S_D(a,b) &= \begin{cases} \max\{a,b\} & \text{if } \min\{a,b\} = 0, \\ 1 & \text{otherwise.} \end{cases} \end{aligned}$$

are t-conorms. In the literature, the t-conorms S_M, S_P, S_L and S_D are often called the *maximum, probabilistic sum, bounded sum* and *drastic sum*, respectively. It may easily be verified that

$$T_M^* = S_M, \; T_P^* = S_P, \; T_L^* = S_L, \; T_D^* = S_D.$$

The following proposition answers the question whether a triangular norm and triangular conorm are determined uniquely by their values on the diagonal of the unit square. In general, this is not the case, but the extremal t-norms and t-conorms T_M, S_M, T_D, S_D are completely determined by their values on the diagonal of the unit square.

1.2 Properties of Triangular Norms and Triangular Conorms

For some important properties of triangular norms and triangular conorms, see also [7, 12].

Proposition 1.1.

(i) The only t-norm T satisfying $T(x,x) = x$ for all $x \in [0,1]$ is the minimum t-norm T_M.
(ii) The only t-conorm S satisfying $S(x,x) = x$ for all $x \in [0,1]$ is the maximum t-conorm S_M.
(iii) The only t-norm T satisfying $T(x,x) = 0$ for all $x \in [0,1]$ is the drastic product T_D.
(iv) The only t-conorm S satisfying $S(x,x) = 0$ for all $x \in [0,1]$ is the drastic sum S_D.

The commutativity and associativity properties allow to extend t-norms and t-conorms, introduced as binary operations, to n-ary operations. Let T be a t-norm. We define its extension to more than two arguments by the formula

$$T^{i+1}(a_1, a_2, \ldots, a_{i+2}) = T(T^i(a_1, a_2, \ldots, a_{i+1}), a_{i+2}), \tag{1.11}$$

where $T^1(a_1, a_2) = T(a_1, a_2)$.

For example, the extensions of T_M, T_P, T_L and T_D to m arguments are

$$T_M^{m-1}(a_1, a_2, \ldots, a_m) = \min\{a_1, a_2, \ldots, a_m\},$$

$$T_P^{m-1}(a_1, a_2, \ldots, a_m) = \prod_{i=1}^{m} a_i,$$

$$T_L^{m-1}(a_1, a_2, \ldots, a_m) = \max\{0, \sum_{i=1}^{m} a_i - (m-1)\},$$

$$T_D^{m-1}(a_1, a_2, \ldots, a_m) = \begin{cases} a_i & \text{if } a_j = 1 \text{ for all } j \neq i, \\ 0 & \text{otherwise.} \end{cases}$$

Given a t-conorm S, we can, in a complete analogy, extend this binary operation to m-tuples $(a_1, a_2, \ldots, a_m) \in [0,1]^m$ by the formula

$$S^{i+1}(a_1, a_2, \ldots, a_{i+2}) = S(S^i(a_1, a_2, \ldots, a_{i+1}), a_{i+2}),$$

where $S^1(a_1, a_2) = S(a_1, a_2)$.

For example, the extensions of S_M, S_P, S_L and S_D to m arguments are

$$S_M^{m-1}(a_1, a_2, \ldots, a_m) = \max\{a_1, a_2, \ldots, a_m\},$$

$$S_P^{m-1}(a_1, a_2, \ldots, a_m) = 1 - \prod_{i=1}^{m}(1 - a_i),$$

$$S_L^{m-1}(a_1, a_2, \ldots, a_m) = \min\{1, \sum_{i=1}^{m} a_i\},$$

$$S_D^{m-1}(a_1, a_2, \ldots, a_m) = \begin{cases} a_i & \text{if } a_j = 1 \text{ for all } j \neq i, \\ 1 & \text{otherwise.} \end{cases}$$

If there is no danger of misunderstanding, the upper index $m-1$ of T or S is omitted.

Now we turn our attention to some algebraic aspects of t-norms and t-conorms, which will be useful later on when T-quasiconcave and T-quasiconvex functions are investigated. Notice that these properties are well-known from the general theory of semigroups.

Definition 1.2. Let T be a t-norm.

(i) An element $a \in [0, 1]$ is called an *idempotent element* of T if $T(a, a) = a$.
(ii) An element $a \in]0, 1[$ is called a *nilpotent element* of T if there exists some positive integer $n \in N$ such that $T^{n-1}(a, \ldots, a) = 0$.
(iii) An element $a \in]0, 1[$ is called a *zero divisor* of T if there exists some $b \in]0, 1[$ such that $T^{n-1}(a, b) = 0$.

By definitions and properties of t-norms we obtain the following proposition summarizing the properties of T_M, T_P, T_L and T_D introduced in Definition 1.2.

Proposition 1.2. *Each $a \in [0, 1]$ is an idempotent element of T_M. Each $a \in (0, 1)$ is both a nilpotent element and zero divisor of T_L, as well as of T_D. The minimum T_M has neither nilpotent elements nor zero divisors, and T_L, as well as T_D possesses only trivial idempotent elements 0 and 1. The product norm T_P has neither non-trivial idempotent elements nor nilpotent elements nor zero divisors.*

Now we add some other usual definitions concerning the properties of triangular norms.

Definition 1.3. A triangular norm T is said to be

(i) *strict* if it is continuous and strictly monotone,
(ii) *Archimedian* if for all $x, y \in]0, 1[$ there exists a positive integer n such that $T^{n-1}(x, \ldots, x) < y$,

(iii) *nilpotent* if it is continuous and if each $a \in]0,1[$ is a nilpotent element of T,
(iv) *idempotent* if each $a \in]0,1[$ is an idempotent element of T.

Notice that if T is strict, then T is Archimedian. The following proposition summarizes the properties of the most popular t-norms in context to the previous definition; see [7].

Proposition 1.3. *The minimum T_M is neither strict, nor Archimedian, nor nilpotent. The product norm T_P is both strict and Archimedian, but not nilpotent. Łukasiewicz t-norm T_L is both strict and Archimedian and nilpotent. The drastic product T_D is Archimedian and nilpotent, but not continuous, thus not strict.*

Strict monotonicity of t-conorms as well as strict, Archimedian and nilpotent t-conorms can be introduced using the duality (1.9), (1.10). Without presenting all technical details, we only mention that it suffices to interchange the words t-norm and t-conorm and the roles of 0 and 1, respectively, in order to obtain the proper definitions and results for t-conorms.

It is obvious that each strict t-norm T is Archimedian since

$$T(x,x) < T(x,1) = x$$

for each $x \in]0,1[)$. The following result can be found in [58].

Proposition 1.4. *A continuous t-norm T is strict if and only if there exists an automorphism φ of the unit interval* $[0,1]$ *such that*

$$T(x,y) = \varphi^{-1}(\varphi(x)\varphi(y)).$$

1.3 Representations of Triangular Norms and Triangular Conorms

Now we show how real-valued functions of one real variable can be used to construct new t-norms and t-conorms, and how new t-norms can be generated from given ones. The construction requires an inverse operation and in order to relax the strong requirement of bijectivity and to replace it by the weaker monotonicity, we first recall some general properties of monotone functions. The following result is crucial for the proper definition of pseudoinverse; see [7].

Proposition 1.5. *Let $f : [a,b] \to [c,d]$ be a non-constant monotone function, where $[a,b]$ and $[c,d]$ are subintervals of the extended real line $\bar{\mathbf{R}} = [-\infty,+\infty]$. Then for each $y \in [c,d] \setminus \mathrm{Ran}(f)$ we have*

$$\begin{aligned}\sup\{x \in [a,b] \mid (f(x)-y)\cdot(f(b)-f(a)) < 0\} \\ = \inf\{x \in [a,b] \mid (f(x)-y)\cdot(f(b)-f(a)) > 0\}.\end{aligned}$$

This result allows us to introduce the following generalization of an inverse function, where we restrict ourselves to the case of non-constant monotone functions.

Definition 1.4. Let $[a, b]$ and $[c, d]$ be subintervals of the extended real line $\bar{\mathbf{R}}$. Let $f : [a, b] \to [c, d]$ be a non-constant monotone function. The *pseudo-inverse* $f^{(-1)} : [c, d] \to [a, b]$ is defined by

$$f^{(-1)}(y) = \sup\{x \in [a, b] \mid (f(x) - y) \cdot (f(b) - f(a)) < 0\}.$$

The following proposition summarizes some evident consequences of Definition 1.4.

Proposition 1.6. *Let $[a, b]$ and $[c, d]$ be subintervals of the extended real line $\bar{\mathbf{R}}$, and let f be a non-constant function mapping $[a, b]$ into $[c, d]$.*

(i) If f is non-decreasing, then for all $y \in [c, d]$ we have

$$f^{(-1)}(y) = \sup\{x \in [a, b] \mid f(x) < y)\}.$$

(ii) If f is non-increasing, then for all $y \in [c, d]$ we have

$$f^{(-1)}(y) = \sup\{x \in [a, b] \mid f(x) > y)\}.$$

(iii) If f is a bijection, then the pseudoinverse $f^{(-1)}$ of f coincides with the inverse function f^{-1} of f.

(iv) If f is a strictly increasing function, then the pseudoinverse $f^{(-1)}$ of f is continuous.

To construct t-norms with the help of functions we start with the best known operations, the usual addition and multiplication of real numbers. The following proposition gives the result known for nearly 200 years and published by N. H. Abel in 1826. We state it here in a bit simplified version.

Proposition 1.7. *Let $f : [a, b] \to [c, d]$ be a continuous strictly monotone function, where $[a, b]$ and $[c, d]$ are subintervals of the extended real line $\bar{\mathbf{R}}$. Suppose that* $\mathrm{Ran}(f) = [c, d]$ *and $\varphi : [c, d] \to [a, b]$ is an inverse function to f. Then the function $F : [a, b] \times [a, b] \to [a, b]$ defined for every $x, y \in [a, b]$ by*

$$F(x, y) = \varphi(f(x) + f(y)) \tag{1.12}$$

is associative.

If we want to obtain a t-norm by means of (1.12), it is obvious that some additional requirements for f are necessary.

Definition 1.5. An *additive generator* of a t-norm T is a strictly decreasing function $g : [0, 1] \to [0, +\infty]$ which is right continuous at 0, satisfies $g(1) = 0$, and is such that for all $x, y \in [0, 1]$ we have

$$g(x) + g(y) \in \mathrm{Ran}(g) \cup [g(0), +\infty], \tag{1.13}$$

$$T(x, y) = g^{(-1)}(g(x) + g(y)). \tag{1.14}$$

A *multiplicative generator* of a t-norm T is a strictly increasing function $\zeta : [0, 1] \to [0, 1]$ which is right continuous at 0, satisfies $\zeta(1) = 1$, and is such that for all $x, y \in [0, 1]$ we have

$$\zeta(x) \cdot \zeta(y) \in \mathrm{Ran}(\zeta) \cup [0, \zeta(0)], \tag{1.15}$$

$$T(x, y) = \zeta^{(-1)}(\zeta(x).\zeta(y)). \tag{1.16}$$

An *additive generator* of a t-conorm S is a strictly increasing function $h : [0, 1] \to [0, +\infty]$ which is left continuous at 1, satisfies $h(0) = 0$, and is such that for all $x, y \in [0, 1]$ we have

$$h(x) + h(y) \in \mathrm{Ran}(g) \cup [h(1), +\infty], \tag{1.17}$$

$$S(x, y) = h^{(-1)}(h(x) + h(y)). \tag{1.18}$$

A *multiplicative generator* of a t-conorm S is a strictly decreasing function ξ : $[0, 1] \to [0, 1]$ which is left continuous at 1, satisfies $\xi(0) = 1$, and is such that for all $x, y \in [0, 1]$ we have

$$\xi(x).\xi(y) \in \mathrm{Ran}(\xi) \cup [0, \xi(1)], \tag{1.19}$$

$$S(x, y) = \xi^{(-1)}(\xi(x).\xi(y)). \tag{1.20}$$

Triangular norms (t-conorms) constructed by means of additive (multiplicative) generators are always Archimedian. This property and some other properties of such t-norms are summarized in the following proposition.

Proposition 1.8. *Let $g : [0, 1] \to [0, +\infty]$ be an additive generator of a t-norm T. Then T is an Archimedian t-norm. Moreover, we have:*

(i) The t-norm T is strictly monotone if and only if $g(0) = +\infty$.
(ii) Each element of $(0, 1)$ is a nilpotent element of T if and only if $g(0) < +\infty$.
(iii) T is continuous if and only if g is continuous.

It was mentioned in Example 1.2 that minimum T_M has no nilpotent elements. Since T_M is not strictly monotone it follows from the preceding theorem that it has no additive generator. If a t-norm T is generated by a continuous generator, then by Proposition 1.8, T is an Archimedian t-norm. The following theorem says that the converse statement is also true; see [7].

Proposition 1.9. *A t-norm T is Archimedian and continuous if and only if there exists a continuous additive generator of T.*

The analogical propositions can be formulated and proved for multiplicative generators and also for t-conorms.

Proposition 1.10. *A t-norm T is Archimedian and continuous if and only if there exists a continuous multiplicative generator ζ of T.*

Proposition 1.11. *A t-conorm S is Archimedian and continuous if and only if there exist both a continuous additive generator h of S, and a multiplicative generator ξ of S.*

Example 1.1 (The Yager t-norms). One of the most popular families in operations research and particularly in fuzzy linear programming is the family of Yager t-norms; see [15]. The results presented here can be found in [7].

Let $\lambda \in [0, +\infty]$. The Yager t-norm T_λ^Y is defined for all $x, y \in [0, 1]$ as follows:

$$T_\lambda^Y(x, y) = \begin{cases} T_D(x, y) & \text{if } \lambda = 0, \\ T_M(x, y) & \text{if } \lambda = +\infty, \\ \max\left\{0, 1 - \left((1-x)^\lambda + (1-y)^\lambda\right)^{1/\lambda}\right\} & \text{otherwise.} \end{cases}$$

The Yager t-conorm S_λ^Y is defined for all $x, y \in [0, 1]$ as:

$$S_\lambda^Y(x, y) = \begin{cases} S_D(x, y) & \text{if } \lambda = 0, \\ S_M(x, y) & \text{if } \lambda = +\infty, \\ \min\left\{1, \left(x^\lambda + y^\lambda\right)^{1/\lambda}\right\} & \text{otherwise.} \end{cases}$$

(i) Obviously, $T_1^Y = T_L$ and $S_1^Y = S_L$.
(ii) For each $\lambda \in]0, +\infty[$, T_λ^Y and T_λ^Y are dual to each other.
(iii) A Yager t-norm T_λ^Y is nilpotent if and only if $\lambda \in]0, +\infty[$.
(iv) A Yager t-conorm S_λ^Y is nilpotent if and only if $\lambda \in]0, +\infty[$.
(v) If $\lambda \in]0, +\infty[$, then the corresponding continuous additive generator $f_\lambda^Y : [0, 1] \to [0, 1]$ of the Yager t-norm T_λ^Y is given for all $x \in [0, 1]$ by

$$f_\lambda^Y(x) = (1-x)^\lambda .$$

(vi) The corresponding continuous additive generator $g_\lambda^Y : [0, 1] \to [0, 1]$ of the Yager t-conorm S_λ^Y is given for all $x \in [0, 1]$ by

$$g_\lambda^Y(x) = x^\lambda .$$

Yager t-norms have been used in several applications of fuzzy set theory. In particular, it was used in extended addition of linear functions with fuzzy parameters. In this context, in [8], it was shown that the sum of piecewise linear function is again piecewise linear, when using Yager t-norm, see also [7].

1.4 Negations and De Morgan Triples

Supplementing t-norms and t-conorms by a special unary function we obtain a triplet which is useful in many-valued logics, fuzzy set theory and their applications.

Definition 1.6. A function $N : [0, 1] \to [0, 1]$ is called a *negation* if it is non-increasing and satisfies the following conditions:

$$N(0) = 1, \ N(1) = 0. \tag{1.21}$$

Moreover, it is called *strict negation*, if it is strictly decreasing and continuous and it is called *strong negation* if it is strict and the following condition of *involution* holds:

$$N(N(x)) = x \quad \text{for all } x \in [0, 1]. \tag{1.22}$$

Since a strict negation N is a strictly decreasing and continuous function, its inverse N^{-1} is also a strict negation, generally different from N. Obviously, $N^{-1} = N$ if and only if (1.22) holds.

Definition 1.7. An *intuitionistic negation* N_I is defined as follows

$$N_I(x) = \begin{cases} 1 & \text{if } x = 0, \\ 0 & \text{otherwise.} \end{cases}$$

A *weak negation* N_W is defined as follows

$$N_W(x) = \begin{cases} 1 & \text{if } x < 1, \\ 0 & \text{if } x = 1. \end{cases}$$

A *standard negation* N is defined by

$$N(x) = 1 - x. \tag{1.23}$$

Strong negations (including the standard one) defined by

$$N_\lambda(x) = \frac{1 - x}{1 + \lambda x},$$

where $\lambda > -1$, are called λ*-complements*.

Notice that N_I is not a strict negation, N_W is a dual operation to N_I, i.e., for all $x \in [0, 1]$, it holds $N_W(x) = 1 - N_I(1 - x)$. The standard negation is a strong negation. An example of strict but not strong negation is the negation N' defined by the formula

$$N'(x) = 1 - x^2.$$

The following proposition characterizing strong negations comes from [38].

Proposition 1.12. *A function* $N : [0, 1] \to [0, 1]$ *is a strong negation if and only if there exists a strictly increasing continuous surjective function* $\varphi : [0, 1] \to [0, 1]$ *such that*

$$N(x) = \varphi^{-1}(1 - \varphi(x)).$$

Definition 1.8. Let T be a t-norm, S be a t-conorm, and N be a strict negation. We say that (T, S, N) is a De Morgan triple if

$$N(S(x, y)) = T(N(x), N(y)).$$

The following proposition is a simple consequence of the above definitions.

Proposition 1.13. *Let* N *be a strict negation,* T *be a t-norm. Let* S *be defined for all* $x, y \in [0, 1]$ *as follows:*

$$S(x, y) = N^{-1}(T(N(x), T(y))).$$

Then (T, S, N) *is a De Morgan triple. Moreover, if* T *is continuous, then* S *is continuous. In addition, if* T *is Archimedian with an additive generator* f*, then* S *is Archimedian with additive generator* $g = f \circ N$ *and* $g(1) = f(0)$.

Example 1.2. A Łukasiewicz-like De Morgan triple (T, S, N) is defined as follows:

$$T(x, y) = \varphi^{-1}(\max\{\varphi(x) + \varphi(y) - 1, 0\}),$$
$$S(x, y) = \varphi^{-1}(\min\{\varphi(x) + \varphi(y), 1\}),$$
$$N(x) = \varphi^{-1}(1 - \varphi(x)),$$

where $\varphi : [0, 1] \to [0, 1]$ is a strictly increasing continuous surjective function.

1.5 Fuzzy Sets

In order to define the concept of a fuzzy subset of a given set X within the framework of standard set theory we are motivated by the concept of upper level set of a function, see also [10]. Throughout this chapter, X is a nonempty set. All propositions are stated without proofs, the reader can find them e.g. in [12].

Definition 1.9. Let X be a nonempty set. A *fuzzy subset A of X* is the family of subsets $A_\alpha \subset X$, where $\alpha \in [0, 1]$, satisfying the following properties:

$$A_0 = X, \tag{1.24}$$

$$A_\beta \subset A_\alpha \quad \text{whenever } 0 \leq \alpha < \beta \leq 1, \tag{1.25}$$

$$A_\beta = \bigcap_{0 \leq \alpha < \beta} A_\alpha. \tag{1.26}$$

A fuzzy subset A of X will be also called a *fuzzy set*. The class of all fuzzy subsets of X is denoted by $\mathcal{F}(X)$.

Definition 1.10. Let $A = \{A_\alpha\}_{\alpha \in [0,1]}$ be a fuzzy subset of X. The $\mu_A : X \to [0, 1]$ defined by

$$\mu_A(x) = \sup\{\alpha \mid \alpha \in [0, 1], x \in A_\alpha\} \tag{1.27}$$

is called the membership function of A, and the value $\mu_A(x)$ is called membership degree of x in the fuzzy set A.

Definition 1.11. Let A be a fuzzy subset of X. The core of A, $\text{Core}(A)$, is defined by

$$\text{Core}(A) = \{x \in X \mid \mu_A(x) = 1\}.$$

If the core of A is nonempty, then A is said to be normalized. The support of A, $\text{Supp}(A)$, is defined by

$$\text{Supp}(A) = \text{Cl}(\{x \in X \mid \mu_A(x) > 0\}).$$

The height of A, $\text{Hgt}(A)$, is defined by

$$\text{Hgt}(A) = \sup\{\mu_A(x) \mid x \in X\}.$$

The upper-level set of the membership function μ_A of A at $\alpha \in [0, 1]$ is denoted by $[A]_\alpha$ and called the α-cut of A, that is,

$$[A]_\alpha = \{x \in X \mid \mu_A(x) \geq \alpha\}. \tag{1.28}$$

Note that if A is normalized, then $\text{Hgt}(A) = 1$, but not vice versa.

In the following two propositions, we show that the family generated by the upper level sets of a function $\mu : X \to [0, 1]$, satisfies conditions (1.24)-(1.26), thus, it generates a fuzzy subset of X and the membership function μ_A defined by (1.27) coincides with μ. Moreover, for a given fuzzy set $A = \{A_\alpha\}_{\alpha \in [0,1]}$, every α-cut $[A]_\alpha$ given by (1.28) coincides with the corresponding A_α.

Proposition 1.14. *Let $\mu : X \to [0, 1]$ be a function and let $A = \{A_\alpha\}_{\alpha\in[0,1]}$ be a family of its upper-level sets, i.e. $A_\alpha = U(\mu, \alpha)$ for all $\alpha \in [0, 1]$. Then A is a fuzzy subset of X and μ is the membership function of A.*

Proposition 1.15. *Let $A = \{A_\alpha\}_{\alpha\in[0,1]}$ be a fuzzy subset of X and let $\mu_A : X \to [0, 1]$ be the membership function of A. Then for each $\alpha \in [0, 1]$ the α-cut $[A]_\alpha$ is equal to A_α.*

These results allow for introducing a natural one-to-one correspondence between fuzzy subsets of X and real-valued functions mapping X to $[0, 1]$. Any fuzzy subset A of X is given by its membership function μ_A and vice-versa, any function $\mu : X \to [0, 1]$ uniquely determines a fuzzy subset A of X, with the property that the membership function μ_A of A is μ.

The notions of inclusion and equality extend to fuzzy subsets as follows. Let $A = \{A_\alpha\}_{\alpha\in[0,1]}$, $B = \{B_\alpha\}_{\alpha\in[0,1]}$ be fuzzy subsets of X. Then

$$A \subset B \text{ if } A_\alpha \subset B_\alpha \quad \text{for each } \alpha \in [0, 1], \tag{1.29}$$

$$A = B \text{ if } A_\alpha = B_\alpha \quad \text{for each } \alpha \in [0, 1]. \tag{1.30}$$

Proposition 1.16. *Let $A = \{A_\alpha\}_{\alpha\in[0,1]}$ and $B = \{B_\alpha\}_{\alpha\in[0,1]}$ be fuzzy subsets of X. Then the following holds:*

$$A \subset B \textit{ if and only if } \mu_A(x) \le \mu_B(x) \quad \textit{for all } x \in X, \tag{1.31}$$

$$A = B \textit{ if and only if } \mu_A(x) = \mu_B(x) \quad \textit{for all } x \in X. \tag{1.32}$$

A subset of X can be considered as a special fuzzy subset of X where all members of its defining a family consist of the same elements. This is formalized in the following definition.

Definition 1.12. Let A be a subset of X. The fuzzy subset $\{A_\alpha\}_{\alpha\in[0,1]}$ of X defined by $A_\alpha = A$ for all $\alpha \in (0, 1]$ is called a *crisp fuzzy subset of X generated by A*. A fuzzy subset of X generated by some $A \subset X$ is called a *crisp fuzzy subset of X* or briefly a crisp subset of X.

Proposition 1.17. *Let $\{A\}_{\alpha\in[0,1]}$ be a crisp subset of X generated by A. Then the membership function of $\{A\}_{\alpha\in[0,1]}$ is equal to the characteristic function of A.*

By Definition 1.12, the set $\mathcal{P}(X)$ of all subsets of X can naturally be embedded into the set of all fuzzy subsets of X and we can write $A = \{A_\alpha\}_{\alpha\in[0,1]}$ if $\{A_\alpha\}_{\alpha\in[0,1]}$ is generated by $A \subset X$. According to Proposition 1.17, we have in this case $\mu_A = \chi_A$. In particular, if A contains only one element a of X, that is, $A = \{a\}$, then we write $a \in \mathcal{F}(X)$ instead of $\{a\} \in \mathcal{F}(X)$ and χ_a instead of $\chi_{\{a\}}$.

Example 1.3. Let $\mu : \mathbf{R} \to [0, 1]$ be defined by $\mu(x) = e^{-x^2}$. Let $A' = \{A'_\alpha\}_{\alpha \in [0,1]}$, $A'' = \{A''_\alpha\}_{\alpha \in [0,1]}$ be two families of subsets in $\mathbf{R}$ defined as follows:

$$A'_\alpha = \{x \mid x \in \mathbf{R},\ \mu(x) > \alpha\},$$
$$A''_\alpha = \{x \mid x \in \mathbf{R},\ \mu(x) \geq \alpha\}.$$

Clearly, A'' is a fuzzy subset of $\mathbf{R}$ and $A' \neq A''$. Observe that (1.24) and (1.25) are satisfied for A' and A''. However, $A'_1 = \emptyset$ and $\bigcap_{0 \leq \alpha < 1} A'_\alpha = \{0\}$, thus (1.26) is not satisfied. Hence A' is not a fuzzy subset of $\mathbf{R}$.

1.6 Operations with Fuzzy Sets

In order to generalize the set operations of intersection, union and complement to fuzzy set operations, it is natural to use triangular norms, triangular conorms and fuzzy negations introduced in Sect. 1.2.

Given a De Morgan triple (T, S, N), i.e., a t-norm T, a t-conorm S and a fuzzy negation N, introduced in Definition 1.8, we define the operations *intersection* $\cap_T$, *union* $\cup_S$ and *complement* $\mathscr{C}_N$ *on* $\mathscr{F}(X)$ as follows: Let A and B be fuzzy subsets of X, and μ_A and μ_B be their membership functions. Then the membership functions of the fuzzy subsets $A \cap_T B$, $A \cup_S B$ and $\mathscr{C}_N A$ of X are defined by

$$\mu_{A \cap_T B}(x) = T(\mu_A(x), \mu_B(x)),$$
$$\mu_{A \cup_S B}(x) = S(\mu_A(x), \mu_B(x)),$$
$$\mu_{\mathscr{C}_N A}(x) = N(\mu_A(x)).$$

The operations introduced by L. Zadeh in [7] have been originally based on $T = T_M = \min$, $S = S_M = \max$ and standard negation N defined in (1.23). The properties of the operations intersection $\cap_T$, union $\cup_S$ and complement $\mathscr{C}_N$ can be derived directly from the corresponding properties of t-norm T, t-conorm S and fuzzy negation N. For brevity, in case of $T = \min$ and $S = \max$, we write only $\cap$ and $\cup$, instead of $\cap_T$ and $\cup_S$.

Notice that for $A \in \mathscr{F}(X)$ we do not necessarily obtain properties which hold for subsets of X. For example,

$$A \cap_T \mathscr{C}_N A = \emptyset, \tag{1.33}$$

$$A \cup_S \mathscr{C}_N A = X, \tag{1.34}$$

may not hold. If the t-norm T in the De Morgan triple (T, S, N) does not have zero divisors, e.g., $T = \min$, then these properties never hold unless A is a crisp set. On the other hand, for the De Morgan triple (T_L, S_L, N) based on Łukaszewicz t-norm $T = T_L$, properties (1.33) and (1.34) are satisfied.

Given a t-norm T and fuzzy subsets A and B of X and Y, respectively, the *Cartesian product* $A \times_T B$ is the fuzzy subset of $X \times Y$ with the following membership function:

$$\mu_{A \times_T B}(x, y) = T(\mu_A(x), \mu_B(y)) \quad \text{for } (x, y) \in X \times Y. \tag{1.35}$$

An interesting and natural question arises, whether the α-cuts of the intersection $A \cap_T B$, union $A \cup_S B$ and Cartesian product $A \times_T B$ of $A, B \in \mathscr{F}(X)$, coincide with the intersection, union and Cartesian product, respectively, of the corresponding α-cuts $[A]_\alpha$ and $[B]_\alpha$.

1.7 Extension Principle

The purpose of the following definition called the *extension principle* (proposed by L. Zadeh in [16] and [17]) is to extend functions or operations having crisp arguments to functions or operations with fuzzy set arguments. Zadeh's methodology can be cast in a more general setting of carrying a membership function via a mapping, see, e.g., [5]. There exist other generalizations for set-to-set mappings; see, e.g., [5, 13]. From now on, X and Y are nonempty sets.

Definition 1.13 (Extension Principle). Let X, Y be sets, $f : X \to Y$ be a mapping. The mapping $\tilde{f} : \mathscr{F}(X) \to \mathscr{F}(Y)$ defined for all $A \in \mathscr{F}(X)$ with $\mu_A : X \to [0, 1]$ and all $y \in Y$ by

$$\mu_{\tilde{f}(A)}(y) = \begin{cases} \sup\{\mu_A(x) \mid x \in X, \ f(x) = y\} & \text{if } f^{-1}(y) \neq \emptyset, \\ 0 & \text{otherwise,} \end{cases} \tag{1.36}$$

is called a *fuzzy extension of* f.

By formula (1.36) we define the membership function of the image of the fuzzy set A by fuzzy extension $\tilde{f}$. A justification of this concept is given in the following theorem stating that the mapping $\tilde{f}$ is a true extension of the mapping f when considering the natural embedding of $\mathscr{P}(X)$ into $\mathscr{F}(X)$ and $\mathscr{P}(Y)$ into $\mathscr{F}(Y)$.

Proposition 1.18. *Let X, Y be sets, $f : X \to Y$ be a mapping, $x_0 \in X$, $y_0 = f(x_0)$. If $\tilde{f} : \mathscr{F}(X) \to \mathscr{F}(Y)$ is defined by (1.36), then*

$$\tilde{f}(x_0) = y_0,$$

and the membership function $\mu_{\tilde{f}(x_0)}$ of the fuzzy set $\tilde{f}(x_0)$ is a characteristic function of y_0, i.e.

$$\mu_{\tilde{f}(x_0)} = \chi_{y_0}. \tag{1.37}$$

A more general form of Proposition 1.18 says that the image of a crisp set by a fuzzy extension of a function is again crisp.

Proposition 1.19. *Let X, Y be sets, $f : X \to Y$ be a mapping, $A \subset X$. Then*

$$\tilde{f}(A) = f(A)$$

and the membership function $\mu_{\tilde{f}(A)}$ of $\tilde{f}(A)$ is a characteristic function of the set $f(A)$, i.e.

$$\mu_{\tilde{f}(A)} = \chi_{f(A)}. \tag{1.38}$$

In the following sections the extension principle will be used in different settings for various sets X and Y, and also for different classes of mappings and relations.

1.8 Binary and Valued Relations

In the classical set theory, a *binary relation* R between the elements of sets X and Y is defined as a subset of the Cartesian product $X \times Y$, that is, $R \subset X \times Y$. A valued relation on $X \times Y$ will be a fuzzy subset of $X \times Y$.

Definition 1.14. A *valued relation R on $X \times Y$* is a fuzzy subset of $X \times Y$. The set of all valued relations on $X \times Y$ is denoted by $\mathcal{F}(X \times Y)$.

The valued relations are sometimes called fuzzy relations, however, we reserve this name for valued relations defined on $\mathcal{F}(X) \times \mathcal{F}(Y)$, which will be defined later.

Every binary relation R, where $R \subset X \times Y$, is embedded into the class of valued relations on $X \times Y$ by its characteristic function χ_R being understood as its membership function μ_R. In this sense, any binary relation is valued.

Particularly, any function $f : X \to Y$ is considered as a binary relation, that is, as a subset R_f of $X \times Y$, where

$$R_f = \{(x, y) \in X \times Y \mid y = f(x)\}. \tag{1.39}$$

Here, R_f may be identified with the valued relation by its characteristic function

$$\mu_{R_f}(x, y) = \chi_{R_f}(x, y) \tag{1.40}$$

for all $(x, y) \in X \times Y$, where

$$\chi_{R_f}(x, y) = \chi_{f(x)}(y). \tag{1.41}$$

In particular, if $Y = X$, then each valued relation R on $X \times X$ is a fuzzy subset of $X \times X$, and it is called a valued relation on X instead of on $X \times X$.

Definition 1.15. Let T be a triangular norm. A valued relation R on X is

(i) *reflexive* if for each $x \in X$

$$\mu_R(x, x) = 1;$$

(ii) *symmetric* if for each $x, y \in X$

$$\mu_R(x, y) = \mu_R(y, x);$$

(iii) T*-transitive* if for each $x, y, z \in X$

$$T(\mu_R(x, y), \mu_R(y, z)) \leq \mu_R(x, z);$$

(iv) *separable* if

$$\mu_R(x, y) = 1 \text{ if and only if } x = y;$$

(v) T*-equivalence* if R is reflexive, symmetric and T-transitive;
(vi) T*-equality* if R is reflexive, symmetric, T-transitive and separable.

Definition 1.16. Let R be a valued relation on $X \times Y$ and let $N : [0, 1] \to [0, 1]$ be a negation.

(i) A valued relation R^{-1} on $Y \times X$ is the inverse of R if $\mu_{R^{-1}}(y, x) = \mu_R(x, y)$ for each $x \in X$ and $y \in Y$.
(ii) A valued relation $\mathscr{C}_N R$ on $X \times Y$ is the complement of R if $\mu_{\mathscr{C}_N R}(x, y) = N(\mu_R(x, y))$ for each $x \in X$ and $y \in Y$. If N is the standard negation, then the index N is omitted.
(iii) If μ_R is upper semicontinuous on $X \times Y$, then R is called closed.

For more information about valued relations, see [38].

Example 1.4. Let $\varphi : \mathbf{R} \to [0, 1]$ be a function. Then R defined by the membership function μ_R for all $x, y \in \mathbf{R}$ by

$$\mu_R(x, y) = \varphi(x - y) \tag{1.42}$$

is a valued relation on $\mathbf{R}$. If

$$\varphi(t) = \begin{cases} 1 & \text{if } t \leq 0, \\ 0 & \text{otherwise,} \end{cases}$$

then R defined by (1.42) is the usual binary relation $\leq$ on $\mathbf{R}$. If

$$\varphi(t) = \begin{cases} 1 & \text{if } t \geq 0, \\ 0 & \text{otherwise,} \end{cases}$$

then R defined by (1.42) is the usual binary relation $\geq$ on $\mathbf{R}$. If

$$\varphi(t) = \begin{cases} 1 & \text{if } t = 0, \\ 0 & \text{otherwise,} \end{cases}$$

then R defined by (1.42) is the usual binary relation $=$ on $\mathbf{R}$.

1.9 Fuzzy Relations

Let X, Y be nonempty sets. Consider a valued relation R on $X \times Y$ given by the membership function $\mu_R : X \times Y \to [0, 1]$. In order to extend this function with crisp arguments to function with fuzzy arguments, we apply the extension principle (1.36) in Definition 1.13. Then we obtain a mapping $\tilde{\mu}_R : \mathcal{F}(X \times Y) \to \mathcal{F}([0, 1])$, that is, values of $\tilde{\mu}_R$ are fuzzy subsets of $[0, 1]$.

Definition 1.17. A fuzzy subset of $\mathcal{F}(X) \times \mathcal{F}(Y)$ is called a *fuzzy relation on* $\mathcal{F}(X) \times \mathcal{F}(Y)$. The set of all fuzzy relations on $\mathcal{F}(X) \times \mathcal{F}(Y)$ is denoted by $\mathcal{F}(\mathcal{F}(X) \times \mathcal{F}(Y))$.

Further on, we shall investigate mappings Ψ assigning to each valued relation R from $\mathcal{F}(X \times Y)$ a fuzzy relation from $\mathcal{F}(\mathcal{F}(X) \times \mathcal{F}(Y))$, that is,

$$\Psi : \mathcal{F}(X \times Y) \to \mathcal{F}(\mathcal{F}(X) \times \mathcal{F}(Y)).$$

Definition 1.18. Let R be a valued relation on $X \times Y$. A fuzzy relation $\tilde{R}$ on $\mathcal{F}(X) \times \mathcal{F}(Y)$ given by the membership function $\mu_{\tilde{R}} : \mathcal{F}(X) \times \mathcal{F}(Y) \to [0, 1]$ is called a *fuzzy extension of relation* R, if, for each $x \in X$, $y \in Y$, it holds

$$\mu_{\tilde{R}}(x, y) = \mu_R(x, y). \tag{1.43}$$

Definition 1.19. Let $\Psi : \mathcal{F}(X \times Y) \to \mathcal{F}(\mathcal{F}(X) \times \mathcal{F}(Y))$ be a mapping. Let for all $R \in \mathcal{F}(X \times Y)$, $\Psi(R)$ be a fuzzy extension of relation R. Then the mapping Ψ is called a *fuzzy extension of valued relations*.

Definition 1.20. Let $\Phi, \Psi : \mathcal{F}(X \times Y) \to \mathcal{F}(\mathcal{F}(X) \times \mathcal{F}(Y))$ be mappings. We say that the *mapping* Φ *is dual to* Ψ, if

$$\Phi(\mathcal{C}R) = \mathcal{C}\Psi(R) \tag{1.44}$$

holds for all $R \in \mathcal{F}(X \times Y)$. For Φ dual to Ψ, $R \in \mathcal{F}(X \times Y)$, the fuzzy relation $\Phi(R)$ is called *dual* to fuzzy relation $\Psi(R)$.

Notice that a mapping Φ is dual to Ψ, if and only if the mapping Ψ is dual to Φ. This fact follows from (1.44) and from the identity

$$\mathscr{C}\mathscr{C}R = R.$$

The analogical statement holds for the dual fuzzy relations $\Phi(R)$ and $\Psi(R)$.

1.10 Fuzzy Quantities and Fuzzy Numbers

In this section, we are concerned with fuzzy subsets of the real line. Therefore we have $X = \mathbf{R}$ and $\mathscr{F}(X) = \mathscr{F}(\mathbf{R})$.

Definition 1.21.

(i) A fuzzy subset $A = \{A_\alpha\}_{\alpha\in[0,1]}$ of $\mathbf{R}$ is called a *fuzzy quantity*. The set of all fuzzy quantities will be denoted by $\mathscr{F}(\mathbf{R})$.
(ii) A fuzzy quantity $A = \{A_\alpha\}_{\alpha\in[0,1]}$ is called a *fuzzy interval* if A_α is nonempty, convex and closed subset of $\mathbf{R}$ for all $\alpha \in [0,1]$. The set of all fuzzy intervals will be denoted by $\mathscr{F}_I(\mathbf{R})$.
(iii) A fuzzy interval A is called a *fuzzy number* if its core is a singleton. The set of all fuzzy numbers will be denoted by $\mathscr{F}_N(\mathbf{R})$.

Notice that the membership function $\mu_A : \mathbf{R} \to [0,1]$ of a fuzzy interval A is quasiconcave on $\mathbf{R}$, that is, for all $x, y \in \mathbf{R}$, $x \neq y$, $\lambda \in]0,1[$, the following inequality holds:

$$\mu_A(\lambda x + (1-\lambda)y) \geq \min\{\mu_A(x), \mu_A(y)\}.$$

By Definition 1.21, each fuzzy interval is normalized, since $\text{Core}(A) = [A]_1$ is nonempty, that is, there exists an element $x_0 \in \mathbf{R}$ with $\mu_A(x_0) = 1$. Then $\text{Hgt}(A) = 1$. Moreover, the restriction of the membership function μ_A to $]-\infty, x_0]$ is non-decreasing and the restriction of μ_A to $[x_0, +\infty[$ is a non-increasing function.

A fuzzy interval A has an upper semicontinuous membership function μ_A or, equivalently, for each $\alpha \in]0,1]$ the α-cut $[A]_\alpha$ is a closed subinterval in $\mathbf{R}$. Such a membership function μ_A, and the corresponding fuzzy interval A, can be fully described by a quadruple $(l; r; F; G)$, where $l, r, \in \mathbf{R}$ with $l \leq r$, and F, G are non-increasing left continuous functions mapping $]0, +\infty[$ into $[0, 1[$, by setting

$$\mu_A(x) = \begin{cases} F(l-x) & \text{if } x \in]-\infty, l[, \\ 1 & \text{if } x \in [l, r], \\ G(x-r) & \text{if } x \in]r, +\infty[. \end{cases} \qquad (1.45)$$

We shall briefly write $A = (l; r; F; G)$. As the ranges of F and G are included in $[0, 1[$, we have $\text{Core}(A) = [l, r]$. We can see that the functions F, G describe the left and right "shape" of μ_A, respectively. Observe also that each crisp number $x_0 \in \mathbf{R}$ and each crisp interval $[a,b] \subset \mathbf{R}$ belongs to $\mathscr{F}_I(\mathbf{R})$, as they may be equivalently expressed by the characteristic functions $\chi_{\{x_0\}}$ and $\chi_{[a,b]}$, respectively. These characteristic functions can be also described in the form (1.45) with $F(x) = G(x) = 0$ for all $x \in]0,+\infty[$.

Example 1.5 (Gaussian fuzzy number). Let $a \in \mathbf{R}$, $\gamma \in]0,+\infty[$, and let $A = (a,a,G,G)$ where

$$G(x) = e^{-\frac{x^2}{\gamma}}.$$

Then the membership function μ_A of A is given by

$$\mu_A(x) = G(x-a) = e^{-\frac{(x-a)^2}{\gamma}}.$$

A class of more specific fuzzy intervals of $\mathscr{F}_I(\mathbf{R})$ is obtained, if the α-cuts are required to be bounded intervals. Let $l, r, \in \mathbf{R}$ with $l \leq r$, let $\gamma, \delta \in [0,+\infty[$ and let L, R be continuous and strictly decreasing functions mapping interval $[0,1]$ into $[0,+\infty[$, i.e., $L, R : [0,1] \to [0,+\infty[$. Moreover, assume that $L(1) = R(1) = 0$, and for each $x \in \mathbf{R}$ let

$$\mu_A(x) = \begin{cases} L^{(-1)}\left(\frac{l-x}{\gamma}\right) & \text{if } x \in]l-\gamma, l[, \gamma > 0, \\ 1 & \text{if } x \in [l,r], \\ R^{(-1)}\left(\frac{x-r}{\delta}\right) & \text{if } x \in]r, r+\delta[, \delta > 0, \\ 0 & \text{otherwise,} \end{cases}$$

where L^{-1}, R^{-1} are inverse functions of L, R, respectively. We shall write $A = (l;r;\gamma;\delta)_{LR}$, and say that A is an *(L,R)-fuzzy interval*. The set of all (L,R)-fuzzy intervals will be denoted by $\mathscr{F}_{LRI}(\mathbf{R})$. The values of γ, δ are called the *left* and the *right spread of* A, respectively. Observe that $\text{Supp}(A) = [l-\gamma, r+\delta]$, $\text{Core}(A) = [l,r]$ and $[A]_\alpha$ is a compact interval for every $\alpha \in]0,1]$. If $r = l$, then A is an *(L,R)-fuzzy number*. The set of all (L,R)-fuzzy numbers will be denoted by $\mathscr{F}_{LRN}(\mathbf{R})$.

Particularly important fuzzy intervals are so called *trapezoidal fuzzy intervals* where $L(x) = R(x) = 1 - x$ for all $x \in [0,1]$. In this case, the subscript LR will be omitted in the notation. If $l = r$, then $A = (r;r;\gamma;\delta)$ is called a *triangular fuzzy number* and the notation is simplified to: $A = (r;\gamma;\delta)$, or, equivalently, $A = (a^L;a^M;a^R)$, see bellow.

In the following section we shall introduce the concept of matrices with fuzzy elements, particularly, with (L,R)-fuzzy numbers.

1.11 Matrices with Fuzzy Elements

From now on, we shall denote any fuzzy set A by the symbol with tilde, i.e. $\tilde{A}$. An $m \times n$ matrix $\tilde{A} = \{\tilde{a}_{ij}\}$, where $\tilde{a}_{ij}$, $i = 1, 2, \ldots, m$, $j = 1, 2, \ldots, n$ are fuzzy quantities, is called a *matrix with fuzzy elements*, i.e.

$$\tilde{A} = \begin{bmatrix} \tilde{a}_{11} & \tilde{a}_{12} & \cdots & \tilde{a}_{1n} \\ \tilde{a}_{21} & \tilde{a}_{22} & \cdots & \tilde{a}_{2n} \\ \vdots & \vdots & \ddots & \vdots \\ \tilde{a}_{m1} & \tilde{a}_{m2} & \cdots & \tilde{a}_{mn} \end{bmatrix} . \tag{1.46}$$

In practice, triangular fuzzy numbers introduced in Sect. 1.10 are suitable for modeling fuzzy quantities by DMs. A *triangular fuzzy number* $\tilde{a} \in \mathcal{F}_{LRN}(\mathbf{R})$ can be equivalently expressed by a triple of real numbers, i.e. $\tilde{a} = (a^L; a^M; a^U)$ where a^L is the *Lower number*, a^M is the *Middle number*, and a^U is the *Upper number*, $a^L \leq a^M \leq a^U$. If $a^L = a^M = a^U$, then $\tilde{a}$ is the crisp number (non-fuzzy number). Evidently, the set of all crisp numbers is isomorphic to the set of real numbers. If $a^L \neq a^M \neq a^U$, then the *membership function* $\mu_{\tilde{a}}$ of $\tilde{a}$ is supposed to be continuous, strictly increasing in the interval $[a^L, a^M]$ and strictly decreasing in $[a^M, a^U]$. Moreover, the *membership grade* $\mu_{\tilde{a}}(x)$ is equal to zero for $x \notin [a^L, a^U]$ and equal to one for $x = a^M$. As usual, the membership function $\mu_{\tilde{a}}$ is assumed to be piece-wise linear, see Fig. 1.1. If $a^L = a^M$ and/or $a^M = a^U$, then the membership function $\mu_{\tilde{a}}$ is discontinuous. The triangular fuzzy numbers $\tilde{a} = (a^L_{ij}; a^M_{ij}; a^U_{ij})$ is *fuzzy positive*, if $a^L_{ij} > 0$.

The arithmetic operations $+, -, \cdot$ and / can be extended to fuzzy numbers by the Extension principle, Sect. 1.6, see also e.g. [4].

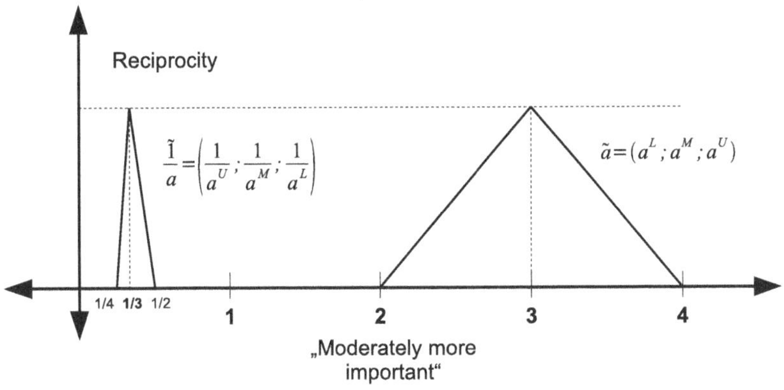

Fig. 1.1 Membership functions of $\tilde{a}$ and $\frac{\tilde{1}}{a}$

Arithmetic operation with fuzzy numbers are defined as follows, see [3], or [15]. Let $\tilde{a} = (a^L; a^M; a^U)$ and $\tilde{b} = (b^L; b^M; b^U)$, where $a^L > 0$, $b^L > 0$, be positive triangular fuzzy numbers.

Addition: $\tilde{a}\tilde{+}\tilde{b} = (a^L + b^L; a^M + b^M; a^U + b^U)$,

Subtraction: $\tilde{a}\tilde{-}\tilde{b} = (a^L - b^U; a^M - b^M; a^U - b^L)$,

Multiplication: $\tilde{a}\tilde{\cdot}\tilde{b} = (a^L \cdot b^L; a^M \cdot b^M; a^U \cdot b^U)$,

Division: $\tilde{a}\tilde{/}\tilde{b} = (a^L/b^U; a^M/b^M; a^U/b^L)$.

Particularly: $\frac{\tilde{1}}{a} = \left(\frac{\tilde{1}}{a^U}; \frac{\tilde{1}}{a^M}; \frac{\tilde{1}}{a^L}\right)$.

For using matrices with triangular fuzzy elements there exist at least following reasons:

- The membership functions of triangular fuzzy elements are usually piece-wise linear, i.e. easy to understand.
- Triangular fuzzy numbers can be easily manipulated, e.g. added, multiplied, see [11].
- Crisp (non-fuzzy) numbers are special cases of triangular fuzzy numbers.
- The reciprocal matrix with triangular fuzzy elements can be considered by the DM as a model for his/her fuzzy pair-wise preference representations concerning n elements (e.g. alternatives). In this model, it is assumed that only $n(n\text{-}1)/2$ judgments are needed, the rest is given by reciprocity condition.
- In practice, when interval-valued matrices are employed, the DM often gives ranges narrower than his or her actual perception would authorize, because he/she might be afraid of expressing information which is too imprecise. On the other hand, triangular fuzzy numbers express rich information because the DM provides both the support set of the fuzzy number as the range that the DM believes to surely contain the unknown ratio of relative importance, and the grades of possibility of occurrence (i.e. membership function) within this range.
- Triangular fuzzy numbers are appropriate in group decision making where a^L can be interpreted as the minimum possible value of DMs judgements, a^U is interpreted as the maximum possible value of DMs judgements, and a^M—the geometric mean of the DMs judgements is interpreted as the mean value, or, the most possible value of DMs judgements, see [10].

1.12 Abelian Linearly Ordered Groups

In this section, we recall some notions and properties related to abelian linearly ordered groups. The matter of this section is based on [1–3], and [4].

Definition 1.22. An *abelian group* is a set, G, together with an operation $\odot$ (read: operation *odot*) that combines any two elements $a, b \in G$ to form another element denoted $a \odot b$. The symbol $\odot$ is a general placeholder for a concretely given operation. The set and operation, $(G, \odot)$, satisfies the following requirements known as the *abelian group axioms*:

- If $a, b \in G$, then $a \odot b \in G$. (*Closure axiom.*)
- If $a, b, c \in G$, then $(a \odot b) \odot c = a \odot (b \odot c)$ holds. (*Associativity axiom.*)
- There exists an element $e \in G$ called the *identity element*, such that for all elements $a \in G$, the equation $e \odot a = a \odot e = a$ holds. (*Identity element axiom.*)
- For each $a \in G$, there exists an element $a^{(-1)} \in G$ called the *inverse element to* a such that $a \odot a^{(-1)} = a^{(-1)} \odot a = e$, where e is the identity element. (*Inverse element axiom.*)
- For all $a, b \in G$, $a \odot b = b \odot a$. (*Commutativity axiom.*)

The *inverse operation* $\div$ to $\odot$ is defined for all $a, b \in G$ as follows

$$a \div b = a \odot b^{(-1)}.$$

In other words, an abelian group is a commutative group. A group in which the group operation is not commutative is called a "non-abelian group" or "non-commutative group".

Definition 1.23. A nonempty set G is *linearly (totally) ordered* under the order relation $\leq$, if the following statements hold for all a, b and c in G:

- *Antisymmetry*:
  If $a \leq b$ and $b \leq a$ then $a = b$;
- *Transitivity*:
  If $a \leq b$ and $b \leq c$ then $a \leq c$;
- *Totality (linearity)*:
  $a \leq b$ or $b \leq a$.
  The *strict order* relation $<$ is defined for $a, b \in G$ as

$$a < b \text{ if } a \leq b \text{ and } a \neq b.$$

Antisymmetry eliminates cases when both a precedes b and b precedes a, and $a \neq b$. A relation having the property of "totality" means that any pair of elements in the set of the relation are comparable under the relation. This also means that the set can be diagrammed as a line of elements, giving it the name "linear". Totality also implies *reflexivity*, i.e., $a \leq a$. Therefore, a total order is also a *partial order*. The partial order has a weaker form of the third condition (it only requires reflexivity, not totality). An extension of a given partial order to a total order is called a linear extension of that partial order.

Definition 1.24. Let $(G, \odot)$ be an abelian group, G be linearly ordered under $\leq$. $(G, \odot, \leq)$ is said to be an *abelian linearly ordered group*, *alo-group* for short, if for all $c \in G$

$$a \leq b \text{ implies } a \odot c \leq b \odot c. \tag{1.47}$$

It is easy to show that for $a \in G$

$$a < e \text{ if and only if } a^{-1} > e, \tag{1.48}$$

$$a > e \text{ if and only if } a^{-1} < e, \tag{1.49}$$

$$a \odot a > a \text{ for all } a > e, \tag{1.50}$$

$$a \odot a < a \text{ for all } a < e. \tag{1.51}$$

If $\mathcal{G} = (G, \odot, \leq)$ is an alo-group, then G is naturally equipped with the order topology induced by $\leq$ and $G \times G$ is equipped with the related product topology. We say that $\mathcal{G}$ is a *continuous alo-group* if $\odot$ is continuous on $G \times G$.

By Definition 1.23, an alo-group $\mathcal{G}$ is a lattice ordered group, see [9]. Hence, there exists $\max\{a, b\}$, for each pair $(a, b) \in G \times G$. Nevertheless, by (1.50), (1.51), a nontrivial alo-group $\mathcal{G} = (G, \odot, \leq)$ has neither the greatest element nor the least element.

Because of the associative property, the operation $\odot$ can be extended by induction to n-ary operation, $n > 2$, by setting

$$\bigodot_{i=1}^{n} a_i = \left(\bigodot_{i=1}^{n-1} a_i\right) \odot a_n. \tag{1.52}$$

Then, for a positive integer n, the (n)-*power* $a^{(n)}$ *of* $a \in G$ is defined by

$$a^{(1)} = a,$$
$$a^{(n)} = \odot_{i=1}^{n} a_i, a_i = a \text{ for all } i = 1, 2, \ldots, n, n \geq 2,$$

and verifies the following properties for $a, b \in G, n \geq 2$:

$$a < b \text{ if and only if } a^{(n)} < b^{(n)}, \tag{1.53}$$

$$a^{(n)} > a \text{ for all } a > e, \tag{1.54}$$

$$a^{(n)} < a \text{ for all } a < e. \tag{1.55}$$

We can extend the meaning of power $a^{(s)}$ to the case that s is a negative integer by setting

$$a^{(0)} = e \text{ and } a^{(-n)} = \left(a^{(n)}\right)^{(-1)} = \left(a^{(-1)}\right)^{(n)}. \tag{1.56}$$

We obtain also

$$(a \div b)^{(n)} = a^{(n)} \div b^{(n)}. \tag{1.57}$$

An *isomorphism* between two alo-groups $\mathcal{G} = (G, \odot, \leq)$ and $\mathcal{G}' = (G', \circ, \preceq)$ is a bijection $h : G \to G'$ that is both a lattice isomorphism and a group isomorphism, i.e.

$$a < b \text{ if and only if } h(a) \preceq h(b) \text{ and } h(a \odot b) = h(a) \circ h(b). \tag{1.58}$$

By the associativity of the operations $\odot$ and $\circ$, the equality in (1.58) can be extended by induction to the n-operation.

Definition 1.25. Let $\mathcal{G} = (G, \odot, \leq)$ be an alo-group. Then $\mathcal{G}$ is *divisible* if for each positive integer n and each $a \in G$ there exists the (n)-th root of a denoted by $a^{(1/n)}$, i.e. $\left(a^{(1/n)}\right)^{(n)} = a$.

The (n)-th root verifies the following property:

$$\text{if } a < b \text{ then } a^{(1/n)} < b^{(1/n)}\,. \tag{1.59}$$

Definition 1.26. Let $\mathcal{G} = (G, \odot, \leq)$ be a divisible alo-group, n is a positive integer. Then the $\odot$*-mean* $m_{\odot}(a_1, a_2, \ldots, a_n)$ *of elements* $a_1, a_2, \ldots, a_n \in G$ is defined as

$$m_{\odot}(a_1, a_2, \ldots, a_n) = \begin{cases} a_1 & \text{for } n = 1, \\ \left(\odot_{i=1}^{n} a_i\right)^{(1/n)} & \text{for } n > 1. \end{cases}$$

Definition 1.27. Let $\mathcal{G} = (G, \odot, \leq)$ be an alo-group. Then the function $\|.\| : G \to G$ defined by $\|a\| = \max\{a, a^{(-1)}\}$ for each $a \in G$ is called a $\mathcal{G}$*-norm*.

The $\mathcal{G}$-norm satisfies the following evident properties, $a, b \in G$, e is the identity element:

(i) $\|a\| = \left\|a^{(-1)}\right\|$;
(ii) $a \leq \|a\|$;
(iii) $\|a\| \geq e$;
(iv) $\|a\| = e$ if and only if $a = e$;
(v) $\left\|a^{(n)}\right\| = \|a\|^{(n)}$;
(vi) $\|a \odot b\| \leq \|a\| \odot \|b\|$ (*triangle inequality*).

Definition 1.28. Let $\mathcal{G} = (G, \odot, \leq)$ be an alo-group. Then the operation $d : G \times G \to G$ defined by $d(a, b) = \|a \div b\|$ for all $a, b \in G$ is called a $\mathcal{G}$*-distance*.

The $\mathcal{G}$-distance d satisfies the following evident properties; $a, b, c \in G$:

(i) $d(a, b) \geq e$;
(ii) $d(a, b) = e$ if and only if $a = b$;
(iii) $d(a, b) = d(b, a)$;
(iv) $d(a, b) \leq d(a, c) \odot d(c, b)$ (*triangle inequality*).

The proof of the following proposition is a straightforward application of the definitions.

Proposition 1.20. *Let* $\mathcal{G} = (G, \odot, \leq)$ *and* $\mathcal{G}' = (G', \circ, \preceq)$ *be alo-groups and* $h : G \to G'$ *be an isomorphism between* $\mathcal{G}$ *and* $\mathcal{G}'$, $a, b \in G$, $a', b' \in G'$. *Then*

$$d_{\mathcal{G}'}(a', b') = h(d_{\mathcal{G}}(h^{-1}(a'), h^{-1}(b')), \tag{1.60}$$

$$d_{\mathcal{G}}(a, b) = h^{-1}(d_{\mathcal{G}'}(h(a), h(b)) \tag{1.61}$$

More definitions and properties of alo-groups of the real line **R** will be presented in Chaps. 2 and 3.

References

1. Bourbaki, N.: Algebra II. Springer, Heidelberg-New York-Berlin (1990)
2. Cavallo, B., D'Apuzzo, L.: A general unified framework for pairwise comparison matrices in multicriterial methods. Int. J. Intell. Syst. **24**(4), 377–398 (2009)
3. Cavallo, B., D'Apuzzo, L., Squillante, M.: About a consistency index for pairwise comparison matrices over a divisible alo-group. Int. J. Intell. Syst. **27**, 153–175 (2012)
4. Cavallo, B., D'Apuzzo, L.: Deriving weights from a pairwise comparison matrix over an alo/group. Soft Comput. **16**, 353–366 (2012)
5. Dubois, D. et al.: Fuzzy interval analysis. Fundamentals of Fuzzy Sets, Series on fuzzy sets, vol. 1. Kluwer Academic, Dordrecht-Boston-London (2000)
6. Frank, H.J.: On the simultaneous associativity of $F(x, y)$ and $x + y - F(x, y)$. *Aequationes Math.* **19**, 194–226 (1979)
7. Klement, E.P., Mesiar, R., Pap, E.: Triangular Norms. Series Trends in Logic. Kluwer Academic, Dordrecht-Boston-London (2000)
8. Kolesárová, A.: Triangular norm-based addition of fuzzy numbers. Tatra Mt. Math. Publ. **6**, 75–81 (1995)
9. Menger, K.: Statistical metrics. Proc. Natl. Acad. Sci. USA **28**, 535–537 (1942)
10. Ralescu, D.: A survey of the representation of fuzzy concepts and its applications. In: Gupta, M.M., Regade, R.K., Yager, R. (eds.) Advances in Fuzzy Sets Theorey and Applications, pp. 77–91. North Holland, Amsterdam (1979)
11. Ramík, J., Římánek, J.: Inequality relation between fuzzy numbers and its use in fuzzy optimization. Fuzzy Sets Syst. **16**, 123–138 (1985)
12. Ramík, J., Vlach, M.: Generalized Concavity in Optimization and Decision Making. Kluwer Publ., Boston-Dordrecht-London (2001)
13. Ramík, J.: Extension principle in fuzzy optimization. Fuzzy Sets Syst. **19**, 29–37 (1986)
14. Schweizer, B., Sklar, A.: Statistical metric spaces. Pacific J. Math. **10**, 313–334 (1960)
15. Yager, R.R.: On a general class of fuzzy connectives. Fuzzy Sets Syst. **4**, 235–242 (1980)
16. Zadeh, L.A.: Fuzzy sets. Inform. Contr **8**, 338–353 (1965)
17. Zadeh, L.A.: The concept of a linguistic variable and its application to approximate reasoning. Inform. Sci. Part I: **8**, 199–249; Part II: **8**, 301–357; Part III: **9**, 43–80 (1975)

Chapter 2
Pairwise Comparison Matrices in Decision Making

Abstract In multicriteria decision making context, a pairwise comparison matrix is a helpful tool to determine the weighted ranking of alternatives or criteria. The entry of the matrix can assume different meanings: it can be a preference ratio (multiplicative case) or a preference difference (additive case), or, it belongs to the unit interval and measures the distance from the indifference that is expressed by 0.5 (fuzzy case). When comparing two elements, the decision maker assigns the value from a scale to any pair of alternatives representing the element of the pairwise preference matrix. Here, we investigate particularly relations between transitivity and consistency of preference matrices being understood differently with respect to the type of preference matrix. By various methods for deriving priorities from various types of preference matrices we obtain the corresponding priority vectors for final ranking of alternatives. The obtained results are also applied to situations where some elements of the fuzzy preference matrix are missing. Finally, a unified framework for pairwise comparison matrices based on abelian linearly ordered groups is presented. Illustrative numerical examples are supplemented.

2.1 Introduction

In various fields of evaluation, selection, and prioritization processes decision maker(s) (DM) try to find the best alternative(s) from a feasible set of alternatives. In many cases, comparison of different alternatives according to their desirability in decision problems cannot be done using only a single criterion or one person. In many decision making problems, procedures have been established to combine opinions about alternatives related to different points of view. These procedures are often based on pairwise comparisons, in the sense that processes are linked to some degree of preference of one alternative over another. According to the nature of the information expressed by the DM, for every pair of alternatives different representation formats can be used to express preferences, e.g. multiplicative preference relations, [39], additive preference relations, [4], fuzzy preference relations, see

---

The chapter was written by "Jaroslav Ramík"

M. Gavalec et al., *Decision Making and Optimization*, Lecture Notes in Economics and Mathematical Systems 677, DOI 10.1007/978-3-319-08323-0_2

[19,38,52], interval-valued preference relations, [70], and also linguistic preference relations, [3].

The decision-making models have been studied for a long time by many authors. A detailed bibliography can be found in the website of the International Society on Multiple Criteria Decision Making, [81]. Many well known methods have been reviewed in [31]: Preference Modeling, Conjoint Measurement, Multi- Attribute Utility Theory (MAUT), Outranking Methods, ELECTRE, PROMETHEE, MACBETH, Fuzzy Multiple Criterion Decision Aid, Multi-objective Programming, Analytic Hierarchy/Network Process (AHP/ANP), and many others.

Usually, experts are characterized by their own personal background and experience of the problem to be solved. Expert opinions may differ substantially, some of them would not be able to efficiently express a preference degree between two or more of the available options, [57]. This may be true due to an expert not possessing a precise or sufficient level of knowledge of part of the problem, or because these experts are unable to discriminate the degree to which some options are better than others. In these situations such an expert will provide an incomplete fuzzy preference matrix, see [3,40,45,70].

Usual procedures for DM problems correct this lack of knowledge of a particular expert using the information provided by the rest of the experts together with aggregation procedures, see [59]. Estimation of missing values in an expert incomplete preference matrix is done using only the preference values provided by these particular experts. By doing this, we assume that the reconstruction of the incomplete preference matrix is compatible with the rest of the information provided by the experts. In this chapter we summarize various approaches to incomplete preference matrix based on different type of degree of preference of one alternative over another, e.g. multiplicative preference relations, additive preference relations, and fuzzy preference relations, see [3,4,20,41,45,46,52,55].

The chapter is organized as follows. Definition of the decision making problem is formulated and multiplicative preference relations and their properties are introduced in Sects. 2.2 and 2.3. In Sect. 2.4. the most popular methods for deriving priorities from multiplicative pairwise comparison matrices are presented. Section 2.5 deals with additive pairwise comparison matrices and in Sect. 2.6 some methods for deriving priorities from additive pairwise comparison matrices are discussed. In Sects. 2.7 and 2.8 we investigate two types of fuzzy pairwise comparison matrices: fuzzy multiplicative and fuzzy additive. We also deal with the problem of deriving priorities for ranking the alternatives. In Sect. 2.9 fuzzy pairwise comparison matrices with missing elements are investigated. Here we deal with the problem of finding the values of missing elements of a given fuzzy pairwise comparison matrix so that the extended matrix is as much fm-consistent or fa-consistent as possible. We consider two important particular cases of fuzzy preference matrix with missing elements where the expert will evaluate only $n-1$ pairwise comparisons of alternatives. Here, we present several yet unpublished results concerning this problem. In Sect. 2.10 we consider pairwise comparison matrices over an abelian linearly (totally) ordered group and, in this way, we provide a general framework for all the above mentioned cases. This section is composed of

4 subsections. By introducing this general setting, we provide a consistency measure that has a natural meaning, it corresponds to the consistency indexes presented in the previous sections and it is easy to calculate it in the additive, multiplicative and fuzzy cases. In some sense we unify a major part of the theory presented in the previous sections. In the last section some concluding considerations and remarks are presented.

2.2 Problem Definition

The *decision making problem* (DM problem) can be formulated as follows. Let $X = \{x_1, x_2, \ldots, x_n\}$ be a finite set of alternatives ($n > 2$). These alternatives have to be ranked from best to worst, using information given by the decision maker (DM) in the form of $n \times n$ *pairwise comparison matrix (PC matrix)* $A = \{a_{ij}\}$, where a_{ij} are elements (components, entries) of the matrix A.

Usually, an ordinal *ranking* of alternatives is required to obtain the best alternative(s), however, it often occurs that the ordinal ranking among alternatives is not a sufficient result and a cardinal ranking called here the *rating* is required.

Some well known limits to our capacity to handle several alternatives at a time (so called cognitive overload, [60]) make it impossible to obtain the rating by a priority weighting vector directly, for instance asking the DM to provide the utility values for the alternatives. Therefore, it is more suitable and also easier to ask the DM for his opinion over the pairs of alternatives and then, once all the necessary information over the pairs is acquired, to derive the rating for the alternatives. The most popular way of eliciting the experts' preferences by pairwise comparisons between the alternatives is the pairwise comparison matrix – a mathematical tool associated with the more general concept: *preference relation*.

The pairwise comparison method was proposed by L. Thurstone in Psychological Review as early as 1927, see [68]. Later on, pairwise comparison matrices have been widely used in many well-known decision making approaches, such as the Analytic Hierarchy Process (AHP), [60], PROMETHEE method, see [31] and many others. A large number of methods deriving a ranking/rating of the alternatives have been proposed in the framework of pairwise comparison matrices in the literature. Two well-known examples are the *eigenvector method (EVM)* in AHP, [60, 61], and the *geometric mean method (GMM)*, being in fact the *Logarithmic Least Squares Method (LLSM)*, see [22].

In one of the most popular MCDM method – the above mentioned Analytic Hierarchy Process (AHP), [59], the decision problem is structured hierarchically at different levels, each level consisting of a finite number of elements. The AHP searches for the priorities representing the relative importance of the decision elements at each particular level. By suitable aggregation it finally calculates the priorities of the alternatives at the bottom level of the hierarchy. Their priorities are interpreted with respect to overall goal at the top of the hierarchy, and elements at upper levels such as criteria, sub-criteria, etc. are used to mediate comparison process. The elicitation process at given level is performed by pairwise comparisons of all elements at given level of the hierarchy with respect to the elements of the upper level. If he/she prefers so, the DM may directly use a numerical value from the scale to express the ratio of elements' relative importance. By inserting numerical values into proper positions a PC matrix is created, and the role of prioritization method is to extract the relative priorities – weights of all compared alternatives, i.e. the rating of alternatives.

The values representing the preferences of the decision elements – alternatives can be also considered as the results of aggregation of pairwise comparisons of a group of decision makers and/or experts. Then the DM problem becomes the *group DM problem (GDM)*. The tournament ranking problem is another well known application of pairwise comparisons, see e.g. [21].

A crucial step in a DM process is the determination of a weighted ranking, i.e. rating, on a set $X = \{x_1, x_2, \ldots, x_n\}$ of alternatives with respect to criteria or experts. A way to determine the rating is to start from a relation represented by the $n \times n$ PC matrix $A = \{a_{ij}\}$; each element of this matrix a_{ij} is a nonnegative real number which expresses how much x_i is preferred to x_j.

The properties of the PC matrix depend on the various meaning given to the number a_{ij}, particularly, the preference matrix $A = \{a_{ij}\}$ becomes: *multiplicative, additive or fuzzy*. In the following sections we shall deal with these cases separately.

2.3 Multiplicative Pairwise Comparison Matrices

In this section, the preferences over the set of alternatives, $X = \{x_1, x_2, \ldots, x_n\}$, will be represented in the following way. Let us assume that the intensities of DM's preferences are given by an $n \times n$ matrix $A = \{a_{ij}\}$ with positive elements in such a way that, for all i and j, the entry a_{ij} indicates the ratio of preference intensity for alternative x_i to that of x_j. In other words, a_{ij} indicates that "x_i is a_{ij} times as good as x_j". T. Saaty in [60] suggests to represent the preference intensities on the absolute scale $\{1/9, 1/8, \ldots, 1/2, 1, 2, \ldots, 8, 9\}$ where $a_{ij} = 1$ indicates equal intensity of preference, whereas $a_{ij} = 9$ indicates extreme intensity of preference for x_i over x_j. Moreover, $a_{ij} \in \{2, 3, \ldots, 8\}$ indicates intermediate evaluations, and the elements of A satisfy the reciprocity condition

$$a_{ji} = \frac{1}{a_{ij}} \text{ for all } i, j \in \{1, 2, \ldots, n\} .$$

If, for example, x_i is 3 times better than x_j, then the goodness of x_j is $1/3$ with respect to the goodness of x_i. The elements of $A = \{a_{ij}\}$ satisfy the following reciprocity condition [60].

A positive $n \times n$ matrix $A = \{a_{ij}\}$ is *multiplicative-reciprocal* (*m-reciprocal*), if

$$a_{ij}.a_{ji} = 1 \text{ for all } i, j \in \{1, 2, \ldots, n\} , \tag{2.1}$$

or, equivalently:

$$a_{ji} = \frac{1}{a_{ij}} \text{ for all } i, j \in \{1, 2, \dots, n\} \ . \tag{2.2}$$

Reciprocity condition (2.1) seems to be natural in many decision situations, however, in some situations the assumption of reciprocity may be too restrictive. For example, in the case when elements under exchange are currencies, a coefficient a_{ij} is a rate of exchange of currency i in relation to currency j, then the presence of transaction costs makes the considered matrix $A = \{a_{ij}\}$ non-reciprocal, and, in particular, (2.1) no longer holds. However, in this chapter, we restrict our consideration only to the most popular reciprocal (particularly, m-reciprocal) preference matrices. The non-reciprocal case has been investigated e.g. in [43].

To establish more important properties to be verified by preference relations is very important for designing good decision making models. One of these properties is the so called consistency property. The lack of consistency in decision making can lead to incompatible conclusions; that is why it is important, if not crucial, to study conditions under which consistency is satisfied, see e.g. [24, 32, 38], or [17]. On the other hand, perfect consistency is difficult to obtain in practice, specially when measuring preferences on a set with a large number of alternatives, e.g. $n > 10$. Clearly, the problem of consistency itself includes two problems: when an expert, considered individually, is said to be consistent and, when the whole group of experts are considered consistent. In this section we will mainly focus on the first problem, assuming that experts' preferences are expressed by means of a preference relation defined over a finite and fixed set of alternatives, expressed appropriately by a square matrix.

A positive $n \times n$ matrix $A = \{a_{ij}\}$ is *multiplicative-consistent* (or, *m-consistent*) [32,60], if

$$a_{ik} = a_{ij}.a_{jk} \text{ for all } i, j, k \in \{1, 2, \dots, n\} \ . \tag{2.3}$$

or, equivalently

$$a_{ij}.a_{jk}.a_{ki} = 1 \text{ for all } i, j, k \in \{1, 2, \dots, n\} \ . \tag{2.4}$$

Here, $a_{ii} = 1$ for all i, and also (2.3) implies (2.1), i.e. an m-consistent matrix is m-reciprocal (however, not vice-versa).

Notice that $a_{ij} > 0$ and m-consistency is not restricted to the Saaty's scale $\{1/9, 1/8, \dots, 1/2, 1, 2, \dots, 8, 9\}$. Here, we extend this scale to the closed interval $[1/\sigma, \sigma]$, where $\sigma > 1$. If σ goes to infinity, we obtain the scale $]0, +\infty[$, i.e. the interval of all positive numbers.

Transitivity is another important property concerning preferences. It represents the idea that the preference intensity obtained by comparing directly two alternatives should be equal to or greater than the preference intensity between those two alternatives x and y obtained using an indirect chain of alternatives x and z, z and y for each $z \in X$.

We say that a positive $n \times n$ matrix $A = \{a_{ij}\}$ is *multiplicative-transitive (m-transitive),20*, if

$$\frac{a_{ik}}{a_{ki}} = \frac{a_{ij}}{a_{ji}}.\frac{a_{jk}}{a_{kj}} \text{ for all } i, j, k \in \{1, 2, \dots, n\} \ , \tag{2.5}$$

or, equivalently:

$$\frac{a_{ij}}{a_{ji}}.\frac{a_{jk}}{a_{kj}}.\frac{a_{ki}}{a_{ik}} = 1 \text{ for all } i, j, k \in \{1, 2, \dots, n\} \ . \tag{2.6}$$

Remark 2.1. Notice that if A is m-consistent, then A is m-transitive. On the other hand, if $A = \{a_{ij}\}$ is m-reciprocal, then A is m-transitive if and only if A is m-consistent. This is not true if A is not m-reciprocal as the following example shows.

Example 2.1. Let $A = \{a_{ij}\}$ be defined as

$$A = \begin{pmatrix} 1 & 2 & 1 \\ 6 & 1 & 2 \\ 9 & 6 & 1 \end{pmatrix} \ .$$

Here, it is easy to verify that A is m-transitive and it is NOT m-consistent (NOT m-reciprocal), as

$$a_{13} = 1 \neq 2.2 = a_{12}.a_{23} \ .$$

In practice, perfect consistency/transitivity is difficult to obtain, particularly when evaluating preferences on a set with a large number of alternatives. If for some positive $n \times n$ matrix $A = \{a_{ij}\}$ and for some $i, j, k = 1, 2, \dots, n$, m-consistency condition (2.3) does not hold, then A is said to be *multiplicative-inconsistent* (or, *m-inconsistent*). Moreover, if for some $n \times n$ positive matrix $A = \{a_{ij}\}$ and for some triple of indexes i, j, k, (2.5) does not hold, then A is said to be *multiplicative-intransitive* (*m-intransitive*). In order to measure the grade of inconsistency/ intransitivity of a given matrix several instruments have been proposed in the literature, see [60].

The following results give a characterization of m-consistent matrix, see also [60].

Proposition 2.1. *Let $A = \{a_{ij}\}$ be a positive $n \times n$ matrix. A is m-consistent if and only if there exists a vector $w = (w_1, w_2, \dots, w_n)$ with $w_i > 0$ for all $i \in \{1, 2, \dots, n\}$, and $\sum_{j=1}^{n} w_j = 1$ such that*

$$a_{ij} = \frac{w_i}{w_j} \text{ for all } i, j \in \{1, 2, \dots, n\} \ . \tag{2.7}$$

Proof. (i) $A = \{a_{ij}\}$ be m-consistent. For $i \in \{1, 2, \ldots, n\}$, set

$$v_i = (a_{i1}a_{i2} \ldots .a_{in})^{\frac{1}{n}} \quad . \tag{2.8}$$

Moreover, set

$$S = \sum_{i=1}^{n} v_i$$

and, finally, define

$$w_i = \frac{v_i}{S} \text{ for } i \in \{1, 2, \ldots, n\} \quad . \tag{2.9}$$

Then, for $i, j = 1, 2, \ldots, n$, by m-reciprocity (2.1) and m-consistency (2.3) we obtain

$$\frac{w_i}{w_j} = (\frac{a_{i1}a_{i2} \ldots a_{in}}{a_{j1}a_{j2} \ldots a_{jn}})^{\frac{1}{n}} = ((a_{i1}a_{1j})(a_{i2}a_{2j}) \ldots (a_{in}a_{nj}))^{\frac{1}{n}}$$
$$= (a_{ij}.a_{ij} \ldots a_{ij})^{\frac{1}{n}} = a_{ij}.$$

Moreover, $\sum_{i=1}^{n} w_i = 1$, consequently, (2.7) is true.

(ii) If (2.7) holds, then evidently (2.3) is satisfied, hence, $A = \{a_{ij}\}$ is m-consistent. □

The following result follows from Remark 2.1.

Proposition 2.2. *Let $A = \{a_{ij}\}$ be a positive m-reciprocal $n \times n$ matrix. $A = \{a_{ij}\}$ is m-consistent if and only if there exists a vector $w = (w_1, w_2, \ldots, w_n)$ with $w_i > 0$ for all $i = 1, 2, \ldots, n$, and*

$$\prod_{j=1}^{n} w_j = 1 \quad , \tag{2.10}$$

such that

$$a_{ij} = \frac{w_i}{w_j} \textit{ for all } i, j \in \{1, 2, \ldots, n\} \quad . \tag{2.11}$$

Proof. (i) Let $A = \{a_{ij}\}$ be m-consistent, for $i \in \{1, 2, \ldots, n\}$, set

$$w_i = (a_{i1}a_{i2} \ldots .a_{in})^{\frac{1}{n}} \quad . \tag{2.12}$$

Then, for $i, j \in \{1,2,\ldots,n\}$, by m-reciprocity (2.1) and m-consistency (2.3) we obtain

$$\frac{w_i}{w_j} = \left(\frac{a_{i1}a_{i2}\ldots a_{in}}{a_{j1}a_{j2}\ldots a_{jn}}\right)^{\frac{1}{n}} = \left((a_{i1}a_{1j})(a_{i2}a_{2j})\ldots(a_{in}a_{nj})\right)^{\frac{1}{n}}$$
$$= (a_{ij}.a_{ij}\ldots a_{ij})^{\frac{1}{n}} = a_{ij}.$$

Moreover,

$$\prod_{i=1}^{n} w_i = ((a_{11}a_{12}\ldots a_{1n})(a_{21}a_{22}\ldots a_{2n})\ldots(a_{n1}a_{n2}\ldots a_{nn}))^{\frac{1}{n}} =$$
$$= ((a_{12}a_{21})(a_{13}a_{31})\ldots(a_{1n}a_{n1})\ldots(a_{2n}a_{n2})\ldots(a_{13}a_{31})$$
$$\ldots(a_{n-1,n})(a_{n,n-1}))^{\frac{1}{n}} = 1\ .$$

Consequently, (2.10) is true.

(ii) If (2.11) holds, then evidently (2.5) is satisfied, hence, $A = \{a_{ij}\}$ is m-consistent. □

Proposition 2.3. *Let $A = \{a_{ij}\}$ be a positive m-reciprocal $n \times n$ matrix. $A = \{a_{ij}\}$ is m-transitive if and only if there exists a vector $w = (w_1, w_2, \ldots, w_n)$ with $w_i > 0$ for all $i \in \{1,2,\ldots,n\}$, and $\sum_{j=1}^{n} w_j = 1$ such that*

$$a_{ij} = \frac{w_i}{w_j} \textit{ for all } i, j \in \{1,2,\ldots,n\}\ . \tag{2.13}$$

Proof. (i) Let $A = \{a_{ij}\}$ be m-transitive and m-reciprocal, then by (2.5) and (2.2)

$$(a_{ij})^2 = (a_{ik})^2.(a_{kj})^2 \text{ for all } i, j, k \in \{1,2,\ldots,n\}\ ,$$

hence $A = \{a_{ij}\}$ is m-consistent. By Proposition 2.1 there exists a vector $w = (w_1, w_2, \ldots, w_n)$ satisfying (2.13).

(ii) If (2.13) holds, then evidently (2.5) is satisfied, hence, $A = \{a_{ij}\}$ is m-transitive. □

2.4 Methods for Deriving Priorities from Multiplicative Pairwise Comparison Matrices

In this section, all matrices are $n \times n$-matrices with positive elements and all vectors are n-dimensional vectors with positive elements.

Let $M^\times$ be the set of all m-reciprocal matrices. Then $M^\times$ is a multiplicative group under component-wise multiplication. If $A = \{a_{ij}\} \in M^\times$ and $B = \{b_{ij}\} \in$

$M^\times$, then $C = A \times B = \{a_{ij}.b_{ij}\} \in M^\times$. Here, by $\times$ we denote the group operation, i.e. the component-wise multiplication of matrices (in contrast to the usual matrix multiplication). Similarly, let $C^\times$ be the set of all m-consistent matrices and $w^\times$ be the set of all vectors $w = (w_1, w_2, \dots, w_n)$ with positive elements such that $\prod_{j=1}^n w_j = 1$. Both $C^\times$ and $w^\times$ are multiplicative groups under component-wise multiplication. Moreover, $C^\times$ is a subgroup of $M^\times$ and by Proposition 2.2, $C^\times$ is isomorphic to $w^\times$.

Let $F^\times$ be the set of all mappings from $M^\times$ into $w^\times$, $f \in F^\times$. Vector $w = (w_1, w_2, \dots, w_n) \in w^\times$ is called the *m-priority vector of* $A = \{a_{ij}\}$ with respect to f, if $w = f(A)$.

If $w = (w_1, w_2, \dots, w_n)$ is the m-priority vector of m-consistent matrix $A = \{a_{ij}\}$ with respect to f satisfying the following condition:

$$a_{ij} = w_i/w_j \text{ for all } i, j \in \{1, 2, \dots, n\} \ , \tag{2.14}$$

then w is called the *m-consistency vector*.

Here, the mapping f defines how the m-priority vector is calculated from the elements a_{ij} of A. If f is defined by (2.12), then (2.14) is satisfied if and only if A is m-consistent.

In the DM problem given by $X = \{x_1, x_2, \dots, x_n\}$ and $A = \{a_{ij}\}$, the rating of the alternatives in X is determined by the priority vector $w = (w_1, w_2, \dots, w_n)$ of A. This vector, if normalized, is called the *vector of weights*. Hence, each element w_i of the m-priority vector w is interpreted as the *relative importance of alternative* x_i. The ranking of alternatives is defined as follows:

$$x_i \succ x_j \text{ if } w_i > w_j \text{ for all } i, j \in \{1, 2, \dots, n\} \ .$$

Therefore, the alternatives $x_1, x_2, \dots, x_n$ in X can be ranked/rated by their relative importance.

Generally, the PC matrix $A = \{a_{ij}\}$ is not m-consistent, hence the priority vector cannot be an m-consistency vector. We face the problem of how to measure the inconsistency of the PC matrix. In order to solve this problem we shall define special consistency indexes based on the corresponding priority vectors.

In the next section, six popular methods for deriving priority vector are discussed. Our choice of methods is based on [20], where 18 different methods for deriving priority vectors from PC matrices under a common framework of effectiveness have been discussed. Later on, in [50], C.-C. Lin published a revised framework for deriving preference values from pairwise comparison matrices together with some new simulation. In [64] B. Srdjevic combined different prioritization methods in AHP. Here, the concept of minimizing aggregated deviation and so called *correctness in error free cases* are presented. Some comparison calculations demonstrate corresponding results of the presented methods. For each method we also derive an associated consistency index – a tool for measuring the grade of inconsistency of the given PC matrix.

2.4.1 Eigenvector Method (EVM)

Historically, one of the oldest method for deriving priorities from multiplicative preference matrix is the eigenvector method (EVM), [15]. This method has an intuitive background but now it is based on Perron-Frobenius theory which is known in several versions, see e.g. [30], where you can find the proof, which is nontrivial. The Perron-Frobenius theorem describes some of the remarkable properties enjoyed by the eigenvalues and eigenvectors of irreducible nonnegative matrices (e.g. positive matrices).

Theorem 2.1 (Perron-Frobenius). *Let $A = \{a_{ij}\}$ be an irreducible nonnegative $n \times n$ matrix. Then the spectral radius of A, $\rho(A)$, is a real eigenvalue, which has a positive (real) eigenvector $w = (w_1, w_2, \ldots, w_n)$. This eigenvalue which is called* the principal eigenvalue *of A is simple (it is not a multiple root of the characteristic equation), and its eigenvector, called* the principal eigenvector *of A is unique up to a multiplicative constant.*

By EVM the priority vector of a positive $n \times n$ matrix $A = \{a_{ij}\}$ is defined as the normalized principal eigenvector $w = F(A)$ given by Perron-Frobenius theorem. Hence, the rating of the alternatives in X is determined by the priority vector $w = (w_1, w_2, \ldots, w_n)$, with $w_i > 0$, for all $i \in \{1, 2, \ldots, n\}$, such that $\sum_{i=1}^{n} w_i = 1$, satisfying $Aw = \rho(A)w$. Since the element of the priority vector w_i is interpreted as the relative importance of alternative x_i, the alternatives $x_1, x_2, \ldots, x_n$ in X are rated by their relative importance.

An m-inconsistency grade of a positive m-reciprocal $n \times n$ matrix A can be measured by the *m-EV-consistency index* $I_{mEV}(A)$ defined in e.g. [60] as

$$I_{mEV}(A) = \frac{\rho(A) - n}{n - 1} \,, \tag{2.15}$$

where $\rho(A)$ is the spectral radius of A (i.e. the principal eigenvalue of A). The proof of the following result can be found in [60].

Proposition 2.4. *If $A = \{a_{ij}\}$ is an $n \times n$ positive m-reciprocal matrix, then $I_{mEV} \geq 0$. Moreover, A is m-consistent if and only if $I_{mEV}(A) = 0$.*

To provide an m-inconsistency measure independently of the dimension n of the matrix A, T. Saaty in [60] proposed the consistency ratio CR_{mEV}. In order to distinguish it here from the other inconsistency measures, we shall call it *m-EV-consistency ratio*. This is obtained by taking the consistency index I_{mEV} to its mean value R_{mEV}, i.e. the mean value of $I_{mEV}(A)$ of positive m-reciprocal matrices of dimension n, whose entries are uniformly distributed random variables on the interval $[1/9, 9]$, i.e.

$$CR_{mEV} = \frac{I_{mEV}}{R_{mEV}} \,. \tag{2.16}$$

The following table relates the dimension n of the positive m-reciprocal matrix in the first row to its corresponding mean value $R_{mEV}(n)$, $n = 3, 4, \ldots, 10$ in the second row.

n	3	4	5	6	7	8	9	10
$R_{mEV}(n)$	0.58	0.90	1.12	1.24	1.32	1.41	1.45	1.49

For this consistency measure it was proposed an estimation of 10% threshold of CR_{mEV}. In other words, a pairwise comparison matrix could be acceptable (in a DM process) if its m-consistency ratio does not exceed 0.1, see [60].

The following theorem is useful for practical computations of the priority vector and also for calculating m-consistency index (2.15), and/or m-consistency ratio (2.16). The proof can be found in [60].

Theorem 2.2. *(Wielandt) Let $A = \{a_{ij}\}$ be a positive $n \times n$ matrix, $e = (1, 1, \ldots, 1)$ be n-vector. Then the principal eigenvector $w = (w_1, w_2, \ldots, w_n)$ corresponding to principal eigenvalue of A is as follows*

$$w = \lim_{k\to\infty} \frac{A^k e}{e^T A^k e} \,. \tag{2.17}$$

Notice that $A^k e$ is the vector of row sums of the k-powered matrix A^k, whereas $e^T A^k e$ is the sum of all elements of A^k. Applying Theorem 2.2 we can easily calculate the priority vector of the matrix A (consequently, the principal eigenvalue and the consistency ratio) by formula (2.17), simply by calculating $2k$-th powers of A, i.e. A, A^2, $(A^2)^2, \ldots$. Here, we apply the usual matrix multiplication, not a component-wise one, as before. Such calculations can be easily performed in a worksheet, e.g. Excel.

It was shown by various researchers, see e.g. [5], that for small deviations of a_{ij} around the consistent ratios w_i/w_j, i.e. for small deviations $a_{ij} - w_i/w_j$, EVM gives reasonably good approximation of the priority vector. However, when the deviations are large, it is generally accepted that the corresponding priority vector is not satisfactory. This is the main reason why we look for another methods for deriving priority vectors.

2.4.2 *Additive Normalization Method (ANM)*

The Additive Normalization Method (ANM) is a heuristic method which is based on Theorem 2.2. To obtain the priority vector $w = (w_1, w_2, \ldots, w_n)$ by this method it is enough to divide the elements of each column of matrix $A = \{a_{ij}\}$ by the sum of that column (i.e. to normalize the column), then add the elements in each resulting row and finally divide this sum by n, the number of elements in the row. ANM gives the first element of the succession (2.17), i.e. for $k = 1$, generated in Theorem 2.2.

The element of the priority vector is given by

$$w_i = \frac{1}{n} \sum_{j=1}^{n} \frac{a_{ij}}{\sum_{k=1}^{n} a_{kj}}, \text{ for all } i = 1, 2, \ldots, n \ . \tag{2.18}$$

Example 2.2. Let $X = \{x_1, x_2, x_3\}$ be the set of alternatives and assume that the intensities of DM's preferences are given by an 3×3 matrix $A = \{a_{ij}\} \in M^{\times}$ defined as

$$A = \begin{pmatrix} 1 & 2 & \frac{1}{16} \\ \frac{1}{2} & 1 & 8 \\ 16 & \frac{1}{8} & 1 \end{pmatrix} .$$

We can compute the priority vector w^* by (2.18) as follows:

$$w^* = (w_1^*, w_2^*, w_3^*) = (0.23, 0.41, 0.35) \ .$$

According to the priority vector w^* we obtain the ranking: $x_2 \succ x_3 \succ x_1$.

In [60] T. Saaty suggested ANM as a simplified and tractable version of EVM. In [50], this method is ranked among the three best ones in simulation experiments, together with EVM and LLSM, see below. Popularity and wide use in practice ANM owes to its extreme simplicity. Although considered inferior it significantly outperforms more sophisticated methods, as is demonstrated in [29, 64] and [50].

2.4.3 *Least Squares Method (LSM)*

The Least Squares (LS) method minimizes L_2 distance function defined for elements of the unknown priority vector $w = (w_1, w_2, \ldots, w_n)$ and known elements a_{ij} of the matrix A by solving the following constrained non-linear optimization problem:

$$\sum_{i,j=1}^{n} \left(a_{ij} - \frac{w_i}{w_j} \right)^2 \longrightarrow \min \tag{2.19}$$

subject to

$$\sum_{j=1}^{n} w_j = 1, w_i \geq \epsilon > 0 \text{ for all } i \in \{1, 2, \ldots, n\} \ .$$

(ϵ is a preselected sufficiently small positive number.)

As the above nonlinear optimization problem may have multiple solutions it is advantageous to convert this problem to the following one

$$\sum_{1 \leq i < j \leq n} \left(a_{ij} w_j - w_i\right)^2 \longrightarrow \min \tag{2.20}$$

subject to

$$\sum_{j=1}^{n} w_j = 1, w_i \geq \epsilon > 0 \text{ for all } i \in \{1, 2, \ldots, n\} \ . \tag{2.21}$$

Optimization problem (2.20), (2.21) which is referred to as *weighted LSM* is transformed into a system of linear equations by differentiating the Lagrangian of (2.20) and equalizing it to zero. It is shown in [8] that in this way the WLSM provides a unique and strictly positive solution $w^* = (w_1^*, w_2^*, \ldots, w_n^*)$. This solution is set as the priority vector of A. If A is m-consistent then it is clear that $A = \{w_i^*/w_j^*\}$, hence, w^* is a consistent vector, see also [23].

An m-inconsistency measure of a positive m-reciprocal $n \times n$ matrix A with respect to LSM is given by the *m-LS-consistency index* $I_{mLS}(A)$ defined as

$$I_{mLS}(A) = \sum_{1 \leq i < j \leq n} \left(a_{ij} w_j^* - w_i^*\right)^2 \ , \tag{2.22}$$

where $w^* = (w_1^*, w_2^*, \ldots, w_n^*)$ is the optimal solution of (2.20), (2.21). The proof of the following proposition which is parallel to Proposition 2.4 is straightforward.

Proposition 2.5. *If $A = \{a_{ij}\}$ is a positive m-reciprocal matrix, then $I_{mLS}(A) \geq 0$. Moreover, A is m-consistent if and only if $I_{mLS}(A) = 0$.*

Example 2.3. Let $X = \{x_1, x_2, x_3\}$ be the set of alternatives and assume that the intensities of DM's preferences are given by an 3×3 matrix $A = \{a_{ij}\} \in M^{\times}$ defined as

$$A = \begin{pmatrix} 1 & 2 & \frac{1}{16} \\ \frac{1}{2} & 1 & 8 \\ 16 & \frac{1}{8} & 1 \end{pmatrix} \ .$$

We can compute the priority vector w^* by solving optimization problem (2.20), (2.21) as follows:

$$w^* = (w_1^*, w_2^*, w_3^*) = (0.2, 0.1, 0.7) \ .$$

By (2.22) we calculate $I_{mLS}(A) = 0.57$. According to the priority vector w^* we obtain the ranking: $x_3 \succ x_1 \succ x_2$.

2.4.4 *Logarithmic Least Squares Method (LLSM)/Geometric Mean Method (GMM)*

The Logarithmic Least Squares Method (LLSM), presented in [5–7, 22], makes use of the multiplicative properties of the PC matrix $A = \{a_{ij}\} \in M^{\times}$. LLSM assumes the minimization of sum of the logarithmic squared deviations from given elements of A, i.e.

$$\sum_{1 \leq i < j \leq n} \left(\ln a_{ij} - \ln \frac{w_i}{w_j} \right)^2 \longrightarrow \min \tag{2.23}$$

subject to

$$\prod_{j=1}^{n} w_j = 1, w_i \geq 0 \text{ for all } i = 1, 2, \ldots, n \ . \tag{2.24}$$

As can be demonstrated, see e.g. [22], the optimal solution of problem (2.23), (2.24) is always unique and can be found simply as the geometric mean of the rows of the reciprocal PC matrix, provided that the set of given pairwise comparisons is complete, i.e. the number of judgments is $n(n-1)/2$. The elements of the priority vector – the optimal solution of problem (2.23), (2.24) $w^* = (w_1^*, w_2^*, \ldots, w_n^*)$ – are defined as the geometric mean of the row elements of A:

$$w_i^* = \left(\prod_{j=1}^{n} a_{ij} \right)^{1/n} \quad \text{for all } i \in \{1, 2, \ldots, n\} \ . \tag{2.25}$$

That is why LLSM is also referred to as the *Geometric Mean Method* (GMM).

Notice that constraint (2.24) can be substituted by the usual normalization constraint as follows:

$$\sum_{j=1}^{n} w_j = 1, w_i \geq 0 \text{ for all } i \in \{1, 2, \ldots, n\} \ . \tag{2.26}$$

Clearly, if the vector w^* is an optimal solution of problem (2.23), (2.24), then $v^* = c.w^*$ is an optimal solution of problem (2.23), (2.26), for a suitable $c > 0$. The opposite assertion is also true: The vector v^* is an optimal solution of problem (2.23), (2.26) if $w^* = d.v^*$ is an optimal solution of problem (2.23), (2.24) for a suitable $d > 0$.

An m-consistency measure of a positive m-reciprocal $n \times n$ matrix $A = \{a_{ij}\}$ with respect to LLSM (GMM) is given by the *m-consistency index* $I_{mGM}(A)$ defined as the minimal value of the objective function (2.23):

$$I_{mGM}(A) = \sum_{1 \leq i < j \leq n} \ln^2(e_{ij}) \ , \tag{2.27}$$

where $w^* = (w_1^*, w_2^*, \ldots, w_n^*)$ given by (2.25) is the optimal solution of (2.23), and

$$e_{ij} = a_{ij}\frac{w_j^*}{w_i^*} \text{ for all } i \in \{1, 2, \ldots, n\} \ .$$

We denote

$$E_A = \{e_{ij}\} = \{a_{ij}\frac{w_j^*}{w_i^*}\}, W_A = \{\frac{w_i^*}{w_j^*}\} \ .$$

Here, $E_A = \{e_{ij}\}$ is called the *error matrix of* $A = \{a_{ij}\}$. By (2.25) we obtain

$$e_{ij} = \left(\prod_{k=1}^{n} a_{ij}a_{jk}a_{ki}\right)^{1/n} \ .$$

Notice that $W_A = \{\frac{w_i^*}{w_j^*}\} \in \mathcal{C}^{\times}$, i.e. W_A is m-consistent. Let us denote $W_A^{(-1)} = \{\frac{w_j^*}{w_i^*}\}$, then evidently

$$A = E_A \times W_A^{(-1)} \ . \tag{2.28}$$

The proof of the following proposition which is parallel to Proposition 2.5 is straightforward.

Proposition 2.6. *If $A = \{a_{ij}\}$ is a positive m-reciprocal matrix, then $I_{mGM}(A) \geq 0$. Moreover, A is m-consistent if and only if $I_{mGM}(A) = 0$.*

Proposition 2.6 enables us to distinguish between m-consistent and m-inconsistent matrices but it is insufficient to determine the degree of m-consistency of m-inconsistent matrices. In fact, a statement of the type: *A is less m-consistent than B if $I_{mGM}(A) > I_{mGM}(B)$* is not meaningful when A and B are of different dimensions. Moreover, a cut-off rule of the type: *A is close enough to being m-consistent if $I_{mGM}(A) \leq \alpha$* for some fixed positive constant α, independent of the dimension n, does not appear to be meaningful.

A natural way to construct a better measure than the above defined m-consistency index I_{mGM} that would preserve the above mentioned properties and addresses its deficiencies is to consider the *relative m-error $RE_m(A)$ of $A = \{a_{ij}\}$* defined as the normalized I_{mGM}, see [5]:

$$RE_m(A) = \frac{\sum_{1 \leq i < j \leq n} \ln^2(e_{ij})}{\sum_{1 \leq i < j \leq n} \ln^2(a_{ij})} \ , \tag{2.29}$$

Proposition 2.7. *If $A = \{a_{ij}\}$ is a positive m-reciprocal matrix, then*

$$0 \leq RE_m(A) \leq 1. \tag{2.30}$$

Proof. As the error component minimizes the sum of squares (2.23) we obtain

$$\sum_{1 \leq i < j \leq n} \ln^2(a_{ij}\frac{w_j^*}{w_i^*}) \leq \sum_{1 \leq i < j \leq n} \ln^2(a_{ij}\frac{1}{1}) = \sum_{1 \leq i < j \leq n} \ln^2(a_{ij}) \ .$$

Hence, (2.30) is satisfied. □

By Proposition 2.7 the values of the relative error belong to the unit interval $[0, 1]$ regardless of n, the dimension of A. Using this measure, we may compare consistency of matrices of different dimensions and justify the use of cut/off rules accepting A as sufficiently consistent if, for example, its relative error satisfies $RE_m(A) \leq 0.1$.

A matrix A is called *totally m-inconsistent* if its relative m-error is maximal, i.e. $RE_m(A) = 1$.

The relative error measures the relative grade of inconsistency of A. Now, we define a new index which will measure the relative grade of consistency of a matrix A by the matrix W_A.

The *relative m-consistency index* $RC_m(A)$ *of* $A = \{a_{ij}\}$ is defined as follows, see [5]:

$$RC_m(A) = \frac{\sum_{1 \leq i < j \leq n} \ln^2(\frac{w_i^*}{w_j^*})}{\sum_{1 \leq i < j \leq n} \ln^2(a_{ij})} \ , \tag{2.31}$$

where $w^* = (w_1^*, w_2^*, \ldots, w_n^*)$ is given by (2.25), the optimal solution of (2.23), (2.24).

Proposition 2.8. *If $A = \{a_{ij}\}$ is a positive m-reciprocal matrix, then*

$$RC_m(A) + RE_m(A) = 1 \ . \tag{2.32}$$

The proof of this proposition will be given later in Sect. 2.5 in connection with additive preference matrices and the logarithmic/exponential isomorphism between multiplicative and additive PC matrices.

Proposition 2.9. *$A = \{a_{ij}\} \in M^{\times}$ is totally m-inconsistent if and only if the row products of A are all ones, i.e.*

$$\prod_{j=1}^{n} a_{ij} = 1 \ for \ all \ i \in \{1, 2, \ldots, n\} \ . \tag{2.33}$$

Proof. (i) Let $A = \{a_{ij}\}$ be totally m-inconsistent, i.e. $RE_m(A) = 1$, hence, by Proposition 2.8 we obtain:

$$RC_m(A) = \frac{\sum_{1 \le i < j \le n} \ln^2(\frac{w_i^*}{w_j^*})}{\sum_{1 \le i < j \le n} \ln^2(a_{ij})} = 0 \ .$$

Therefore, $w_i^* = w_j^*$ for all $i, j = 1, 2, \ldots, n$ and

$$\sum_{1 \le i < j \le n} \ln^2(\frac{w_i^*}{w_j^*}) = 0 \ ,$$

consequently, (2.33) holds.

(ii) Let (2.33) be satisfied. Then by (2.25) $w_i^* = w_j^* = 1$ for all $i, j \in \{1, 2, \ldots, n\}$ and then

$$\sum_{1 \le i < j \le n} \ln^2(a_{ij}.\frac{w_j^*}{w_i^*}) = \sum_{1 \le i < j \le n} \ln^2(a_{ij}) \ ,$$

hence, $RE_m(A) = 1$. □

Example 2.4. Let $X = \{x_1, x_2, x_3\}$ be the set of alternatives and assume that the intensities of preferences are given by an 3×3 matrix $A = \{a_{ij}\} \in M^{\times}$ defined as (see also Example 2.2 and Example 2.3):

$$A = \begin{pmatrix} 1 & 2 & \frac{1}{16} \\ \frac{1}{2} & 1 & 8 \\ 16 & \frac{1}{8} & 1 \end{pmatrix} \ .$$

We compute $w^* = (w_1^*, w_2^*, w_3^*)$ as the row geometric averages of A, then we compute $B = \{\ln^2(\frac{w_i^*}{w_j^*})\}$ and $C = \{\ln^2(a_{ij})\}$ as follows:

$$w^* = (w_1^*, w_2^*, w_3^*) = ((1 \cdot 2 \cdot \frac{1}{16})^{\frac{1}{3}}, (\frac{1}{2} \cdot 1 \cdot 8)^{\frac{1}{3}}, (16 \cdot \frac{1}{8} \cdot 1)^{\frac{1}{3}}) = (0.50, 1.59, 1.26) \ ,$$

the normalized priority vector $v^* = \frac{1}{3.35}.w^* = (0.15, 0.47, 0.38)$,

$$B = \begin{pmatrix} 0 & 1.33 & 0.85 \\ 0.48 & 0 & 0.05 \\ 0.85 & 0.05 & 0 \end{pmatrix} \ ,$$

$$C = \begin{pmatrix} 0 & 0.48 & 7.69 \\ 0.48 & 0 & 4.32 \\ 7.69 & 4.32 & 0 \end{pmatrix} \ .$$

Then we calculate

$$RC_m(A) = \frac{\sum_{1\le i<j\le 3} \ln^2(\frac{w_i^*}{w_j^*})}{\sum_{1\le i<j\le 3} \ln^2(a_{ij})} = \frac{2.24}{12.49} = 0.18 \ .$$

Moreover, by (2.32) we have $RE_m(A) = 1 - RC_m(A) = 1 - 0.18 = 0.82$. We conclude that the relative m-consistency of A is 18 percent and relative m-error of A is 82 percent. Hence, the m-inconsistency of A is relatively high.

According to the priority vector v^* we obtain the ranking: $x_2 \succ x_3 \succ x_1$. When comparing this result with Example 2.3 we obtain a different ranking of the variants.

2.4.5 *Fuzzy Programming Method (FPM) and Other Methods for Deriving Priorities*

The FP method proposed by Mikhailov in [53] firstly states that if reciprocal matrix $A = \{a_{ij}\}$ is m-consistent, then the system of $m = n(n-1)/2$ linear equations:

$$a_{ij}w_j - w_i = 0 \text{ for all } i, j \in \{1, 2, \ldots, n\}, i < j \ ,$$

can be represented as follows:

$$C_j w = 0, \ j \in \{1, 2, \ldots, m\} \ , \tag{2.34}$$

where by C_j we denote the j-th row vector of coefficients of the matrix of the system (2.34).

If A is inconsistent, it is desirable to find such values of vector w, so that (2.34) is approximately satisfied, i.e. $Cw \approx 0$.

The FPM represents (2.34) geometrically as an intersection of fuzzy hyperlines and transforms the prioritization problem to optimization one, determining the values of the priorities that correspond to the point with the highest *measure of intersection*. In this way the prioritization problem is reduced to a fuzzy programming problem that can be solved as a standard linear program:

$$\mu \longrightarrow \max$$

subject to

$$\mu.d_j^+ + C_j.w \le d_j^+, \ j \in \{1, \ldots, m\} \ ,$$

$$\mu.d_j^- - C_j.w \le d_j^-, \ j \in \{1, \ldots, m\} \ ,$$

$$\sum_{i=1}^{n} w_i = 1, w_i \ge 0 \text{ for all } i \in \{1, 2, \ldots, n\} \ .$$

Here, the positive values of the left and right tolerance parameters d_j^+ and d_j^- represent the admissible interval of approximate satisfaction of the crisp equality (2.34).

Other methods for deriving priorities from multiplicative PC matrices can be found in the literature, see e.g. [2, 20, 64, 74–80], or [50].

2.5 Additive Pairwise Comparison Matrices

The above mentioned interpretation of preferences on the set of alternatives $X = \{x_1, x_2, \ldots, x_n\}$ described by a multiplicative pairwise comparison matrix A (e.g. in AHP, see e.g. [59, 62]) is, however, not always appropriate for a DM. Evaluating the preference of two elements of a pair, say, x_i and x_j with respect to e.g. "design of products", or "hight of persons" might cause a problem. Here, saying e.g. that product x_i is 3 times nicer than x_j, or, John is 5 times higher than Peter, might be peculiar. It appears that Saaty's ratio scale for pairwise comparisons do not represent reality and often produces an exaggerated results beyond our common sense, although mathematically the operation of the m-reciprocal matrix seems to be useful. The perception of interval, or difference, of two alternatives is relatively much simpler than the perception of the ratio of both.

In this section, the preferences over the set of alternatives X will be represented in the following way. Let us assume that the intensities of DM's preferences are given by an $n \times n$ matrix $A = \{a_{ij}\}$ with real number elements in such a way that, for all i and j, the entry a_{ij} indicates the difference of preference intensity for alternative x_i to that of x_j. In other words, a_{ij} means that the difference between intensities "x_i and x_j is a_{ij}". Hence the element a_{ij} can be positive, zero or negative. The perception of interval (or difference) of two objects is relatively much simpler than the perception of the ratio of both. The reasons are that operations of addition and subtraction are easier than the operations of multiplication and division, which, in fact, are based on addition and subtraction, respectively. Therefore, addition and subtraction are straightforward for the comparison of two objects.

K.T.F. Yuen in [9] suggests to represent the preference intensities by the relative scale $\{-\kappa, -7\kappa/8, -6\kappa/8, \ldots, -\kappa/8, 0, \kappa/8, 2\kappa/8, \ldots, 7\kappa/8, \kappa\}$ in contrast to the Saaty's absolute scale $\{1/9, 1/8, \ldots, 1/2, 1, 2, \ldots, 8, 9\}$.

Here, $a_{ij} = 0$ indicates equal intensity of both alternatives and $a_{ij} = \kappa$, is the maximal difference (i.e. utility) of intensities of the compared alternatives, [9]. Here, κ indicates how people perceive the value of the difference of paired objects in different scenarios. Here, $2.8 + 1 = 17$ is the number of the intervals of the scale schema which corresponds to 7 ± 2 stages in a single rating process. In general, the scale for measuring preference intensities is the interval of all real numbers, i.e. $\mathbf{R} =]-\infty, +\infty[$.

The elements of $A = \{a_{ij}\}$ satisfy the following reciprocity condition. A real $n \times n$ matrix $A = \{a_{ij}\}$ is *additive-reciprocal* (*a-reciprocal*), if

$$a_{ji} = -a_{ij} \text{ for all } i, j \in \{1, 2, \ldots, n\} \ . \tag{2.35}$$

A real $n \times n$ matrix $A = \{a_{ij}\}$ is *additive-consistent* (or, *a-consistent*) [32, 60], if

$$a_{ik} = a_{ij} + a_{jk} \text{ for all } i, j, k \in \{1, 2, \ldots, n\} \ . \tag{2.36}$$

From (2.36) we obtain $a_{ii} = 0$ for all i, and also (2.36) implies (2.35), i.e. an a-consistent matrix is a-reciprocal (however, not vice-versa).

Here, all matrices are $n \times n$-matrices with real number elements and all vectors are n-dimensional vectors with real number elements.

Let M^+ be the set of all a-reciprocal matrices. Then M^+ is an additive group under component-wise addition. If $A = \{a_{ij}\} \in M^+$ and $B = \{b_{ij}\} \in M^+$, then $C = A + B = \{a_{ij} + b_{ij}\} \in M^+$. Here, by "+" we denote the group operation, i.e. the component-wise addition of matrices. Similarly, let C^+ be the set of all a-consistent matrices and v^+ be the set of all vectors $v = (v_1, v_2, \ldots, v_n)$ with real number elements such that $\sum_{j=1}^{n} v_j = 0$. Both C^+ and v^+ are additive groups under component-wise addition. Moreover, C^+ is a subgroup of M^+ and, as we will see by Proposition 2.10, C^+ is isomorphic to v^+.

Let F^+ be the set of all mappings from M^+ into v^+, $f \in F^+$. Vector $v = (v_1, v_2, \ldots, v_n) \in v^+$ is called the *a-priority vector of* $A = \{a_{ij}\}$ with respect to f, if $v = f(A)$.

If $v = (v_1, v_2, \ldots, v_n)$ is the a-priority vector of a-consistent matrix $A = \{a_{ij}\}$ with respect to f satisfying the following condition:

$$a_{ij} = v_i - v_j \text{ for all } i, j \in \{1, 2, \ldots, n\} \ , \tag{2.37}$$

then v is called the *a-consistency vector of A*.

Here, the mapping f defines how the a-priority vector is calculated from the elements a_{ij} of A.

Proposition 2.10. *An matrix $A = \{a_{ij}\}$ is a-consistent if and only if there exists a vector $v = (v_1, v_2, \ldots, v_n)$ with $\sum_{j=1}^{n} v_j = 0$ such that*

$$a_{ij} = v_i - v_j \textit{ for all } i, j \in \{1, 2, \ldots, n\} \ . \tag{2.38}$$

Proof. (i) Let let $A = \{a_{ij}\}$ be a-consistent, then by setting

$$v_i = \frac{1}{n} \sum_{k=1}^{n} a_{ik} \text{ for all } i \in \{1, 2, \ldots, n\} \ ,$$

we obtain

$$v_i - v_j = \frac{1}{n} \sum_{k=1}^{n} (a_{ik} - a_{jk}) = a_{ij} \text{ for all } i \in \{1, 2, \ldots, n\} \ .$$

Considering a-reciprocity (2.36) we obtain

$$\sum_{i=1}^{n} v_i = \frac{1}{n} \sum_{i=1}^{n} \sum_{k=1}^{n} a_{ik} = 0 \text{ for all } i \in \{1, 2, \ldots, n\} .$$

(ii) Let (2.38) be true, then easily (2.35) is satisfied. □

In the DM problem given by $X = \{x_1, x_2, \ldots, x_n\}$ and $A = \{a_{ij}\}$, the rating of the alternatives in X is determined by the a-priority vector $v = (v_1, v_2, \ldots, v_n)$ of A. The ranking of alternatives is defined as follows:

$$x_i \succ x_j \text{ if } v_i > v_j \text{ for all } i, j \in \{1, 2, \ldots, n\} .$$

Therefore, the alternatives $x_1, x_2, \ldots, x_n$ in X can be ranked by the natural ranking of the components of the corresponding a-priority vector.

Generally, the PC matrix $A = \{a_{ij}\}$ is not a-consistent, hence the a-priority vector cannot be an a-consistency vector. We face the problem of how to measure the inconsistency of the PC matrix. To solve this problem we shall define special a-consistency indexes based on the corresponding a-priority vectors.

2.6 Methods for Deriving Priorities from Additive Pairwise Comparison Matrices

The Least Squares Method (LSM), presented in Sect. 2.4.3 makes use also of the additive properties of the PC matrix $A = \{a_{ij}\} \in M^+$, see [35]. Here, LSM assumes the minimization of sum of the squared deviations from given elements of A, i.e.

$$\sum_{1 \le i < j \le n} \left(a_{ij} - (v_i - v_j)\right)^2 \longrightarrow \min \tag{2.39}$$

subject to

$$\sum_{j=1}^{n} v_j = 0 . \tag{2.40}$$

As is demonstrated e.g. in [22] the optimal solution of problem (2.39), (2.40) can be found simply as the arithmetic mean of the rows of an a-reciprocal PC matrix. The elements of the a-priority vector – the optimal solution of problem (2.39), (2.40) $v^* = (v_1^*, v_2^*, \ldots, v_n^*)$ are defined as the arithmetic mean of the row elements of A:

$$v_i^* = \frac{1}{n} \sum_{j=1}^{n} a_{ij} \text{ for all } i \in \{1, 2, \ldots, n\} . \tag{2.41}$$

An a-consistency measure of a matrix $A = \{a_{ij}\}$ with respect to LSM is given by the *a-consistency index of* A, $I_a(A)$, defined as the minimal value of the objective function (2.39):

$$I_{aLS}(A) = \sum_{1 \le i < j \le n} e_{ij}^2 \ , \tag{2.42}$$

where $v^* = (v_1^*, v_2^*, \ldots, v_n^*)$ is given by (2.41), the optimal solution of (2.39), (2.40), moreover, $E_A^a = \{e_{ij}\} = \{a_{ij} - (v_j^* - v_i^*)\}$, $V_A = \{v_i^* - v_j^*\}$.

Here, $E_A^a = \{e_{ij}\}$ is called the *a-error matrix of* $A = \{a_{ij}\}$. By (2.41) we obtain $e_{ij} = \frac{1}{n}\sum_{k=1}^{n}(a_{ij} + a_{jk} + a_{ki})$.

Notice that $V_A = \{v_i^* - v_j^*\} \in C^+$, i.e. the matrix V_A is a-consistent. Then evidently

$$A = E_A^a - V_A \ . \tag{2.43}$$

The proof of the following proposition which is parallel to Proposition 2.6 is straightforward.

Proposition 2.11. *If $A = \{a_{ij}\}$ is an a-reciprocal matrix, then $I_{aLS}(A) \ge 0$. Moreover, A is a-consistent if and only if $I_{aLS}(A) = 0$.*

Proposition 2.11 enables us to distinguish between a-consistent and a-inconsistent matrices but it is insufficient to determine the degree of a-consistency from a-inconsistent matrices. In fact, a statement of the type: *A is less a-consistent than B if $I_{aLS}(A) > I_{aLS}(B)$* is not meaningful when A and B are of different dimensions. Moreover, a cut-off rule of the type: *A is close enough to being a-consistent if $I_{aLS}(A) \le \alpha$* for some fixed positive constant α, independent of the dimension n, does not appear to be meaningful.

We want to construct a better measure than the above defined a-consistency index I_a. A natural way that would preserve the above mentioned properties and addresses its deficiencies is to consider the *relative a-error $RE_a(A)$ of $A = \{a_{ij}\}$* defined as the "normalized", I_{aLS}, see [5]:

$$RE_a(A) = \frac{\sum_{1 \le i < j \le n} e_{ij}^2}{\sum_{1 \le i < j \le n} a_{ij}^2} \ , \tag{2.44}$$

Proposition 2.12. *If $A = \{a_{ij}\}$ is an a-reciprocal matrix, then*

$$0 \le RE_a(A) \le 1 \ . \tag{2.45}$$

Proof. Since the error component minimizes the sum of squares (2.39) we have

$$\sum_{1 \le i < j \le n} (a_{ij} - (v_i^* - v_j^*))^2 \le \sum_{1 \le i < j \le n} (a_{ij} - (v_i - v_j))^2 \le \sum_{1 \le i < j \le n} a_{ij}^2 \ .$$

Hence, (2.45) is satisfied. □

By Proposition 2.12 the values of the relative a-error belong to the unit interval $[0, 1]$ regardless of n, the dimension of A. Using this measure, we may compare consistency of matrices of different dimensions and justify the use of cut/off rules accepting A as sufficiently consistent if, for example, its relative a-error satisfies $RE_a(A) \leq 0.1$.

A matrix A is called *totally a-inconsistent* if its relative a-error is maximal, i.e. $RE_a(A) = 1$.

The relative a-error measures the relative grade of a-inconsistency of A. Now, we define a new index which will measure the relative grade of a-consistency of a matrix A by the matrix V_A.

The *relative a-consistency index* $RC_a(A)$ *of* $A = \{a_{ij}\}$ is defined as follows, see [5]:

$$RC_a(A) = \frac{\sum_{1 \leq i < j \leq n} (v_i^* - v_j^*)^2}{\sum_{1 \leq i < j \leq n} a_{ij}^2} , \tag{2.46}$$

where $v^* = (v_1^*, v_2^*, \ldots, v_n^*)$ is given by (2.41), the optimal solution of (2.39), (2.40).

The main result of this subsection is based on the classical projection theorem, see [33]:

Theorem 2.3. *Let A be a vector in $\mathbf{R}^N$ and S be a subspace of $\mathbf{R}^N$. The vector A can be uniquely represented in the form:*

$$A = C + E , \tag{2.47}$$

where $C \in S$ and $E \perp S$. Moreover, C is a unique vector satisfying:

$$\|A - C\| \leq \|A - X\| \textit{ for all } X \in S , \tag{2.48}$$

where $\|A\|$ is the Euclidean norm of A.

Now, we formulate the main result:

Proposition 2.13. *If $A = \{a_{ij}\}$ is an a-reciprocal matrix, then*

$$RC_a(A) + RE_a(A) = 1 . \tag{2.49}$$

Proof. We may rewrite the $n \times n$ matrix A as an $N = n^2$-dimensional vector by ordering its elements by rows. In the N-dimensional Euclidean space $\mathbf{R}^N$ the vectors corresponding to the a-consistent matrices form a subspace S. Combining the projection theorem with the solution of the minimization problem which is characterized by the arithmetic mean, we see that the decomposition of A into its a-consistent and a-inconsistent components is an orthogonal decomposition, i.e. $C \perp E$, or,

$$\sum_{1 \leq i < j \leq n} c_{ij} e_{ij} = 0 .$$

Using this equation with $a_{ij} = c_{ij} + e_{ij}$, we obtain

$$\sum_{1 \le i < j \le n} a_{ij}^2 = \sum_{1 \le i < j \le n} c_{ij}^2 + 2 \sum_{1 \le i < j \le n} c_{ij} e_{ij} + \sum_{1 \le i < j \le n} e_{ij}^2 ,$$

and therefore

$$\sum_{1 \le i < j \le n} a_{ij}^2 = \sum_{1 \le i < j \le n} c_{ij}^2 + \sum_{1 \le i < j \le n} e_{ij}^2 .$$

Hence, for $A \neq 0$ we obtain (2.49). □

Now, we derive the parallel result to Proposition 2.9.

Proposition 2.14. $A = \{a_{ij}\} \in M^+$ *is totally a-inconsistent if and only if the row sums of* A *are all zeros, i.e.*

$$\sum_{j=1}^{n} a_{ij} = 0 \, \textit{for all } i \in \{1, 2, \ldots, n\} . \tag{2.50}$$

Proof. (i) Let $A = \{a_{ij}\}$ be totally a-inconsistent, i.e. $RE_a(A) = 1$, hence, by Proposition 2.13 we obtain:

$$RC_a(A) = \frac{\sum_{1 \le i < j \le n} (v_i^* - v_j^*)^2}{\sum_{1 \le i < j \le n} a_{ij}^2} = 0 .$$

Therefore, $v_i^* = v_j^*$ for all $i, j = 1, 2, \ldots, n$ and by (2.40) we have $v_i^* = 0$ for all $i \in \{1, 2, \ldots, n\}$. Then by (2.41) we obtain

$$\frac{1}{n} \sum_{j=1}^{n} a_{ij} = v_i^* = 0 \text{ for all } i \in \{1, 2, \ldots, n\} .$$

Hence, (2.50) holds.

(ii) Let (2.50) be satisfied. Then by (2.41) $v_i^* = \frac{1}{n} \sum_{j=1}^{n} a_{ij} = 0$ for all $i, j \in \{1, 2, \ldots, n\}$. Consequently, $e_{ij} = a_{ij}$ for all $i, j \in \{1, 2, \ldots, n\}$ and

$$\sum_{1 \le i < j \le n} e_{ij}^2 = \sum_{1 \le i < j \le n} a_{ij}^2 ,$$

hence, $RE_a(A) = 1$, i.e. A is totally a-inconsistent. □

We illustrate the theory by a simple example.

Example 2.5. Let $X = \{x_1, x_2, x_3\}$ be the set of alternatives and assume that the intensities of preferences are given by an 3×3 matrix $A = \{a_{ij}\} \in M^+$ defined as, see [5]:

$$A = \begin{pmatrix} 0 & 1 & -4 \\ -1 & 0 & 7 \\ 4 & -7 & 0 \end{pmatrix} .$$

We can now compute the priority vector v^* as follows:

$$v^* = (v_1^*, v_2^*, v_3^*) = (\frac{1}{3}(0+1-4), \frac{1}{3}(-1+0+7), \frac{1}{3}(4-7+0)) = (-1, 2, -1) .$$

Then

$$V_A = \{v_i^* - v_j^*\} = \begin{pmatrix} 0 & -3 & 0 \\ 3 & 0 & 3 \\ 0 & -3 & 0 \end{pmatrix} ,$$

$$RC_a(A) = \frac{\sum_{1 \le i < j \le 3}(v_i^* - v_j^*)^2}{\sum_{1 \le i < j \le 3} a_{ij}^2} = \frac{36}{132} = 0.27.$$

Moreover, by (2.32) we have $RE_a(A) = 1 - RC_a(A) = 1 - 0.27 = 0.73$. We conclude that the relative a-consistency of A is 27 percent and relative a-error of A is 73 percent. Hence, the a-inconsistency of A is relatively high.

According to the priority vector v^* we obtain the ranking: $x_2 \succ x_3 = x_1$.

2.7 Fuzzy Pairwise Comparison Matrices

Sometimes it is more natural, when comparing x_i to x_j, that the decision maker assigns the positive value b_{ij} to x_i and b_{ji} to x_j, where $b_{ij} + b_{ji} = 1$. Here, the preferences on $X = \{x_1, x_2, \dots, x_n\}$ can be understood as a *fuzzy preference relation*, or, *valued relation*, with membership function $\mu : X \times X \to [0, 1]$, and $\mu(x_i, x_j) = b_{ij}$ denotes the preference of the alternative x_i over x_j [32,47,55,60,65, 66,72]. Therefore, the fuzzy preference relation on X can be represented as a *fuzzy pairwise comparison matrix (FPC matrix)*. Here, $b_{ij} = \frac{1}{2}$ indicates indifference between x_i and x_j, $b_{ij} > \frac{1}{2}$ indicates that x_i is preferred to x_j, $b_{ij} = 1$ indicates that x_i is absolutely preferred to x_j, $b_{ij} = 0$ indicates that x_j is absolutely preferred to x_i, [18, 25–27]. Other important properties of fuzzy pairwise comparison matrix $B = \{b_{ij}\}$, can be presented as follows.

An $n \times n$ FPC matrix $B = \{b_{ij}\}$ with $0 \le b_{ij} \le 1$ for all $i, j \in \{1, 2, \dots, n\}$ is *fuzzy-reciprocal* (*f-reciprocal*) [17], if

$$b_{ij} + b_{ji} = 1 \text{ for all } i, j \in \{1, 2, \dots, n\} , \quad (2.51)$$

or, equivalently

$$b_{ji} = 1 - b_{ij} \text{ for all } i, j \in \{1, 2, \ldots, n\} . \tag{2.52}$$

Evidently, if (2.51) holds, then $b_{ii} = \frac{1}{2}$ for all $i \in \{1, 2, \ldots, n\}$.

For making a "coherent" choice (when assuming FPC matrices) some properties have been suggested in the literature [32, 67].

The nomenclature of properties of fuzzy PC matrices has, however, not been stabilized yet, compare e.g. [17, 52, 67, 71]. Here, we use the usual nomenclature which is as close as possible to the one used in the literature.

We say that FPC matrix $B = \{b_{ij}\}$ is *fuzzy-additive-consistent* (*fa-consistent*) [17], if

$$b_{ik} - \frac{1}{2} = (b_{ij} - \frac{1}{2}) + (b_{jk} - \frac{1}{2}) \text{ for all } i, j, k \in \{1, 2, \ldots, n\} . \tag{2.53}$$

Equation (2.53) can be equivalently rewritten as (see [67]):

$$b_{ik} = \frac{1}{2} + b_{ij} - b_{kj} \text{ for all } i, j, k \in \{1, 2, \ldots, n\} . \tag{2.54}$$

or, see [17],

$$b_{ij} + b_{jk} + b_{ki} = \frac{3}{2} \text{ for all } i, j, k \in \{1, 2, \ldots, n\} . \tag{2.55}$$

Notice that if $B = \{b_{ij}\}$ is fa-consistent, then B is f-reciprocal.

Let FPC matrix $B = \{b_{ij}\}$ be fuzzy reciprocal according to (2.51). We say that $B = \{b_{ij}\}$ is *fuzzy-multiplicative-consistent* (*fm-consistent*), [13], if B is multiplicatively transitive by (2.5), i.e. if

$$\frac{b_{ik}}{b_{ki}} = \frac{b_{ij}}{b_{ji}} . \frac{b_{jk}}{b_{kj}} \text{ for all } i, j, k \in \{1, 2, \ldots, n\}. \tag{2.56}$$

This property is sometimes called "transitivity", see e.g. [52], however, here we reserve this name for different concept, see below.

It is not difficult to prove some relationships among individual consistency properties of FPC matrices. In Fig. 2.1 a scheme of properties of PC matrices is depicted. Moreover, we present some examples of matrices with particular properties showing that the corresponding inclusions depicted in Fig. 2.1 are strict.

(1) Let $A_1 = \{a_{1ij}\}$ be defined as

$$A_1 = \begin{pmatrix} 0.5 & 0.6 & 0.5 \\ 0.4 & 0.5 & 0.4 \\ 0.5 & 0.6 & 0.5 \end{pmatrix} .$$

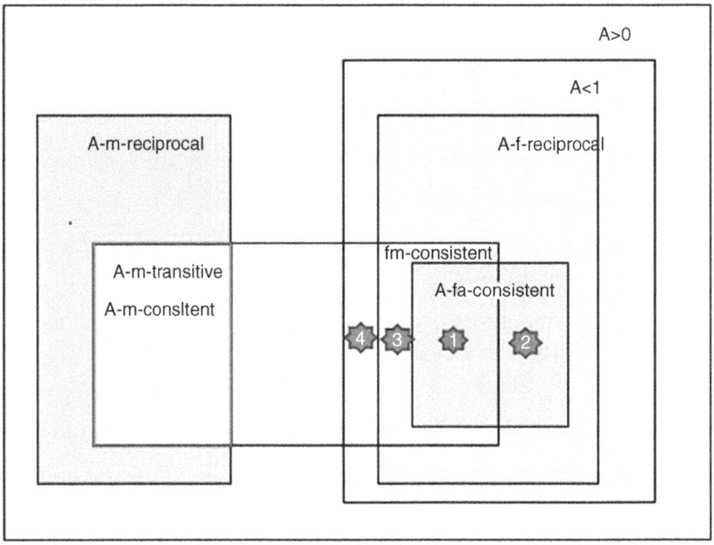

Fig. 2.1 The scheme of properties

Here, A_1 is f-reciprocal and m-transitive defined by (2.5), i.e. fm-consistent, and fa-consistent.

(2) Let $A_2 = \{a_{2ij}\}$ be defined as

$$A_2 = \begin{pmatrix} 0.5 & 0.3 & 0.2 \\ 0.7 & 0.5 & 0.4 \\ 0.8 & 0.6 & 0.5 \end{pmatrix}.$$

Here, A_2 is fa-consistent and NOT fm-consistent.

(3) Let $A_3 = \{a_{3ij}\}$ be defined as

$$A_3 = \begin{pmatrix} 0.5 & 0.25 & 0.1 \\ 0.75 & 0.5 & 0.25 \\ 0.9 & 0.75 & 0.5 \end{pmatrix}.$$

Here, A_3 is fm-consistent and NOT fa-consistent.

(4) Let $A_4 = \{a_{4ij}\}$ be defined as

$$A_4 = \begin{pmatrix} 0.5 & 0.9 & 0.2 \\ 0.9 & 0.5 & 0.3 \\ 0.2 & 0.3 & 0.5 \end{pmatrix}.$$

Here, A_4 is m-transitive and NOT f-reciprocal.

In what follows, we shall investigate some relationships between f-reciprocal and m-reciprocal pairwise comparison matrices. We start with extension of the result published by E. Herrera-Viedma et al. [17]. For this purpose, given $\sigma > 1$, we define the following function φ_σ and its inverse function φ_σ^{-1} as

$$\varphi_\sigma(t) = \frac{1}{2}(1 + \frac{\ln t}{\ln \sigma}) \text{ for } t \in [\frac{1}{\sigma}, \sigma] \ , \tag{2.57}$$

$$\varphi_\sigma^{-1}(t) = \sigma^{2t-1} \text{ for } t \in [0, 1] \ . \tag{2.58}$$

We obtain the following results, characterizing fa-consistent and m-consistent matrices, see [17,55]. By $\sigma > 1$ evaluation scale $[1/\sigma; \sigma]$ is defined.

Proposition 2.15. *Let $\sigma > 1$, $A = \{a_{ij}\}$ be an $n \times n$ matrix with $\frac{1}{\sigma} \leq a_{ij} \leq \sigma$ for all $i, j \in \{1, 2, \ldots, n\}$. If $A = \{a_{ij}\}$ is m-consistent then $B = \{\varphi_\sigma(a_{ij})\}$ is fa-consistent.*

Proof. Let let $A = \{a_{ij}\}$ be an $n \times n$ matrix with $\frac{1}{\sigma} \leq a_{ij} \leq \sigma$ for all i and j. Suppose that A is m-consistent and set $b_{ij} = \varphi_\sigma(a_{ij})$ for all i and j. Then by (2.55) and (2.57)

$$b_{ij} + b_{jk} + b_{ki} = \frac{1}{2}(3 + \frac{\ln a_{ij}a_{jk}a_{ki}}{\ln \sigma}) \text{ for all } i, j, k \in \{1, 2, \ldots, n\} \ .$$

Hence, by (2.3) we have $a_{ij}a_{jk}a_{ki} = 1$ for all $i, j, k \in \{1, 2, \ldots, n\}$ and then

$$b_{ij} + b_{jk} + b_{ki} = \frac{3}{2} \text{ for all } i, j, k \in \{1, 2, \ldots, n\} \ .$$

and $B = \{\varphi_\sigma(a_{ij})\}$ is fa-consistent. □

Proposition 2.16. *Let $\sigma > 1$, $B = \{b_{ij}\}$ be an $n \times n$ matrix with $0 \leq b_{ij} \leq 1$ for all $i, j \in \{1, 2, \ldots, n\}$. If $B = \{b_{ij}\}$ is fa-consistent then $A = \{\varphi_\sigma^{-1}(b_{ij})\}$ is m-consistent.*

Proof. Let $B = \{b_{ij}\}$ be fa-consistent. Then, by setting $a_{ij} = \varphi_\sigma^{-1}(b_{ij})$, we obtain

$$a_{ij}a_{jk}a_{ki} = \sigma^{2b_{ij}-1}\sigma^{2b_{jk}-1}\sigma^{2b_{ki}-1} = \sigma^{2(b_{ij}+b_{jk}+b_{ki})-3} = \sigma^{2(\frac{3}{2})-3} = 1$$

for all $i, j, k \in \{1, 2, \ldots, n\}$. Therefore, $A = \{a_{ij}\}$ is m-consistent. □

Now, let us define the function ϕ and its inverse function ϕ^{-1} as follows

$$\phi(t) = \frac{t}{1 + t} \text{ for } t > 0 \ , \tag{2.59}$$

$$\phi^{-1}(t) = \frac{t}{1 - t} \text{ for } 0 < t < 1 \ . \tag{2.60}$$

We obtain the following results, characterizing fm-consistent and m-consistent matrices, see [55].

Proposition 2.17. *Let $A = \{a_{ij}\}$ be an $n \times n$ matrix with $0 < a_{ij}$ for all i and j. If $A = \{a_{ij}\}$ is m-consistent then $B = \{b_{ij}\} = \{\phi(a_{ij})\}$ is fm-consistent.*

Proof. By (2.59) and m-reciprocity (2.1) we obtain for all i, j:

$$b_{ij} = \frac{a_{ij}}{1+a_{ij}} = \frac{\frac{1}{a_{ji}}}{1+\frac{1}{a_{ji}}} = \frac{1}{1+a_{ji}} = \frac{a_{ji}}{1+a_{ji}} a_{ij} = b_{ji} a_{ij} \ .$$

Then $b_{ij}b_{jk}b_{ki} = b_{ji}b_{kj}b_{ik}a_{ij}a_{jk}a_{ki}$, for all i, j and k. Hence, by (2.3) and by (2.6), B is m-transitive. Moreover,

$$b_{ij} + b_{ji} = \frac{1}{1+a_{ji}} + \frac{a_{ji}}{1+a_{ji}} = 1 \text{ for all } i, j \ .$$

Hence, by (2.51), B is f-reciprocal, consequently, B is fm-consistent. □

Proposition 2.18. *Let $B = \{b_{ij}\}$ be an fm-reciprocal $n \times n$ matrix with $0 < b_{ij} < 1$ for all i and j. If $B = \{b_{ij}\}$ is fm-consistent then $A = \{a_{ij}\} = \{\phi^{-1}(b_{ij})\}$ is m-consistent.*

Proof. Let B be f-reciprocal. Then

$$a_{ij} = \frac{b_{ij}}{1-b_{ij}} = \frac{b_{ij}}{b_{ji}} \ ,$$

Then by m-transitivity (2.5), we obtain

$$a_{ij}a_{jk}a_{ki} = \frac{b_{ij}}{b_{ji}}\frac{b_{jk}}{b_{kj}}\frac{b_{ki}}{b_{ik}} = 1 \ ,$$

hence, $A = \{a_{ij}\}$ is m-consistent. □

From Proposition 2.16 it is clear that the concept of m-transitivity plays a similar role for f-reciprocal fuzzy preference matrices as the concept of m-consistency does for m-reciprocal matrices.

2.8 Methods for Deriving Priorities from Fuzzy Pairwise Comparison Matrices

A parallel result to Proposition 2.1 can be derived for f-reciprocal matrices.

Proposition 2.19. *Let $A = \{a_{ij}\}$ be an f-reciprocal $n \times n$ matrix with $0 < a_{ij} < 1$ for all i and j. $A = \{a_{ij}\}$ is fm-consistent if and only if there exists a vector*

$v = (v_1, v_2, \ldots, v_n)$ *with* $v_i > 0$ *for all* $i = 1, 2, \ldots, n$, *and* $\sum_{j=1}^{n} v_j = 1$ *such that*

$$a_{ij} = \frac{v_i}{v_i + v_j} \textit{ for all } i, j = 1, 2, \ldots, n \ . \tag{2.61}$$

Proof. By Proposition 2.18, $B = \{b_{ij}\}$ is fm-consistent if and only if $A = \{a_{ij}\} = \{\frac{b_{ij}}{1-b_{ij}}\}$ is m-consistent. By Proposition 2.4, this result is equivalent to the existence of a vector $v = (v_1, v_2, \ldots, v_n)$ with $v_i > 0$ for all $i \in \{1, 2, \ldots, n\}$, such that $\frac{b_{ij}}{1-b_{ij}} = \frac{v_i}{v_j}$ for all $i, j \in \{1, 2, \ldots, n\}$, or, equivalently, $b_{ij} = \frac{v_i}{v_i + v_j}$ for all $i, j \in \{1, 2, \ldots, n\}$, i.e. (2.61) is true. □

Here, the vector $v = (v_1, v_2, \ldots, v_n)$ with $v_i > 0$ for all $i = 1, 2, \ldots, n$, and $\sum_{j=1}^{n} v_j = 1$ is called the *fm-priority vector of* A.

The previous result is a representation theorem for fm-consistent matrices, the next two results concern fa-consistent matrices.

Proposition 2.20. *Let* $A = \{a_{ij}\}$ *be an* $n \times n$ *matrix with* $0 < a_{ij} < 1$ *for all* $i, j \in \{1, 2, \ldots, n\}$. $A = \{a_{ij}\}$ *is fa-consistent if and only if*

$$a_{ij} = \frac{1}{2}(1 + nu_i - nu_j) \textit{ for all } i, j \in \{1, 2, \ldots, n\} \ , \tag{2.62}$$

where

$$u_i = \frac{2}{n^2} \sum_{k=1}^{n} a_{ik} \textit{ for all } i \in \{1, 2, \ldots, n\} \ , \tag{2.63}$$

and

$$\sum_{i=1}^{n} u_i = 1 \ . \tag{2.64}$$

Proof. (i) Let let $A = \{a_{ij}\}$ be fa-consistent, then A is f-reciprocal and by (2.51) we obtain

$$\sum_{i,j=1}^{n} a_{ij} = \frac{1}{2} n^2 \ . \tag{2.65}$$

From (2.63), (2.65) we obtain

$$\sum_{i=1}^{n} u_i = \frac{2}{n^2} \sum_{i,j=1}^{n} a_{ij} = 1 \ . \tag{2.66}$$

From (2.63), (2.66) we also have

$$\frac{1}{2} + \frac{n}{2}(u_i - u_j) = \frac{1}{2} + \frac{1}{n}\sum_{k=1}^{n}(a_{ik} - a_{jk}) \text{ for all } i, j \in \{1, 2, \ldots, n\} \ .$$

By fa-consistency of B, (2.54), we obtain

$$a_{ij} - \frac{1}{2} = a_{ik} - a_{jk} \text{ for all } i, j \in \{1, 2, \ldots, n\} \ .$$

If (2.9) is satisfied for all $i, j \in \{1, 2, \ldots, n\}$, then

$$a_{ij} = \frac{1}{2}(1 + nu_i - nu_j) \text{ for all } i, j \in \{1, 2, \ldots, n\} \ .$$

(ii) Let (2.62) be true, then easily (2.53) is satisfied. □

Here, the vector $u = (u_1, u_2, \ldots, u_n)$ with $u_i > 0$ for all i=1,2,...,n, and $\sum_{j=1}^{n} u_j = 1$ is called the *fa-priority vector of A*.

By the symbol I_{mY} we denote the m-consistency index that has been defined for various methods for deriving priorities of multiplicative pairwise comparison matrices, where $Y \in \{EV, LS, GM\}$, i.e. I_{mEV}, I_{mLS}, I_{mGM}.

Now, we shall investigate inconsistency property also for f-reciprocal matrices. For this purpose we use relations between m-consistent and fm-consistent/fa-consistent matrices derived in Propositions 2.15 - 2.18.

Let $B = \{b_{ij}\}$ be an f-reciprocal $n \times n$ matrix with $0 < b_{ij} < 1$ for all $i, j \in \{1, 2, \ldots, n\}$, and let $Y \in \{EV, LS, GM\}$. We define the *faY-consistency index* $I_{faY}(B)$ of $B = \{b_{ij}\}$ as

$$I_{faY}(B) = \phi(I_{mY}(A)), \text{ where } A = \{\phi^{-1}(b_{ij})\} \ . \tag{2.67}$$

From (2.67) we obtain the following result which is parallel to Proposition 2.4.

Proposition 2.21. *If $B = \{b_{ij}\}$ is an f-reciprocal $n \times n$ fuzzy matrix with $0 < b_{ij} < 1$ for all $i, j = 1, 2, \ldots, n$, $Y \in \{EV, LS, MG\}$. Then $I_{faY}(B) \geq 0$. Moreover, B is fa-consistent if and only if $I_{faY}(B) = 0$.*

Proof. The proof follows from the corresponding proofs of the propositions in Sect. 2.4. □

Further, we shall be dealing with inconsistency of f-multiplicative matrices. Recall transformation functions φ_σ and φ_σ^{-1} defined by (2.57), (2.58), where $\sigma > 1$ is a given scale value. Let $B = \{b_{ij}\}$ be an f-reciprocal $n \times n$ matrix with $0 < b_{ij} < 1$ for all $i, j = 1, 2, \ldots, n$, $Y \in \{EV, LS, MG\}$. We define the *fmY-consistency index* $I^{\sigma}_{fmY}(B)$ of $B = \{b_{ij}\}$ as

$$I^{\sigma}_{fmY}(B) = \varphi_\sigma(I_{mY}(A_\sigma)), \text{ where } A_\sigma = \{\varphi_\sigma^{-1}(b_{ij})\} \ . \tag{2.68}$$

From (2.58), (2.68) we obtain the following result which is parallel to Propositions 2.4 and 2.21.

Proposition 2.22. *If $B = \{b_{ij}\}$ is an f-reciprocal $n \times n$ matrix with $0 < b_{ij} < 1$ for all $i, j \in \{1, 2, \ldots, n\}$, $Y \in \{EV, LS, MG\}$, then $I^{\sigma}_{fmY}(B) \geq 0$. Moreover, B is fm-consistent if and only if $I^{\sigma}_{fmY}(B) = 0$.*

Proof. The proof follows from the corresponding proofs of the propositions in Sect. 2.4. □

Example 2.6. Let $X = \{x_1, x_2, x_3, x_4\}$ be the set of 4 alternatives. The preferences on X are described by the PC matrix $B = \{b_{ij}\}$, where

$$B = \begin{pmatrix} 0.5 & 0.6 & 0.6 & 0.9 \\ 0.4 & 0.5 & 0.6 & 0.7 \\ 0.4 & 0.4 & 0.5 & 0.5 \\ 0.1 & 0.3 & 0.5 & 0.5 \end{pmatrix}. \tag{2.69}$$

Here, $B = \{b_{ij}\}$ is f-reciprocal and fm-inconsistent, as it may be directly verified by (2.5), e.g. $b_{12}.b_{23}.b_{31} \neq b_{32}.b_{21}.b_{13}$. At the same time, B is fa-inconsistent as $b_{12} + b_{23} + b_{31} = 1.9 \neq 1.5$. Now, we consider the scale $[1/\sigma, \sigma]$ with $\sigma = 9$. Then we calculate

$$E = \{\phi^{-1}(b_{ij})\} = \begin{pmatrix} 1 & 1.50 & 1.50 & 9.00 \\ 0.67 & 1 & 1.5 & 2.33 \\ 0.67 & 0.67 & 1 & 1 \\ 0.11 & 0.43 & 1 & 1 \end{pmatrix},$$

$$F = \{\varphi_9^{-1}(b_{ij})\} = \begin{pmatrix} 1 & 1.55 & 1.55 & 5.80 \\ 0.64 & 1 & 1.55 & 2.41 \\ 0.64 & 0.64 & 1 & 1 \\ 0.17 & 0.42 & 1 & 1 \end{pmatrix}.$$

Further, we calculate the maximal eigenvalues $\rho(E) = 4.29$ and $\rho(F) = 4.15$.

By (2.15), (2.68) we obtain $I_{fmEV}(B) = 0.11$ with the priority vector $w^{fmEV} = (0.47, 0.25, 0.18, 0.10)$, which gives the ranking of the alternatives as $x_1 \succ x_2 \succ x_3 \succ x_4$. Similarly, $I^{9}_{faEV}(B) = 0.056$ with the priority vector $w^{faEV} = (0.44, 0.27, 0.18, 0.12)$, giving the same ranking of alternatives $x_1 \succ x_2 \succ x_3 \succ x_4$.

In order to investigate a relationship between the fa-consistency and fm-consistency indexes of f-reciprocal matrices, we performed a simulation experiment with randomly generated 1000 f-reciprocal matrices, ($n = 4$ and $n = 10$), then we calculated corresponding indexes I_{faEV} and I^{σ}_{fmEV}, with $\sigma = 9$.

Example 2.7. **Case study**

In the previous example we could see that the values of $I_{faEV}(A)$ and $I_{fmEV}(A)$ are different for a f-reciprocal matrix A. In order to investigate a relationship

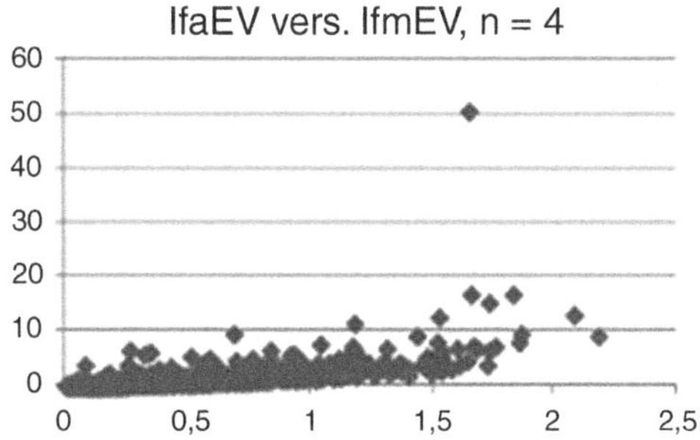

Fig. 2.2 $n = 4$

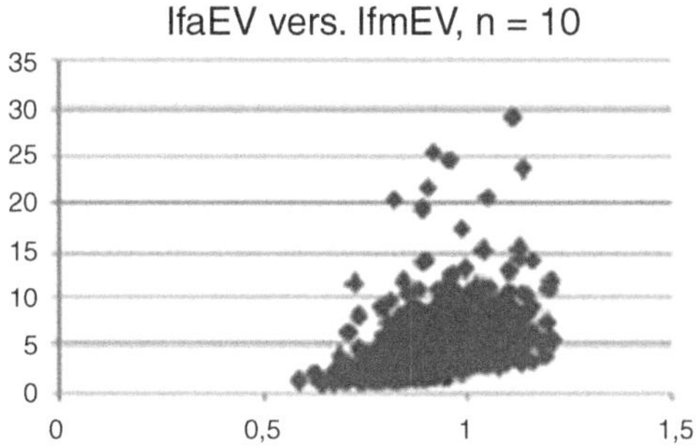

Fig. 2.3 $n = 10$

between the fa-inconsistency and fm-inconsistency of A, we randomly generated 1000 f-reciprocal matrices, (with $n = 4$ and $n = 10$), then we calculated corresponding indexes $I_{faEV}(A)$ and $I^{\sigma}_{fmEV}(A)$, with $\sigma = 9$, and depicted the points $(I_{fmEV}(A), I^{\sigma}_{faEV}(A))$ in Figs. 2.2 and 2.3, respectively. Here, each simulated matrix is represented by unique point in the plane given by two indexes.

Numerical experiments have shown that there is no particular relationship between fm-consistency and fa-consistency.

2.9 Fuzzy Pairwise Comparison Matrices with Missing Elements

In many decision-making processes we assume that experts are capable of providing preference degrees between any pair of possible alternatives. However, this may not be always true, which makes a missing information problem. A missing value in the fuzzy pairwise comparison matrix is not equivalent to a lack of preference of one alternative over another. A missing value can be the result of the incapacity of an expert to quantify the degree of preference of one alternative over another, see [48, 49, 51, 69]. In this case he/she may decide not to guess the preference degree between some pairs of alternatives. It must be clear that when an expert is not able to express a particular value b_{ij}, because he/she does not have a clear idea of how the alternative x_i is better than alternative x_j, this does not mean that he/she prefers both options with the same intensity. To model these situations, in the following we introduce the incomplete fuzzy pairwise comparison matrix, or, in other words, the fuzzy pairwise comparison matrix with missing elements, [34].

For the sake of simplicity of presentation, from now on we identify the alternatives $x_1, x_2, \ldots, x_n$ with integers $1, 2, \ldots, n$, i.e. $X = \{1, 2, \ldots, n\}$ is the set of alternatives, $n > 1$. Moreover, let, $X \times X = X^2$ be the Cartesian product of X, i.e. $X^2 = \{(i, j)|i, j \in X\}$. Let $K \subset X^2$, $K \neq X^2$ and $\mathscr{B}$ be the fuzzy preference relation on K given by the membership function $\mu_{\mathscr{B}} : K \to [0, 1]$. The fuzzy preference relation $\mathscr{B}$ is represented by the $n \times n$ *fuzzy pairwise comparison matrix (FPC matrix)* $B(K) = \{b_{ij}\}_K$ *with missing elements depending on* K as follows

$$b_{ij} = \begin{cases} \mu_{\mathscr{B}}(i, j) \text{ if } (i, j) \in K \ , \\ - \qquad\quad \text{ if } (i, j) \notin K \ . \end{cases}$$

In what follows we shall assume that each FPC matrix $B(K) = \{b_{ij}\}_K$ with missing elements is f-reciprocal with respect to K, i.e.

$$b_{ij} + b_{ji} = 1 \text{ for all } (i, j) \in K \ .$$

Let $L \subset X^2$, and $L = \{(i_1, j_1), (i_2, j_2), \ldots, (i_q, j_q)\}$ be a set of couples (i, j) of alternatives such that there exist b_{ij}, with $0 \leq b_{ij} \leq 1$ for all $(i, j) \in L$. We assume that the symmetric subset L' to L, i.e. $L' = \{(j_1, i_1), (j_2, i_2), \ldots, (j_q, i_q)\}$ also belongs to the subset of K, i.e. $L' \subset K$. Each subset K of X^2 can be represented as follows: $K = L \cup L' \cup D$, where L is the set of couples of alternatives (i, j) given by the DM/expert, L' is given by the reciprocity property and D is the diagonal of this matrix, i.e. $D = \{(1, 1), (2, 2), \ldots, (n, n)\}$, where $b_{ii} = 0.5$ for all $i \in X$. The reciprocity property means that if the expert is able to quantify b_{ij}, then he/she is able to quantify b_{ji} for some i, j. The elements b_{ij} with $(i, j) \in X^2$ - K are called *the missing elements of matrix* $B(K)$. The missing elements of $B(K)$ are denoted by symbol "-" ("dash"). On the other hand, the preference degrees given by the experts are denoted by b_{ij} where $(i, j) \in K$. By f-reciprocity it is sufficient that the expert will quantify only the elements b_{ij}, where $(i, j) \in L$, such that $K = L \cup L' \cup D$.

The problem of missing elements in a FPC matrix is well known and has been solved differently by many authors, e.g. Alonso, S. et al. [3], Herrera-Viedma, E. et al. [19, 41], Chiclana, F. et al. [42], Fedrizzi, M. and Giove, S. [28], or Ramik, J. [56].

The first approach, [3], is based on calculating the missing values of an incomplete fuzzy additive PC matrix. This calculation is done by using only the known elements of the matrix by reconstruction of the incomplete matrix being compatible with the rest of the information provided by the DM/expert. The primary purpose of this method is to maintain or maximize the global measure of fa-consistency. The drawback of this approach is that it does not allow to deal with non-fuzzy PC matrices. The other approaches to the problem of missing elements in fuzzy PC matrices are based on optimization procedures associated with calculating the priority vectors. These approaches are not limited to FPC matrices, they can be applied also to non-fuzzy PC matrices, see Sects. 2.4 and 2.6. In the next two subsections we shall study the second approach based on the optimization procedures, then, in the next subsection we shall deal with the first method.

2.9.1 Problem of Missing Elements in FPC Matrices Based on Optimization

Now, we shall deal with the problem of finding the values of missing elements of a given fuzzy pairwise comparison matrix so that the extended matrix is as much fm-consistent/fa-consistent as possible. In the ideal case the extended matrix would become fm-consistent/fa-consistent.

Let $K \subset X^2$, let $B(K) = \{b_{ij}\}_K$ be a fuzzy pairwise comparison matrix with missing elements, let $K = L \cup L' \cup D$, $L = \{(i_1, j_1), (i_2, j_2), \ldots, (i_q, j_q)\}$ be a set of couples given by the DM. We assume that

$$\{1, 2, \ldots, n\} \subset \{i_1, j_1, i_2, j_2, \ldots, i_q, j_q\} \; . \tag{2.70}$$

Remark 2.2. In the terminology of *graph theory* condition (2.70) means that K is the *covering of* $\{1, 2, \ldots, n\}$.

The matrix $B^{fm}(K) = \{b_{ij}^{fm}\}_K$ called the *fm-extension of* $B(K)$ is defined as follows

$$b_{ij}^{fm} = \begin{cases} b_{ij} & \text{if } (i,j) \in K, \\ \frac{v_i^*}{v_i^* + v_j^*} & \text{if } (i,j) \notin K \; . \end{cases} \tag{2.71}$$

Here, $v^* = (v_1^*, v_2^*, \ldots, v_n^*)$ called the *fm-priority vector with respect to* K is the optimal solution of the following problem

$$(P_{fm}) \qquad d_{fm}(v, K) = \sum_{(i,j)\in K} \left(b_{ij} - \frac{v_i}{v_i + v_j} \right)^2 \longrightarrow \min$$

subject to

$$\sum_{j=1}^{n} v_j = 1, \; v_i \geq \epsilon > 0 \text{ for all } i \in \{1, 2, \ldots, n\} \; .$$

(ϵ is a preselected sufficiently small positive number)

Notice, that fm-consistency index of the matrix $B^{fm}(K) = \{b_{ij}^{fm}\}_K$ is defined by (2.67) as $I_{fmY}(B^{fm}(K))$, where $Y \in \{EV, LS, GM\}$.

Assumption (2.70) is natural as otherwise, if e.g. $k \in \{1, 2, \dots, n\}$ does not belong to $\{i_1, j_1, i_2, j_2, \dots, i_q, j_q\}$, then v_k^* is not defined by problem (P_{fm}) and therefore the fm-extension of $B(K)$ cannot be defined by (2.71).

The proof of the following proposition follows directly from Proposition 2.2.

Proposition 2.23. *$B^{fm}(K) = \{b_{ij}^{fm}\}_K$ is fm-consistent, (i.e. $I_{fm}(B^{fm}(K)) = 0$) if and only if $d_{fm}(v^*, K) = 0$.*

Now, we would like to find the values of missing elements of a given fuzzy pairwise comparison matrix so that the extended matrix is as much fa-consistent as possible. In the ideal case the extended matrix would become fa-consistent.

Again, let $K \subset I^2$, let $B(K) = \{b_{ij}\}_K$ be a fuzzy pairwise comparison matrix with missing elements with (2.70), $K = L \cup L' \cup D$ as before. The matrix $B^{fa}(K) = \{b_{ij}^{fa}\}_K$ called an *fa-extension of $B(K)$ with respect to K* is defined as follows

$$b_{ij}^{fa} = \begin{cases} b_{ij} & \text{if } (i,j) \in K\ , \\ max\{0, min\{1, \frac{1}{2}(1 + nu_i^* - nu_j^*)\}\} & \text{if } (i,j) \notin K\ . \end{cases}$$

Here, $u^* = (u_1^*, u_2^*, \dots, u_n^*)$ called the *fa-priority vector with respect to K* is the optimal solution of the following problem

$$(P_{fa}) \qquad d_{fa}(v, K) = \sum_{(i,j)\in K} \left(b_{ij} - \frac{1}{2}(1 + nu_i - nu_j)\right)^2 \longrightarrow \min$$

subject to

$$\sum_{j=1}^{n} u_j = 1, u_i \geq \epsilon > 0 \text{ for all } i=1,2,\dots,n.$$

Now, let fa-consistency index $I_{faLS}^{\sigma}(B^{ft}(K))$ of the matrix $B^{ft}(K) = \{b_{ij}^{fa}\}_K$ be defined by (2.68) with a given $\sigma > 0$. The next proposition follows directly from Proposition 2.3.

Proposition 2.24. *If $B^{fa}(K) = \{b_{ij}^{fa}\}_K$ is fa-consistent (i.e. $I_{faLS}^{\sigma}(B^{fa}(K)) = 0$), then $d_{fa}(u^*, K) = 0$.*

2.9.2 Particular Cases of FPC Matrices with Missing Elements Based on Optimization

For a complete definition of an $n \times n$ reciprocal fuzzy pairwise comparison matrix we need $N = n(n-1)/2$ pairs of elements to be evaluated by an expert. For example, if $n = 10$, then $N = 45$, which is a considerable amount of pairwise comparisons. We ask that the expert would evaluate only 'around n' pairwise comparisons of alternatives which seems to be a reasonable amount. In this section we shall deal with two important particular cases of fuzzy preference matrix with missing elements where the expert will evaluate only $n-1$ pairwise comparisons of alternatives so that condition (2.70) is satisfied. Let $K \subset I^2$ be a set of indexes given by an expert, $B(K) = \{b_{ij}\}_K$ be a fuzzy preference matrix with missing elements. Moreover, let $K = L \cup L' \cup D$. In fact, it is sufficient to assume that the expert will evaluate for instance the matrix elements of L: $b_{12}, b_{23}, b_{34}, \ldots, b_{n-1,n}$.

2.9.3 Case $L = \{(1,2), (2,3), \ldots, (n-1, n)\}$

Here, we assume that the expert evaluates $n-1$ elements of the fuzzy preference matrix $B(K)$, $b_{12}, b_{23}, b_{34}, \ldots, b_{n-1,n}$. First, we investigate the fm-extension of $B^{fm}(K)$. We derive the following result.

Proposition 2.25. *Let $L = \{(1,2), (2,3), \ldots, (n-1, n)\}$, $0 < b_{ij} < 1$ with $b_{ij}+b_{ji} = 1$ for all $(i,j) \in K$, $K = L \cup L' \cup D$, and $L' = \{(2,1), (3,2), \ldots, (n, n-1)\}$, $D = \{(1,1), \ldots, (n,n)\}$. Then fm-priority vector $v^* = (v_1^*, v_2^*, \ldots, v_n^*)$ with respect to K is given as*

$$v_1^* = \frac{1}{S} \text{ and } v_{i+1}^* = a_{i,i+1} v_i^* \text{ for } i = 1, 2, \ldots, n-1, \tag{2.72}$$

where

$$S = 1 + \sum_{i=1}^{n-1} a_{i,i+1} a_{i+1,i+2} \ldots a_{n-1,n} \text{ and } a_{ij} = \frac{1 - b_{ij}}{b_{ij}} \text{ for all } (i,j) \in K \ . \tag{2.73}$$

Proof. If (2.72) and (2.73) is satisfied, then

$$v_{i+1}^* = a_{i,i+1} a_{i-1,i} \ldots a_{1,2} v_1^* \text{ for } i \in \{1, 2, \ldots, n-1\} \ ,$$

hence

$$\sum_{j=1}^{n} v_j^* = 1, v_j^* > 0 \text{ for all } j \in \{1, 2, \ldots, n\} \ ,$$

and also

$$b_{i,i+1} = \frac{v_i^*}{v_i^* + v_{i+1}^*} \text{ for } i \in \{1, 2, \ldots, n-1\} \ .$$

Then $v = (v_1^*, \ldots, v_1^*)$ is an optimal solution of (P_{fm}). □

By (2.67) it follows that $B^{fm}(K) = \{b_{ij}^{fm}\}_K$ is fm-consistent.

Now, we investigate the fa-extension $B^{fa}(K)$ of $B(K)$. We obtain the following result.

Proposition 2.26. *Let* $L = \{(1,2),(2,3),\ldots,(n-1,n)\}$, $0 < b_{ij} < 1$ *with* $b_{ij} + b_{ji} = 1$ *for all* $(i,j) \in K$, $K = L \cup L' \cup D$, *and* $L' = \{(2,1),(3,2),\ldots,(n,n-1)\}$, $D = \{(1,1),\ldots,(n,n)\}$. *Let* $u^* = (u_1^*, u_2^*, \ldots, u_n^*)$ *be defined as*

$$u_i^* = \frac{2}{n^2}\sum_{j=1}^{n-1}\alpha_j - \frac{2}{n}\alpha_{i-1} - \frac{n-i-1}{n} \text{ for } i \in \{1,2,\ldots,n\} \ , \tag{2.74}$$

where

$$\alpha_0 = 0, \alpha_j = \sum_{i=1}^{j} b_{i,i+1} \text{ for } j \in \{1,2,\ldots,n\} \ . \tag{2.75}$$

If $u^* = (u_1^*, \ldots, u_n^*)$ *is a vector with positive elements, then* u^* *is an fa-priority vector with respect to* K.

Proof. If (2.74) and (2.75) is satisfied, then

$$\sum_{j=1}^{n} u_j^* = nu_1^* + \frac{2}{n}\sum_{j=2}^{n}\alpha_{j-1} + n - 1 = 1,$$

and

$$u_j^* > 0 \text{ for all } j \in \{1,2,\ldots,n\} \ .$$

We get also

$$\frac{1}{2}(1 + nu_i^* - nu_{i+1}^*) = b_{i,i+1} \text{ for } i \in \{1,2,\ldots,n-1\} \ .$$

Then $u^* = (u_1^*, \ldots, u_n^*)$ is an optimal solution of (P_{fa}). □

Remark 2.3. In general, the optimal solution $u^* = (u_1^*, u_2^*, \ldots, u_n^*)$ of (P_{fa}) does not satisfy condition

$$0 \le \frac{1}{2}(1 + nu_i^* - nu_j^*) \le 1, \text{ for all } i,j \in \{1,2,\ldots,n\} \ , \tag{2.76}$$

i.e. $B = \{b_{ij}\} = \{\frac{1}{2}(1 + nu_i^* - nu_j^*)\}$ is not a fuzzy preference matrix. We can easily prove the necessary and sufficient condition for satisfying (2.76) based on evaluations $b_{i,i+1}$.

Proposition 2.27. *Let* $L = \{(1,2),(2,3),\ldots,(n-1,n)\}$, $0 \le b_{ij} \le 1$ *with* $b_{ij} + b_{ji} = 1$ *for all* $(i,j) \in K$, $K = L \cup L' \cup D$, *and* $L' = \{(2,1),(3,2),\ldots,$

$(n, n-1)\}$, $D = \{(1,1), \ldots, (n,n)\}$. *Then the fa-extension* $B^{at}(K) = \{b_{ij}^{at}\}_K$ *is fa-consistent if and only if*

$$\left| \sum_{k=i}^{j-1} b_{k,k+1} - \frac{j-i}{2} \right| \leq \frac{1}{2} \text{ for all } i, j, k \in \{1, 2, \ldots, n\} . \tag{2.77}$$

Proof. By Proposition 2.3 and Proposition 2.26 it is sufficient to show that $B^{fa}(K) = \{b_{ij}^{fa}\}_K$ is fa-consistent if and only if

$$0 \leq \frac{1}{2}(1 + nu_i^* - nu_{i+1}^*) \leq 1, \text{ for all } i \in \{1, 2, \ldots, n-1\} . \tag{2.78}$$

where $u^* = (u_1^*, \ldots, u_n^*)$ is defined by (2.74) and (2.75).

By (2.74) and (2.75) we get

$$\frac{1}{2}(1 + nu_i^* - nu_j^*) = \alpha_{j-1} - \alpha_{i-1} + \frac{i-j+1}{2} = \sum_{k=i}^{j-1} b_{k,k+1} - \frac{j-i-1}{2} . \tag{2.79}$$

Hence, (2.78) is equivalent to

$$0 \leq \sum_{k=i}^{j-1} b_{k,k+1} - \frac{j-i-1}{2} \leq 1 \text{ for } i \in \{1, 2, \ldots, n-1\}, j \in \{i+1, \ldots, n\} ,$$

which, after some rearrangement, is the same as in (2.77). □

Example 2.8. Let $L = \{(1,2), (2,3), (3,4)\}$, the expert evaluations be $b_{12} = 0.9, b_{23} = 0.8, b_{34} = 0.6$, with $b_{ij} + b_{ji} = 1$ for all $(i,j) \in L$, let $K = L \cup L' \cup D$. Hence $B(K) = \{b_{ij}\}_K$ is a fuzzy preference matrix with missing elements as follows

$$B(K) = \begin{pmatrix} 0.5 & 0.9 & - & - \\ 0.1 & 0.5 & 0.8 & - \\ - & 0.2 & 0.5 & 0.5 \\ - & - & 0.4 & 0.5 \end{pmatrix} . \tag{2.80}$$

Solving (P_{fm}) we obtain ac-priority vector v^* with respect to K, particularly, $v^* = (0.864, 0.096, 0.024, 0.016)$. By (2.74) we obtain $B^{fm}(K)$ - fm-extension of $B(K)$ as follows

$$B^{fm}(K) = \begin{pmatrix} 0.5 & 0.9 & 0.97 & 0.98 \\ 0.1 & 0.5 & 0.8 & 0.86 \\ 0.03 & 0.2 & 0.5 & 0.6 \\ 0.02 & 0.14 & 0.4 & 0.5 \end{pmatrix} , \tag{2.81}$$

where, $B^{fm}(K)$ is fm-consistent, as $d_{fm}(v, B(K)) = 0$, hence $I_{fmLS}(B^{fm}(K)) = 0$.

Solving (P_{fa}) we obtain fa-priority vector u^* with respect to K,

$u^* = (0.487, 0.287, 0.137, 0.088)$. Moreover, $B^{fa}(K)$ is an fa-extension of $B(K)$ as

$$B^{fa}(K) = \begin{pmatrix} 0.5 & 0.9 & 1.0 & 1.0 \\ 0.1 & 0.5 & 0.8 & 0.9 \\ 0.0 & 0.2 & 0.5 & 0.6 \\ 0.0 & 0.1 & 0.4 & 0.5 \end{pmatrix}, \tag{2.82}$$

where, $B^{fa}(K)$ is not fa-consistent, as $d_{fa}(v, B(K)) > 0$. This fact can be easily verified as $I^9_{faLS}(B^{fa}(K)) = 0.057$.

2.9.4 Case $L = \{(1,2), (1,3), \ldots, (1,n)\}$

Now, we assume that the expert evaluates the pairs of a fixed element $x \in X$ with the remaining $n - 1$ elements, i.e. the fuzzy preference matrix $B(K)$ is given by $b_{12}, b_{13}, \ldots, b_{1n}$. We investigate the fm-extension of $B(K)$ and obtain the following result.

Proposition 2.28. *Let* $L = \{(1,2), (1,3), \ldots, (1,n)\}$, $0 < b_{ij} < 1$ *with* $b_{ij} + b_{ji} = 1$ *for all* $(i, j) \in K$, $K = L \cup L' \cup D$, *and* $L' = \{(2,1), (3,2), \ldots, (n, n-1)\}$, $D = \{(1,1), \ldots, (n,n)\}$. *Then fm-priority vector* $v^* = (v_1^*, v_2^*, \ldots, v_n^*)$ *with respect to K is given as*

$$v_1^* = \frac{1}{V} \text{ and } v_{i+1}^* = a_{1,i+1} v_i^* \text{ for } i = 1, 2, \ldots, n-1 \ , \tag{2.83}$$

where

$$V = 1 + \sum_{i=1}^{n-1} a_{1,i+1} \text{ and } a_{ij} = \frac{1 - b_{ij}}{b_{ij}} \text{ for all } (i, j) \in K \ . \tag{2.84}$$

Proof. If (2.82) and (2.84) is satisfied, then

$$v_{i+1}^* = \frac{1 - b_{1,i+1}}{b_{1,i+1}} v_1^* \text{ for } i \in \{1, 2, \ldots, n-1\} \ ,$$

hence

$$b_{1,i+1} = \frac{v_i^*}{v_i^* + v_{i+1}^*} \text{ for } i \in \{1, 2, \ldots, n-1\} \ .$$

and also

$$\sum_{j=1}^{n} v_j^* = v_1^* + a_{1,2}v_1^* + \ldots + a_{1,n}v_1^* = 1, v_j^* > 0 \text{ for all } j \in \{1,2,\ldots,n\} .$$

Consequently, $v = (v_1^*, \ldots, v_n^*)$ is an optimal solution of (P_{fm}). □

We conclude that the fm-extension of $B(K)$, i.e. matrix $B^{fm}(K) = \{b_{ij}^{fm}\}_K$ is fm-consistent.

Now, we investigate the fa-extension matrix $B^{fa}(K)$ of $B(K)$. We derive the following result.

Proposition 2.29. *Let* $L = \{(1,2),(1,3),\ldots,(1,n)\}$, $0 < b_{ij} < 1$ *with* $b_{ij}+b_{ji} = 1$ *for all* $(i,j) \in K$, $K = L \cup L' \cup D$, *and* $L' = \{(2,1),(3,2),\ldots,(n,n-1)\}$, $D = \{(1,1),\ldots,(n,n)\}$. *Let* $u^* = (u_1^*, u_2^*, \ldots, u_n^*)$ *be defined as follows*

$$u_1^* = \frac{2}{n^2}\sum_{j=1}^{n-1} b_{1,j+1} + \frac{1}{n^2} \text{ and } u_{i+1}^* = u_1^* + \frac{1-2b_{1,i+1}}{n} \text{ for } i = 1,2,\ldots,n-1 . \tag{2.85}$$

If $u^* = (u_1^*, \ldots, u_n^*)$ *is a vector with positive elements, then* u^* *is an fa-priority vector with respect to* K.

Proof. If (2.85) is satisfied, then

$$\sum_{j=1}^{n} u_j^* = nu_1^* + \frac{2}{n}\sum_{j=2}^{n} b_{1,i+1} + \frac{n-1}{n} + \frac{1}{n} = 1 ,$$

and

$$u_j^* > 0 \text{ for all } j \in \{1,2,\ldots,n\} .$$

We get also

$$\frac{1}{2}(1 + nu_i^* - nu_{i+1}^*) = 0.5 + b_{i,1} - b_{i+1,1} = b_{i,i+1} \text{ for } i \in \{1,2,\ldots,n-1\} .$$

Hence, $u^* = (u_1^*, \ldots, u_n^*)$ is an optimal solution of (P_{fa}). □

Remark 2.4. In general, the optimal solution $u^* = (u_1^*, u_2^*, \ldots, u_n^*)$ of (P_{fa}) does not satisfy condition (2.76), i.e. $B = \{b_{ij}^*\} = \{\frac{1}{2}(1 + nu_i^* - nu_j^*)\}$ is not a fuzzy PC matrix. By a similar way we can prove the result parallel to Proposition 2.27.

Proposition 2.30. *Let* $L = \{(1,2),(2,3),\ldots,(n-1,n)\}$, $0 \le b_{ij} \le 1$ *with* $b_{ij}+b_{ji} = 1$ *for all* $(i,j) \in K$, $K = L \cup L' \cup D$, *and* $L' = \{(2,1),(3,2),\ldots,(n,n-1)\}$, $D = \{(1,1),\ldots,(n,n)\}$. *Then the fa-extension* $B^{fa}(K) = \{b_{ij}^{fa}\}_K$ *is* *fa-consistent if and only if*

$$\left|b_{1j} - b_{1i}\right| \leq \frac{1}{2} \textit{ for all } i, j \in \{1, 2, \ldots, n\} . \tag{2.86}$$

Proof. By Proposition 2.3 and Proposition 2.26 it is sufficient to show that $B^{fa}(K) = \{b_{ij}^{fa}\}_K$ is fa-consistent if and only if

$$0 \leq \frac{1}{2}(1 + nu_i^* - nu_{i+1}^*) \leq 1, \text{ for all } i \in \{1, 2, \ldots, n-1\} . \tag{2.87}$$

where $u^* = (u_1^*, \ldots, u_n^*)$ is defined by (2.85).

However, by (2.85) we obtain

$$\frac{1}{2}(1 + nu_i^* - nu_j^*) = 0.5 + b_{i1} - b_{j1} \text{ for } i, j \in \{1, 2, \ldots, n\} . \tag{2.88}$$

Hence, (2.87) is equivalent to

$$0 \leq \frac{1}{2}(1 + nu_i^* - nu_j^*) = 0.5 + b_{i1} - b_{j1} \leq 1 \text{ for } i, j \in \{1, 2, \ldots, n\} ,$$

which, after some rearrangements, is equal to (2.86). □

Example 2.9. Let $L = \{(1,2), (1,3), (1,4)\}$, the expert evaluations be $b_{12} = 0.9, b_{13} = 0.8, b_{14} = 0.3$, with $b_{ij} + b_{ji} = 1$ for all $(i, j) \in L$, let $K = L \cup L' \cup D$. Hence $B(K) = \{b_{ij}\}_K$ is a fuzzy preference matrix with missing elements as follows

$$B(K) = \begin{pmatrix} 0.5 & 0.9 & 0.8 & 0.3 \\ 0.1 & 0.5 & - & - \\ 0.2 & - & 0.5 & - \\ 0.7 & - & - & 0.5 \end{pmatrix} . \tag{2.89}$$

Solving (P_{fm}) we obtain fm-priority vector v^* with respect to K, particularly, $v^* = (0.271, 0.030, 0.068, 0.632)$. Then we obtain $B^{fm}(K)$ – fm-extension of $B(K)$ as

$$B^{fm}(K) = \begin{pmatrix} 0.5 & 0.9 & 0.80 & 0.30 \\ 0.10 & 0.5 & 0.30 & 0.04 \\ 0.20 & 0.70 & 0.5 & 0.10 \\ 0.70 & 0.96 & 0.90 & 0.5 \end{pmatrix} , \tag{2.90}$$

where, $B^{fm}(K)$ is fm-consistent, as $d_{fm}(v, B(K)) = 0$, hence $I_{fmLS}(B^{fm}(K)) = 0$. Moreover, solving (P_{fa}) we obtain fa-priority vector u^* with respect to K, $u^* = (0.312, 0.113, 0.162, 0.412)$. Then $B^{fa}(K)$ is an fa-extension of $B(K)$ as

$$B^{fa}(K) = \begin{pmatrix} 0.5 & 0.90 & 0.80 & 0.30 \\ 0.10 & 0.5 & 0.40 & 0.00 \\ 0.20 & 1.00 & 0.5 & 0.00 \\ 0.70 & 1.00 & 1.00 & 0.5 \end{pmatrix}, \tag{2.91}$$

where $B^{fa}(K)$ is not fa-consistent, as $|b_{12} - b_{14}| > 0.6 > \frac{1}{2}$.

2.9.5 *Problem of Missing Elements in FPC Matrices: Other Methods*

The problem of missing elements in a FPC matrix has been solved by many authors, e.g. Alonso, S. et al. [3], Herrera-Viedma, E. et al. [19,41], Chiclana, F. et al. [42], or Fedrizzi, M. and Giove, S. [28].

The first approach, see [3, 19, 41, 42], is based on calculating the missing values of an incomplete fuzzy additive PC matrix. This calculation is done by using only the known elements of the matrix, therefore by assuring that the reconstruction of the incomplete matrix is compatible with the rest of the information provided by the DM/expert. The primary purpose of this method is to maintain or maximize the global fa-consistency. The drawback is that it does not allow to deal with fuzzy multiplicative PC matrices and non-fuzzy PC matrices.

Let us now assume that one or more comparisons are missing. As a consequence, the PC matrix is incomplete and it is no longer possible to derive the priorities for the alternatives using the well known methods of the EV, or GM, to mention only the most popular ones. Some methods have been proposed in the literature to solve the incomplete comparison problem. Most of these methods are formulated in the multiplicative framework [16, 36, 37, 54, 63], some other in the additive framework [70, 73]. Let us very briefly describe the most important ideas presented in the above mentioned literature.

Two methods have been proposed by P.T. Harker. The first one [36], based on GM method, is based on the concept of 'connecting path'. If alternatives x_i and x_j are not compared with each other, let us denote by $(x_i; x_j)$ the *missing comparison* (MC) and let x_{ij} be the corresponding numerical value to be estimated; a connecting path of size r has the following form

$$x_{ij} = a_{ik_1}.a_{k_1k_2}\dots a_{k_rj} \ , \tag{2.92}$$

where the comparison values at the right hand side of (2.92) are known. The connecting path of size two, also called elementary connecting path, corresponds to the more familiar expression

$$x_{ij} = a_{ik}a_{kj} \ . \tag{2.93}$$

Note that each connecting path corresponds to an indirect comparison between x_i and x_j. Harker proposes that the value x_{ij} of the MC should be equal to the geometric mean of all connecting paths related to this MC. The drawback of this method is that the number of connecting paths grows with the number n of the alternatives in such a way that the method becomes computationally intractable for many real world problems.

The second Harker's method [37] is based on the following idea. The missing element $(x_i; x_j)$ is set to be equal to w_i/w_j, where the components of the vector w are not known and are to be calculated. In other words, the missing entries are completed by setting them equal to the value they should approximate. The matrix obtained with the described substitution is denoted by C. An auxiliary nonnegative matrix A is then associated to C satisfying $Aw = Cw$. The matrix A is nonnegative and *quasi reciprocal*, in the sense that all its positive entries are reciprocal, but it contains entries equal to zero. In this way, Harker transforms the original problem in that of computing the principal eigenvector of a nonnegative quasi reciprocal matrix. In order to justify his method, Harker develops a theory for such type of matrices, following the Saaty's one for positive reciprocal matrices.

Shiraishi et al. in [63] propose a heuristic method which is based on a property of a coefficient of the characteristic polynomial of a PC matrix A. More precisely, the coefficient c_3 of λ^{n-3} is viewed as an index of consistency for A. Therefore, in order to maximize the consistency of the PC matrix, the authors consider the missing entries in the PC matrix as variables $z_1, \ldots, z_m$ and propose to maximize $c_3(z_1, \ldots, z_m)$ as a function of these variables. In [16] a Least Squares type method is proposed. Instead of first calculating the missing entries of a PC matrix, the priority vector w is directly calculated as the solution of a constrained optimization problem. Here the variables are the n components w_i of w and only the known entries a_{ij} are approximated by w_i/w_j. The corresponding error is minimized as a function of $w_1, \ldots, w_n$. In [70], Xu proposes to calculate the priority vector w of incomplete fuzzy preference matrix by a goal programming approach. This method minimizes the errors

$$\epsilon_{ij} = |a_{ij} - 0.5 + 0.5(u_i - u_j)| \tag{2.94}$$

for all missing elements $(x_i; x_j)$. He also proposes his goal programming approach with another type of consistency.

In his second proposal, Xu in [73] develops a method, for incomplete fuzzy preference relations, similar to that introduced by Harker [36] for incomplete multiplicative PC matrix. In [73] the priority vector w is calculated by solving a system of equations which corresponds to the Harker's auxiliary eigenproblem.

In [28], Fedrizzi and Giove proposed a method which is based on the resolution of an optimization problem with an objective function measuring the additive consistency of the incomplete fuzzy preference matrix. Indeed, for each triplet of alternatives (x_i, x_j, x_k), Fedrizzi and Giove define its associated inconsistency contribution as

$$L_{ijk} = (a_{ik} + a_{kj} - a_{ij} - 0.5)^2 \ . \tag{2.95}$$

It is worth noting that the error between a value a_{ij} and its local estimated one obtained by using the intermediate alternative x_k, denoted as ep_k^{ij}, is the square root of L_{ijk}. The *global inconsistency index of a fuzzy PC matrix A* is defined as follows:

$$\rho = 6 \sum_{i<j<k} L_{ijk} \ . \tag{2.96}$$

The missing values in an incomplete fuzzy PC matrix are treated as variables in the global consistency index. The stationary vector that minimizes the global inconsistency function is taken as the estimated values for the unknown preference values. Obviously, these estimated values are the most consistent with the available preference values.

Under reciprocity, if a value a_{ij} is missing, then the value a_{ji} is also missing. Therefore, it makes sense in this context to denote these two missing preference values as the missing comparison $(x_i; x_j)$. When a single comparison $(x_i; x_j)$ is missing, Fedrizzi–Giove's method produces the following linear equation:

$$(n-2)a_{ij} - \sum_{k=1,k\neq j}^{n} a_{ki} - \sum_{k=1,k\neq i}^{n} a_{kj} - \frac{n}{2} = 0 \ . \tag{2.97}$$

Example 2.10. Let us assume the following incomplete PC matrix, (see [28])

$$A = \begin{pmatrix} 0.5 & 0.2 & 0.6 & 0.4 \\ 0.8 & 0.5 & x & y \\ 0.4 & 1-x & 0.5 & z \\ 0.6 & 1-y & 1-z & 0.5 \end{pmatrix}, \tag{2.98}$$

where $a_{23} = x$, $a_{24} = y$, and $a_{34} = z$. The global inconsistency index of A is

$$\rho = 6(0.9 - x)^2 + (0.7 - y)^2 + (0.3 - z)^2 + (0.5 - x + y - z)^2 \ . \tag{2.99}$$

The optimal solution corresponds to $x = 0.9, y = 0.7, z = 0.3$. These values coincide with the estimated values obtained by Herrera–Viedma in [19].

To establish the condition under which this method can guarantee the successful reconstruction of an incomplete fuzzy preference relation, the authors introduce the concept of an independent/dependent set of missing comparisons.

(1) A set of missing comparisons is called *independent* when no alternative is shared between any two of their missing comparisons.
(2) A set of missing comparisons is called *dependent* when for every partition of it into two subsets, there exists at least one alternative belongs to the both subsets.

Each set of missing comparisons can be expressed as a disjoint union of independent and/or dependent sets of missing comparisons.

The maximum cardinality of an independent set of missing comparisons is calculated for an even or odd number of alternatives, respectively. Fedrizzi and Giove show that the optimal values of an independent set of missing comparisons exist and are computed by solving each one of the corresponding linear equations independently of the rest. However, in the case of a dependent set of missing comparisons, the optimal values exist and are unique if its cardinality is lower than $n-1$. When this is not the case, i.e. if the cardinality of a set of missing comparisons is greater than or equal to $n-1$, Fedrizzi–Giove's method does not guarantee the existence nor the uniqueness of estimated values for the missing comparisons.

2.10 Unified Framework for Pairwise Comparison Matrices over ALO Groups

2.10.1 Unified Framework

As it was mentioned before, a pairwise comparison matrix $A = \{a_{ij}\}$ is a helpful tool to determine the ranking on a set $X = \{x_1, x_2, \ldots, x_n\}$ of alternatives (or, criteria). The entry a_{ij} of the matrix assumes three different meanings: a_{ij} can be a preference ratio (multiplicative case), or a preference difference (additive case), or a_{ij} belongs to the unit interval [0, 1] and measures the distance from the indifference that is expressed by 0.5 (fuzzy case).

In this section we consider pairwise comparison matrices over an*abelian linearly (totally) ordered group* and, in this way, we provide a general framework for all the above mentioned cases. By introducing this more general setting, we provide a consistency measure that has a natural meaning, it corresponds to the consistency indices presented in the previous sections and it is easy to compute it in the additive, multiplicative and fuzzy cases. In some sense we unify major part of the theory presented in the previous sections. The matter of this section is based on the textbook by N. Bourbaki [9] and works of B. Cavallo and L. D'Apuzzo [11–14]. Some elements of abelian linearly (totally) ordered groups are summarized in Chap. 1, Sect. 1.12.

2.10.2 Continuous Alo-Groups over a Real Interval

Let G be a subset of the real line $\mathbf{R}$, $\leq$ be the total order on G inherited from the usual order on $\mathbf{R}$ and $\mathcal{G} = (G, \odot, \leq)$ be a real alo-group. Let G be a proper interval of $\mathbf{R}$. By (1.50) and (1.51), G is an open interval.

By $\mathbf{Q}$ we denote the set of the rational numbers, $\mathbf{Q}^+$ is the set of all positive rational numbers, operation $+$ is the usual addition and $\cdot$ is the usual multiplication on $\mathbf{R}$. We obtain the following examples of real alo-groups, see [11, 13].

Example 2.11. Let $\mathscr{R} = (\mathbf{R}, +, \leq)$ and $\mathscr{Q} = (\mathbf{Q}, +, \leq)$ be two alo-groups. Both alo-groups are continuous with: $e = 0$, $a^{(-1)} = -a$, $a^{(n)} = na$, $a \div b = a - b$; the norm $\|a\| = |a| = a \vee (-a)$ generates the usual distance over $\mathbf{R}$ (resp. $\mathbf{Q}$): $|a-b| = (a-b) \vee (b-a)$. Both $\mathbf{R}$ and $\mathbf{Q}$ are divisible, the (n)-root of $x^{(n)} = a$ is the solution of $nx = a$, usually indicated by the symbol a/n. The mean $m_{\odot}(a_1, a_2, \ldots, a_n)$ is the usual arithmetic average $\frac{1}{n}\sum a_i$.

Example 2.12. $\mathscr{R}^+ = (]0, +\infty[, \cdot, \leq)$ and $\mathscr{Q}^+ = (\mathbf{Q}^+, \cdot, \leq)$ are two continuous alo-groups with: $e = 1, x^{(-1)} = x^{-1} = 1/x$, $x^{(n)} = x^n$, $x \div y = \frac{x}{y}$, $\|a\| = a \vee a^{-1}$; and both $d_{\mathscr{R}^+}(a,b)$ and $d_{\mathscr{Q}^+}(a,b)$ are given by $\|a \div b\| = \|\frac{a}{b}\| = \frac{a}{b} \vee \frac{b}{a} \in [1, +\infty[$. The alo-group $\mathscr{R}^+$ is divisible and the (n)-root of a is $x = a^{1/n}$. The mean $m_{\cdot}(a_1, a_2, \ldots, a_n)$ is the geometric mean $\left(\prod_{i=1}^{n} a_i\right)^{1/n}$. The alo-group $\mathscr{Q}^+$ is not divisible as e.g. $x^2 = 2$ has no solution in Q^+.

Example 2.13. $]\mathbf{0}, \mathbf{1}[_{\mathbf{m}}$=$(]0, 1[, \cdot_f, \leq)$ is *fuzzy-multiplicative alo-group* with:

$$a \cdot_f b = \frac{ab}{ab + (1-a)(1-b)}, \ e = 0,5, \ a^{(-1)} = 1 - a \ ,$$

$$\|a\| = max\{a, 1-a\} \ .$$

Here,

$$\|a \div_f b\| = max\{\frac{a(1-b)}{a(1-b) + (1-a)b}, \frac{(1-a)b}{a(1-b) + (1-a)b}\} \in]0, 1[\ . \tag{2.100}$$

We obtain also the mean

$$m_{\cdot_f}(a_i; i \in \{1, \ldots, n\}) = \frac{(\prod_{i=1}^{n} a_i)^{1/n}}{(\prod_{i=1}^{n} a_i)^{1/n} + (\prod_{i=1}^{n}(1-a_i))^{1/n}} \ . \tag{2.101}$$

Fuzzy multiplicative alo-group $]\mathbf{0}, \mathbf{1}[_{\mathbf{m}}$ is divisible and continuous. For more details and properties, see [14].

Example 2.14. Fuzzy additive alo-group $\mathscr{R}_a = (]-\infty, +\infty[, +_f, \leq)$ is a continuous alo-group with:

$$a +_f b = a + b - 0.5, \ e = 0.5, \ a^{(-1)} = 1 - a,$$

$$a^{(n)} = n.a - \frac{n-1}{2}, \ a -_f b = a - b + 0.5, \ \|a\| = max\{a, 1-a\} \ ,$$

and

$$d_{\mathscr{R}_a}(a,b) = \|a -_f b\| = max\{a - b + 0.5, b - a - 0.5\} \ .$$

The fuzzy additive alo-group $\mathscr{R}_a$ is divisible and the (n)-root of a is $x = a/n$. The mean $m_{+_f}(a_1, a_2, \ldots, a_n) = \frac{1}{n}\sum_{i=1}^{n} a_i$. The entries of a PC matrix $A = \{a_{ij}\}$ are taken from the unit interval [0, 1] with a fuzzy interpretation. A is *fa-consistent* if

$$a_{ik} = a_{ij} +_f a_{jk} = a_{ij} + a_{jk} - 0.5 \text{ for all } i, j, k \in \{1, 2, \ldots, n\} \ .$$

The following result by Aczel [1] will be helpful. It shows that if G is a proper open interval of **R** and $\leq$ is the total order on G inherited from the usual order on **R**, then a continuous real alo-group can be built based on the real alo-group $\mathscr{R}$, or the real alo-group $\mathscr{R}^+$. For the proof of the following theorem, see [1].

Theorem 2.4. *Let G be a proper open interval of* **R** *and $\leq$ the total order on G inherited from the usual order on* **R**, *let $\odot$ be a binary operation on G. Then $\odot$ is a continuous, associative and cancellative operation if and only if there exists a continuous and strictly monotonic function $\phi : J \to G$ such that or each $x, y \in J$:*

$$x \odot y = \phi(\phi^{-1}(x) + \phi^{-1}(y)) \ . \tag{2.102}$$

and J is **R** *or one of the real intervals $]-\infty, \gamma[$, $]-\infty, \gamma]$, $]\delta, +\infty[$, $[\delta, +\infty[$, where γ and δ are suitable constants. The function ϕ in (2.102) is unique up to a linear transformation of the variable x.*

Notice that by (2.102) $\odot$ is commutative and strictly increasing in both variables. By Theorem 2.4 we obtain the following consequence, see [11].

Theorem 2.5. *Let G be a proper open interval of* **R** *and $\leq$ the total order on G inherited from the usual order on* **R**. *The following assertions are equivalent:*

(i) *$\mathscr{G} = (G, \odot, \leq)$ is a continuous alo-group;*
(ii) *there exists a continuous and strictly increasing function $\phi : R \to G$ verifying the equality in (2.102);*
(iii) *there exists a continuous and strictly increasing function $\psi :]0, +\infty[\to G$ verifying the equality*

$$x \odot y = \psi(\psi^{-1}(x) \cdot \psi^{-1}(y)) \text{ for each } x, y \in]0, +\infty[\ . \tag{2.103}$$

Proof. First, we prove the equivalence of *(i)* and *(ii)*. By Theorem 2.4 and (1.47), $\mathscr{G}$ is a continuous alo-group if and only if there exists a continuous and strictly monotonic function $\phi : J \to G$ defined on a proper interval J of **R** and verifying (2.102). This function can be chosen strictly increasing as it is unique up to the linear transformation of the variable x. In order to prove the required equivalence it is enough to prove that J, the domain of ϕ, coincides with **R**. By (2.102) observe that for $x \in J$:

$$x = x \odot e \Leftrightarrow \phi^{-1}(x) + \phi^{-1}(e) = 0 \ ,$$

hence $0 \in J$ and

$$x \odot x^{(-1)} = e \Leftrightarrow \phi^{-1}(x) + \phi^{-1}(x^{(-1)}) = \phi^{-1}(e) = 0 \Leftrightarrow \phi^{-1}(x^{(-1)}) = -\phi^{-1}(x) \ .$$

Therefore, if $a = \phi^{-1}(x) \in J$ then $-a = \phi^{-1}(x) \in J$. The rest follows by Theorem 2.4.

Second, we prove the equivalence of *(ii)* and *(iii)*. Let *(ii)* be true. Setting ψ as a composition of ϕ and logarithmic function log, i.e. $\psi(x) = \phi(\log x)$ for $x \in]0, +\infty[$, we obtain

$$\psi^{-1}(y) = \exp(\phi^{-1}(y)) \text{ and } \psi(\psi^{-1}(x)\psi^{-1}(y)) =$$

$$= \phi(\log(\exp(\phi^{-1}(x))\exp(\phi^{-1}(y)))) = x \odot y \ .$$

Therefore, *(iii)* is true. The reverse implication can be proven by an analogous way. □

Applying Theorem 2.5, we provide, in the following propositions, two examples of continuous real alo-groups over a limited interval of **R**.

Proposition 2.31. *Let* $+_f$ *be the binary operation defined for all* $x, y \in]-\infty, +\infty[$ *by*

$$x +_f y = x + y - 0.5 \textit{ for all } x, y \in] - \infty, +\infty[\ . \tag{2.104}$$

and $\leq$ *be the order inherited by the usual order in* **R**. *Then* $\mathbf{]0, 1[_a} = (] - \infty, +\infty[, +_f, \leq\ )$ *is a continuous alo-group with* $e = 0.5$, $x^{(-1)} = 1 - x$ *for each* $x \in] - \infty, +\infty[$.

Proof. The function $g :]0, +\infty[\to] - \infty, +\infty[$ defined by

$$g(t) = 0.5(1 + \log t) \text{ for each } t \in]0, +\infty[\ , \tag{2.105}$$

is a bijection between $]0, +\infty[$ and $] - \infty, +\infty[$ being continuous and strictly increasing. For $a, b \in]0, +\infty[$ and $x = g(a)$, $y = g(b)$, applying (2.104) we get

$$\begin{aligned} g(a) +_f g(b) &= 0.5(1 + \log a) + 0.5(1 + \log b) - 0.5 \\ &= 0.5(1 + \log a.b) = g(a \cdot b) \ . \end{aligned} \tag{2.106}$$

Hence, $x +_f y = g(g^{-1}(x) \cdot g^{-1}(y))$, and (2.103) is verified with $\psi = g$. Finally, it is easy to verify that $x +_f 0.5 = x$ and $x +_f (1 - x) = 0.5$. □

Proposition 2.32. *Let* $\cdot_f :]0,1[^2 \to]0,1[$ *be the operation defined for all* $x, y \in]0,1[$ *by*

$$x \cdot_f y = \frac{xy}{xy + (1-x)(1-y)} \tag{2.107}$$

and $\leq$ *be the order inherited by the usual order in* $\mathbf{R}$. *Then* $\mathbf{]0,1[_m} = (]0,1[, \cdot_f, \leq)$ *is a continuous alo-group with* $e = 0,5$ *and* $x^{(-1)} = 1 - x$ *for each* $x \in]0,1[$.

Proof. The function $v :]0,+\infty[\to]0,1[$ defined for all $t \in]0,+\infty[$ by

$$v(t) = \frac{t}{t+1}$$

is a bijection between $]0,+\infty[$ and $]0,1[$ being continuous and strictly increasing.
For $a, b \in]0,+\infty[$ and $x = v(a), y = v(b)$, we get

$$v(a) \cdot_f v(b) = \frac{ab}{ab+1} = v(a \cdot b)\ . \tag{2.108}$$

Hence, $x \cdot_f y = v(v^{-1}(x) \cdot v^{-1}(y))$, and (2.102) is verified with $\psi = v$. Finally, it is easy to verify that $x \cdot_f 0,5 = x$ and $x \cdot_f (1-x) = 0,5$. □

By Examples 2.11 and 2.12 and Proposition 2.32, we shall call:

- $\mathcal{R} = (\mathbf{R}, +, \leq)$ the *additive alo-group*, or, *a-alo-group*;
- $\mathcal{R}^+ = (]0,+\infty[, \cdot, \leq)$ the *multiplicative alo-group*, or, *m-alo-group*;
- $\mathcal{R}_a = (\mathbf{R}, +_f, \leq)$ the *fuzzy-additive alo-group*, or, *fa-alo-group*;
- $\mathbf{]0,1[_m} = (]0,1[, \cdot_f, \leq)$ the *fuzzy-multiplicative alo-group*, or, *fm-alo-group*.

The isomorphism between $\mathcal{R}^+$ and $\mathcal{R}$ is

$$h : x \in]0,+\infty[\to \log x \in \mathbf{R}, h^{-1} : y \in \mathbf{R} \to \exp(y) \in]0,+\infty[\ . \tag{2.109}$$

The isomorphism between $\mathcal{R}^+$ and $\mathcal{R}_a$ is the function g in (2.105) and its inverse, i.e.

$$g : x \in]0,+\infty[\to 0.5(1 + \log x) \in \mathbf{R},\ g^{-1} : y \in \mathbf{R} \to \exp(2y-1) \in]0,+\infty[\ . \tag{2.110}$$

The isomorphism between $\mathcal{R}^+$ and $\mathbf{]0,1[_m}$ is the function v in (2.102) and its inverse, i.e.

$$v : x \in]0,+\infty[\to \frac{1-x}{1+x} \in]0,1[,\ v^{-1} : y \in]0,1[\to \frac{y}{1-y} \in]0,+\infty[\ . \tag{2.111}$$

2.10.3 Pairwise Comparison Matrices over a Divisible Alo-Group

Let $\mathcal{G} = (G, \odot, \leq)$ be a divisible alo-group. A pairwise comparison system over G is a pair $(X, \mathcal{A})$ constituted by a set $X = \{x_1, x_2, \ldots, x_n\}$ and a relation $\mathcal{A} : X^2 \to G$, where $\mathcal{A}(x_i, x_j) = a_{ij} \in G$, and $A = \{a_{ij}\}$ is a pairwise comparison matrix (PC matrix). In the context of an evaluation problem, the element a_{ij} can be interpreted as a measure on G of the preference of x_i over x_j. Here, $a_{ij} > e$ implies that x_i is strictly preferred to x_j, whereas $a_{ij} < e$ expresses the opposite and $a_{ij} = e$ means that x_i and x_j are indifferent. Moreover, $A = \{a_{ij}\}$ is assumed to be reciprocal with respect to the operation $\odot$, that is,

$$a_{ji} = a_{ij}^{(-1)} \text{ for all } i, j \in \{1, 2, \ldots, n\} \ . \quad \odot\textit{-reciprocity} \tag{2.112}$$

Hence, by definition $a_{ii} = e$ for all $i \in \{1, 2, \ldots, n\}$, and $a_{ij} \odot a_{ji} = e$ for all $i, j \in \{1, 2, \ldots, n\}$.

In the sequel, by $PC_n(\mathcal{G})$, the set of all $\odot$–reciprocal PC matrices of order $n \geq 3$ over $\mathcal{G}$ will be denoted. Particularly, a matrix of $PC_n(\mathcal{R})$ is an *additive* PC matrix, a matrix of $PC_n(\mathcal{R}^+)$ is a *multiplicative* PC matrix, whereas a matrix of $PC_n(\mathcal{R}_a)$, or, $PC_n(]\mathbf{0}, \mathbf{1}[_{\mathbf{m}})$, is a *fuzzy* PC matrix.

Let $A = \{a_{ij}\} \in PC_n(\mathcal{G})$. The following notation will be used.

- A_i is the i-th row of A, i.e. $A_i = (a_{i1}, a_{i2}, \ldots, a_{in})$;
- A^j is the j-th column of A, i.e. $A^j = (a_{1j}, a_{2j}, \ldots, a_{nj})$;
- $m_\odot(A_i)$ is the $\odot$-mean of row $A_i = (a_{i1}, a_{i2}, \ldots, a_{in})$;
- $w_{m_\odot}(A) = (m_\odot(A_1), m_\odot(A_2), \ldots, m_\odot(A_n))$ is the vector of $\odot$-means of rows of A;
- $\rho_{ijk} = a_{ik} \div (a_{ij} \odot a_{jk})$ is the element of G.

By the definition of $\mathcal{G}$-distance, see Chap. 1, Sect. 1.12, we obtain

$$d_{\mathcal{G}}(a_{ik} \div (a_{ij} \odot a_{jk})) = \|\rho_{ijk}\|. \tag{2.113}$$

Because of the assumption $\odot$-reciprocity the equality $a_{ik} = a_{ij} \odot a_{jk}$ does not depend on the considered order of the indexes i, j, k, that is,

$$a_{ik} = a_{ij} \odot a_{jk} \Leftrightarrow a_{ij} = a_{ik} \odot a_{kj} \Leftrightarrow a_{jk} = a_{ji} \odot a_{ik} \Leftrightarrow \tag{2.114}$$

$$a_{jk} = a_{ji} \odot a_{ik} \Leftrightarrow a_{ji} = a_{jk} \odot a_{ki} \Leftrightarrow \ldots . \tag{2.115}$$

Definition 2.1. Let $A = \{a_{ij}\} \in PC_n(\mathcal{G})$.

- Let $x_i, x_j, x_k \in X$. A PC matrix $A = \{a_{ij}\}$ is *$\odot$-consistent with respect to the 3-subset*$\{x_i, x_j, x_k\}$ of X if $a_{ik} = a_{ij} \odot a_{jk}$;

- A PC matrix $A = \{a_{ij}\}$ is $\odot$-*consistent* if it is $\odot$-consistent with respect to each 3 – subset of X, i.e.

$$a_{ik} = a_{ij} \odot a_{jk} \text{ for all } i, j, k \in \{1, 2, \ldots, n\} \text{ - } \odot\text{-}consistency \ . \quad (2.116)$$

In the above mentioned context, a fuzzy PC matrix is defined over $]\mathbf{0}, \mathbf{1}[_{\mathbf{m}}$ and the condition of $\cdot$-consistency becomes $\cdot_f$-consistency, i.e. by (2.107) we obtain:

$$a_{ik} = a_{ij} \cdot_f a_{jk} = \frac{a_{ij}a_{jk}}{a_{ij}a_{jk} + (1 - a_{ij})(1 - a_{jk})} \text{ for all } i, j, k \in \{1, 2, \ldots, n\} \ . \quad (2.117)$$

It is clear that for each $A = \{a_{ij}\} \in PC_n(\mathcal{G})$ the property of $\odot$-consistency is equivalent to the one of the following conditions:

$$a_{ij} = a_{ik} \div a_{jk} \text{ for all } i, j, k \in \{1, 2, \ldots, n\} \ , \quad (2.118)$$

$$\rho_{ijk} = e \text{ for all } i, j, k \in \{1, 2, \ldots, n\} \ , \quad (2.119)$$

$$d_{\mathcal{G}}(a_{ik} \div (a_{ij} \odot a_{jk})) = e \text{ for all } i, j, k \in \{1, 2, \ldots, n\} \ . \quad (2.120)$$

Because of the equivalences in (2.115), when checking equations (2.118)–(2.120) we may restrict only to the cases $i < j < k, i, j, k \in \{1, 2, \ldots, n\}$.

Definition 2.2. A vector $w = (w_1, w_2, \ldots, w_n)$, with $w_i \in G$ for all $i \in \{1, 2, \ldots, n\}$ is a $\odot$-*consistent vector with respect to* $A = \{a_{ij}\} \in PC_n(\mathcal{G})$ if

$$w_i \div w_j = a_{ij} \text{ for all } i, j \in \{1, 2, \ldots, n\} \ . \quad (2.121)$$

By (2.121) and the equivalences in changing the indexes (2.115) we obtain for $i, j \in \{1, 2, \ldots, n\}$:

$$w_i > w_j \iff a_{ij} > e \text{ and } w_i = w_j \iff a_{ij} = e \ . \quad (2.122)$$

Thus, the elements of the $\odot$-consistent vector of $A = \{a_{ij}\}$ correspond with the preferences expressed by the entries a_{ij} of the PC matrix.

Proposition 2.33. *$A = \{a_{ij}\} \in PC_n(\mathcal{G})$ is $\odot$-consistent if and only if there exists a $\odot$-consistent vector $w = (w_1, w_2, \ldots, w_n), w_i \in G$.*

Proof. Let $A = \{a_{ij}\}$ be consistent. Then by (2.118), $a_{ik} \div a_{jk} = a_{ij}$ for all $i, j, k \in \{1, 2, \ldots, n\}$. Hence, the equalities (2.121) are verified for each column vector $w = A^k, k \in \{1, 2, \ldots, n\}$. On the other hand, if w is an $\odot$-consistent vector, then

$$a_{ij} \odot a_{jk} = (w_i \div w_j) \odot (w_j \div w_k) = w_i \odot w_j^{(-1)} \odot w_j \odot w_k^{(-1)}$$

$$= w_i \odot w_k^{(-1)} = a_{ik} \ .$$

□

Proposition 2.34. *Let $A = \{a_{ij}\} \in PC_n(\mathcal{G})$. The following assertions are equivalent:*

(i) $A = \{a_{ij}\}$ *is* $\odot$*-consistent;*
(ii) *each column* A^k *is an* $\odot$*-consistent vector;*
(iii) $w_{m_\odot}(A) = (m_\odot(A_1), m_\odot(A_2), \ldots, m_\odot(A_n))$ *- the vector of* $\odot$*-means of rows of* A *is an* $\odot$*-consistent vector.*

Proof. The equivalence of *(i)* and *(ii)* follows directly from Proposition 2.32.

Let *(ii)* be true, then by (1.57) we get

$$(m_\odot(A_i) \div m_\odot(A_j))^{(n)} = (m_\odot(A_i))^{(n)} \div (m_\odot(A_j))^{(n)} =$$

$$= (a_{i1} \odot a_{i2} \odot \ldots \odot a_{in}) \odot (a_{j1} \odot a_{j2} \odot \ldots \odot a_{jn})^{(-1)} =$$

$$= (a_{i1} \odot a_{j1}) \odot \ldots \odot (a_{in} \odot a_{jn}) = a_{ij}^{(n)} \ .$$

Therefore, (2.121) is verified and *(iii)* is true. The opposite implication is similar.

□

Proposition 2.35. *Let $\mathcal{G} = (G, \odot, \leq)$ and $\mathcal{G}' = (G', \circ, \preceq)$ be divisible alo-groups and $h : G \to G'$ be an isomorphism between $\mathcal{G}$ and $\mathcal{G}'$, $a, b \in G$, $a', b' \in G'$. Then*

$$H : A = \{a_{ij}\} \in PC_n(\mathcal{G}) \to H(A) = \{h(a_{ij})\} \in PC_n(\mathcal{G}')$$

is a bijection between $PC_n(\mathcal{G})$ and $PC_n(\mathcal{G}')$ that preserves the $\odot$-consistency, that is A is $\odot$-consistent if and only if A' is $\circ$-consistent.

Proof. The mapping H is an injection because h is an injective function. By applying h to the entries of the matrix $A = \{a_{ij}\}$, we get the matrix $A = \{h(a_{ij})\}$, that is reciprocal too. Therefore, $H(A) = \{h(a_{ij})\} \in PC_n(\mathcal{G}')$. The rest of the proof is straightforward. □

2.10.4 Consistency Index in Alo-Groups

Let $\mathcal{G} = (G, \odot, \leq)$ be a divisible alo-group and $A = \{a_{ij}\} \in PC_n(\mathcal{G})$. By Definition 2.2, A is $\odot$-inconsistent if at least one triple $\{x_i, x_j, x_k\}$ is $\odot$-inconsistent. The closeness to the consistency depends on the degree of consistency with respect to each 3 – subset $\{x_i, x_j, x_k\}$ and it can be measured by an *average* of these degrees. In order to define a consistency index for $A = \{a_{ij}\}$, we first consider the case that X has only 3 elements.

Let $X = \{x_1, x_2, x_3\}$ be the set of 3 elements and A be a PC matrix

$$A = \begin{pmatrix} a_{11} & a_{12} & a_{13} \\ a_{21} & a_{22} & a_{23} \\ a_{31} & a_{32} & a_{33} \end{pmatrix} \in PC_3(\mathcal{G}). \tag{2.123}$$

By (2.120), $A = \{a_{ij}\}$ is inconsistent if and only if $d_{\mathcal{G}}(a_{13}, a_{12} \odot a_{23}) > e$. It is natural to say that the more A is inconsistent the more $d_{\mathcal{G}}(a_{13}, a_{12} \odot a_{23})$ is far from e. We formulate the following definition.

Definition 2.3. The $\odot$-*consistency index* of the matrix in (2.123) is given by

$$I_{\mathcal{G}}(A) = \|\rho_{123}\| = d_{\mathcal{G}}(a_{13}, a_{12} \odot a_{23}) \ . \tag{2.124}$$

As particular cases, we obtain:

- If $A \in PC_3(\mathcal{R}^+)$ then

$$I_{\mathcal{R}^+}(A) = \frac{a_{13}}{a_{12} \cdot a_{23}} \vee \frac{a_{12} \cdot a_{23}}{a_{13}} \in [1, +\infty[\ , \tag{2.125}$$

  and A is $\cdot$-consistent if and only if $I_{\mathcal{R}^+}(A) = 1$;
- If $A \in PC_3(\mathcal{R})$, then

$$I_{\mathcal{R}}(A) = |a_{13} - a_{12} - a_{23}| = (a_{13} - a_{12} - a_{23}) \vee (a_{12} + a_{23} - a_{13}) \in [0, +\infty[\ ,$$

  and A is $+$-consistent if and only if $I_{\mathcal{R}}(A) = 0$;
- If $A \in PC_3(\mathcal{R}_a)$, then

$$I_{]0,1[_m}(A) = \frac{a_{13}(1-a_{12})(1-a_{23})}{a_{13}(1-a_{12})(1-a_{23}) + a_{12}(1-a_{13})a_{23}} \vee \frac{a_{12}(1-a_{13})a_{23}}{(1-a_{13})a_{12}a_{23} + (1-a_{12})(1-a_{23})a_{13}}. \tag{2.126}$$

  and A is an $\cdot_f$-consistent if and only if $I_{]0,1[_m}(A) = 0.5$.
- If $A \in PC_3(\mathcal{R}_a)$, then

$$I_{]0,1[_a}(A) = (a_{13} - a_{12} - a_{23} + 0.5) \vee (a_{12} + a_{23} - a_{13} + 0.5) \tag{2.127}$$

  and A is $+_f$-consistent if and only if $I_{\mathcal{R}_a}(A) = 0.5$.

The following proposition shows that the more $I_{\mathcal{G}}(A)$ is close to e the more the mean vector $w_{m_\odot}$ is close to be an $\odot$-consistent vector.

Proposition 2.36. *Let $w_{m\odot} = (w_1, w_2, w_3)$ be the mean vector associated to the matrix in (2.123) and $\rho = \rho_{123}$. Then*

$$d_{\mathcal{G}}(w_i \div w_j, a_{ij}) = \|\rho\|^{\frac{1}{3}} \text{ for all } i \neq j.$$

Proof. By definition of ρ we obtain $a_{13} = \rho \odot a_{12} \odot a_{23}$, $a_{21} = \rho \odot a_{23} \odot a_{31}$ and $a_{32} = \rho \odot a_{12} \odot a_{31}$. By the above equalities and the equality $a_{ii} = e$, we get

- $w_1 = (a_{11} \odot a_{12} \odot a_{13})^{(\frac{1}{3})} = (a_{12}^{(2)} \odot \rho \odot a_{23})^{(\frac{1}{3})}$;
- $w_2 = (a_{21} \odot a_{22} \odot a_{23})^{(\frac{1}{3})} = (a_{23}^{(2)} \odot \rho \odot a_{31})^{(\frac{1}{3})}$;
- $w_3 = (a_{31} \odot a_{32} \odot a_{33})^{(\frac{1}{3})} = (a_{31}^{(2)} \odot \rho \odot a_{12})^{(\frac{1}{3})}$.

Then, we get $(w_1 \div w_2) \div a_{12} = \rho^{(\frac{1}{3})}$ and $a_{12} \div (w_1 \div w_2) = (\rho^{(-1)})^{(\frac{1}{3})}$, thus $d_{\mathcal{G}}(w_1 \div w_2, a_{12}) = ||\rho||^{\frac{1}{3}}$.

Similarly, we obtain $d_{\mathcal{G}}(w_2 \div w_3, a_{23}) = \|\rho\|^{\frac{1}{3}}$ and $d_{\mathcal{G}}(w_3 \div w_1, a_{31}) = \|\rho\|^{\frac{1}{3}}$.

□

Proposition 2.37. *Let $\mathcal{G} = (G, \odot, \leq)$ and $\mathcal{G}' = (G', \circ, \preceq)$ be divisible alo-groups, $h : G \to G'$ be an isomorphism between $\mathcal{G}$ and $\mathcal{G}'$ and $A' = \{h(a_{ij})\} \in PC_3(\mathcal{G}')$. Then $I_{\mathcal{G}'}(A') = h(I_{\mathcal{G}}(A))$.*

Proof. By the equality $h(a_{12}) \odot h(a_{23}) = h(a_{12} \odot a_{23})$, we obtain
$I_{\mathcal{G}'}(A') = d_{\mathcal{G}'}(h(a_{13}), h(a_{12}) \odot h(a_{23})) = h(d_{\mathcal{G}}(a_{13}, a_{12} \odot a_{23})) = h(I_{\mathcal{G}}(A))$.

□

Finally, let G be a proper open interval of **R** and $\leq$ the total order on G inherited from the usual order on **R**, $\mathcal{G} = (G, \odot, \leq)$ be a continuous alo-group and $\phi : \mathbf{R} \to G$ and $\psi :]0, +\infty[\to G$ the continuous strictly monotonic functions satisfying (2.102) and (2.103), respectively. Then the $\odot$-consistency index of the matrix A in (2.123) is given by

$$I_{\mathcal{G}}(A) = \phi(I_{\mathcal{R}}(\phi^{-1}(A))) = \psi(I_{\mathcal{R}^+}(\psi^{-1}(A))). \tag{2.128}$$

Let $v :]0, +\infty[\to]0, 1[, v(t) = \frac{t}{1+t}$ for all $t \in]0, +\infty[$ be the isomorphism, $v^{-1} :]0, 1[\to]0, +\infty[, v^{-1}(s) = \frac{s}{1-s}$ for all $s \in]0, 1[$, the inverse isomorphism. Then

$$A' = \{v^{-1}(a_{ij})\} = \{\frac{a_{ij}}{1 - a_{ij}}\}, \tag{2.129}$$

and

$$I_{]\mathbf{0},\mathbf{1}[_m}(A) = v(I_{\mathcal{R}^+}(v^{-1}(A))) = \frac{I_{\mathcal{R}^+}(\{\frac{a_{ij}}{1-a_{ij}}\})}{1 + I_{\mathcal{R}^+}(\{\frac{a_{ij}}{1-a_{ij}}\})}. \tag{2.130}$$

Now, let $g :]0, +\infty[\to] - \infty, +\infty[$, $g(t) = 0.5(1 + \log t)$ for all $t \in]0, +\infty[$ be the isomorphism, $g^{-1} :] - \infty, +\infty[\to]0, +\infty[$, $g^{-1}(s) = \exp(2s - 1)$ for all

$s \in]-\infty, +\infty[$, the inverse isomorphism. Then

$$A'' = \{g^{-1}(a_{ij})\} = \{\exp(2a_{ij} - 1)\}, \tag{2.131}$$

and

$$I_{]0,1[_a}(A) = g(I_{\mathscr{R}^+}(g^{-1}(A))) = 0.5(1 + \log I_{\mathscr{R}^+}(\{\exp(2a_{ij} - 1)\}). \tag{2.132}$$

Example 2.15. Consider the matrix

$$A = \begin{pmatrix} 0.5 & 0.3 & 0.2 \\ 0.7 & 0.5 & 0.1 \\ 0.8 & 0.9 & 0.5 \end{pmatrix} \in PC_3(]\mathbf{0}, \mathbf{1}[_m) \ .$$

Then by (2.126) we obtain

$$I_{]0,1[_m}(A) = \frac{0.2 \cdot 0.7 \cdot 0.9}{0.2 \cdot 0.7 \cdot 0.9 + 0.3 \cdot 0.1 \cdot 0.8} \vee \frac{0.3 \cdot 0.8 \cdot 0.9}{0.2 \cdot 0.7 \cdot 0.9 + 0.3 \cdot 0.1 \cdot 0.8} =$$

$$= 0.84 \vee 0.16 = 0.84 \ .$$

Applying v^{-1} by (2.129) we obtain

$$A' = \begin{pmatrix} 1 & \frac{3}{7} & \frac{1}{4} \\ \frac{7}{3} & 1 & \frac{1}{9} \\ 4 & 9 & 1 \end{pmatrix} \in PC_3(\mathscr{R}^+) \ .$$

By (2.125), the $\cdot$-consistency index is $I_{\mathscr{R}^+}(A') = \frac{21}{4} \vee \frac{4}{21} = \frac{21}{4}$.

In accordance with (2.130) we obtain $I_{]0,1[_m}(A) = v(I_{\mathscr{R}^+}(A')) = \frac{\frac{21}{4}}{1+\frac{21}{4}} = 0.84$.

Now, we extend our approach to the $n \times n$ PC matrices with $n > 3$. Let $\mathscr{G} = (G, \odot, \leq)$, $A = \{a_{ij}\} \in PC_n(\mathscr{G})$. By $T(A)$ we denote the set of 3-element subsets $\{a_{ij}, a_{jk}, a_{ik}\}$ of A with $i < j < k$. Clearly, $n_T = \frac{n!}{3!(n-3)!}$ is the cardinality of the set $T(A)$. Denote

$$A_{ijk} = \begin{pmatrix} a_{ii} & a_{ij} & a_{ik} \\ a_{ji} & a_{jj} & a_{jk} \\ a_{ki} & a_{kj} & a_{kk} \end{pmatrix} \ . \tag{2.133}$$

Here, A_{ijk} is a submatrix of A related to the 3 – subset $\{x_i, x_j, x_k\}$ and $I_{\mathscr{G}}(A_{ijk}) = ||\rho_{ijk}|| = d_{\mathscr{G}}(a_{ik}, a_{ij} \odot a_{jk})$ is its $\odot$-consistency index. By definition (2.116), a $\odot$-consistency index of A should be expressed in terms of the $\odot$-consistency indexes $I_{\mathscr{G}}(A_{ijk})$. Hence, we formulate the following definition.

Definition 2.4. Let $\mathcal{G} = (G, \odot, \leq)$ be an alo-group, $n \geq 3$. The *$\odot$-consistency index* of $A = \{a_{ij}\} \in PC_n(\mathcal{G})$ is given by

$$I_{\mathcal{G}}(A) = \left(\bigodot_{i<j<k} I_{\mathcal{G}}(A_{ijk}) \right)^{(1/n_T)} = \left(\bigodot_{i<j<k} d_{\mathcal{G}}(a_{ik}, a_{ij} \odot a_{jk}) \right)^{(1/n_T)} . \quad (2.134)$$

By the above definition (2.134) it follows that $I_{\mathcal{G}}(A_{ijk}) \geq e$ for all $1 \leq i < j < k \leq n$ and $A = \{a_{ij}\}$ is $\odot$-consistent if and only if $I_{\mathcal{G}}(A_{ijk}) = e$ for all $1 \leq i < j < k \leq n$.

As particular cases, we have

- if $A \in PC_n(\mathcal{R}^+)$, then $I_{\mathcal{R}^+}(A) = \left(\prod_{i<j<k} I_{\mathcal{R}^+}(A_{ijk}) \right)^{\frac{1}{n_T}} \geq 1$ and A is m -consistent if and only if $I_{\mathcal{R}^+}(A) = 1$;
- if $A \in PC_n(\mathcal{R})$, then $I_{\mathcal{R}}(A) = \frac{1}{n_T} \sum_{i<j<k} I_{\mathcal{R}}(A_{ijk}) \geq 0$, and A is a-consistent if and only if $I_{\mathcal{R}}(A_{ijk}) = 0$ for all $1 \leq i < j < k \leq n$;
- if $A \in PC_n(]\mathbf{0}, \mathbf{1}[_\mathbf{m})$, then $I_{]\mathbf{0},\mathbf{1}[_\mathbf{m}}(A) = m_{.f}(I_{]\mathbf{0},\mathbf{1}[_\mathbf{m}}(A_{ijk}); i < j < k) \in [0.5, 1[$ and A is fm-consistent if and only if $I_{]\mathbf{0},\mathbf{1}[_\mathbf{m}}(A) = 0.5$.
- if $A \in PC_n(\mathcal{R}_a)$, then $I_{\mathcal{R}_a}(A) = \frac{1}{n_T}(\sum_{i<j<k} I_{\mathcal{R}_a}(A_{ijk})) \in [0.5, 1[$ and A is fa-consistent if and only if $I_{]\mathbf{0},\mathbf{1}[_\mathbf{a}}(A) = 0.5$.

Proposition 2.36 holds also for $n > 3$ as follows:

Let $\mathcal{G} = (G, \odot, \leq)$ and $\mathcal{G}' = (G', \circ, \preceq)$be divisible alo-groups, $h : G \to G'$ be an isomorphism between $\mathcal{G}$ and $\mathcal{G}'$ and $A' = \{h(a_{ij})) \in PC_n(\mathcal{G}')$. Then

$$I_{\mathcal{G}'}(A') = h(I_{\mathcal{G}}(A)) \ .$$

Therefore, (2.128) and (2.129) are also valid for $\mathcal{G} = (G, \odot, \leq)$, and $A = \{a_{ij}\} \in PC_n(\mathcal{G})$. Particularly, if $A' = \{\log a_{ij}\}$, $A = \{\exp a'_{ij}\}$, then

$$I_{\mathcal{R}}(A') = \log(I_{\mathcal{R}^+}(A)), I_{\mathcal{R}^+}(A) = \exp(I_{\mathcal{R}}(A')) \ . \quad (2.135)$$

Example 2.16. [11] Consider the matrix

$$A = \begin{pmatrix} 1 & \frac{1}{7} & \frac{1}{7} & \frac{1}{5} \\ 7 & 1 & \frac{1}{2} & \frac{1}{3} \\ 7 & 2 & 1 & \frac{1}{9} \\ 5 & 3 & 9 & 1 \end{pmatrix} \in PC_4(\mathcal{R}^+) \ ;$$

then

$$I_{\mathcal{R}^+}(A) = \sqrt[4]{I_{\mathcal{R}^+}(A_{234}) \cdot I_{\mathcal{R}^+}(A_{134}) \cdot I_{\mathcal{R}^+}(A_{124}) \cdot I_{\mathcal{R}^+}(A_{123})} =$$
$$= \sqrt[4]{6 \cdot 12.6 \cdot 4.2 \cdot 2} = 5.02 \ .$$

Let $h(t) = \log t$ be the isomorphism between $\mathscr{R}^+$ and $\mathscr{R}$. By applying h to the entries of A, we get

$$A' = \begin{pmatrix} 0 & -\log 7 & -\log 7 & -\log 5 \\ \log 7 & 0 & -\log 2 & -\log 3 \\ \log 7 & \log 2 & 0 & -\log 9 \\ \log 5 & \log 3 & \log 9 & 0 \end{pmatrix} \in PC_4(\mathscr{R}) \ ;$$

and the a-consistency index is

$$I_{\mathscr{R}}(A') = \frac{I_{\mathscr{R}}(A'_{234}) + I_{\mathscr{R}}(A'_{134}) + I_{\mathscr{R}}(A'_{124}) + I_{\mathscr{R}}(A'_{123})}{4} =$$

$$= \frac{1.792 + 2.534 + 1.435 + 0.693}{4} = 1.613 \ .$$

In accordance with (2.128), we obtain

$$I_{\mathscr{R}}(A') = \log(I_{\mathscr{R}^+}(A)) = \log(5.02) = 1.613 \ .$$

2.11 Conclusions

In this chapter we investigated PC matrices where relations among the entries are treated in four ways depending on a particular interpretation: multiplicatively, additively, fuzzy multiplicatively and fuzzy additively. By various methods for deriving priorities from various types of preference matrices we obtained the corresponding priority vectors for a final ranking of alternatives. Moreover, we derived some new results for situations where some elements of the fuzzy preference matrix are missing. Finally, a unified framework for pairwise comparison matrices based on abelian linearly ordered groups were presented. Four basic particular cases of PC matrices were discussed and illustrative numerical examples were presented.

References

1. Aczel, J.: Lectures on Functional Equation and Their Applications. Academic Press, New York and London (1966)
2. Aguaron, J., Moreno-Jimenez, J.M.: The geometric consistency index: Approximated thresholds. Eur. J. Oper. Res. **147**(1), 137–145 (2003)
3. Alonso, S., Chiclana, F., Herrera. F., Herrera-Viedma, E., Alcala-Fdes, J., Porcel, C.: A consistency-based procedure to estimate missing pairwise preference values. Int. J. Intell. Syst. **23**, 155–175 (2008)

4. Barzilai, J.: Deriving weights from pairwise comparison matrices. J. Oper. Res. Soc. **48**(12), 1226–1232 (1997)
5. Barzilai, J.: Consistency measures for pairwise comparison matrices. J. Multicrit. Dec. Anal. **7**, 123–132 (1998)
6. Barzilai, J.: Notes on the analytic hierarchy process. In: Proceedings of the NSF Design and Manufacturing Research Conference, pp. 1–6. Tampa, Florida (2001)
7. Barzilai, J.: Preference function modeling: The mathematical foundations of decision theory. In: Ehrgott, M. et al. (eds.) Trends in Multiple Criteria Decision Analysis, pp. 57–86. Springer, Berlin-Heidelberg-New York (2003)
8. Blankmeyer, E.: Approaches to consistency adjustments. J. Optim. Theory Appl. **154**, 479–488 (1987)
9. Bourbaki, N.: Algebra II. Springer, Heidelberg-New York-Berlin (1990)
10. Brin, S., Page. L.: The anatomy of a large-scale hypertextualweb search engine. Comput. Networks ISDN Syst. **30**, 107–117 (1998)
11. Cavallo, B., D'Apuzzo, L.: A general unified framework for pairwise comparison matrices in multicriterial methods. Int. J. Intell. Syst. **24**(4), 377–398 (2009)
12. Cavallo, B., D'Apuzzo, L.: Characterizations of consistent pairwise comparison matrices over abelian linearly ordered groups. Int. J. Intell. Syst. **25**, 1035–1059 (2010)
13. Cavallo, B., D'Apuzzo, L., Squillante, M.: About a consistency index for pairwise comparison matrices over a divisible alo-group. Int. J. Intell. Syst. **27**, 153–175 (2012)
14. Cavallo, B., D'Apuzzo, L.: Deriving weights from a pairwise comparison matrix over an alo/group. Soft Comput. **16**, 353–366 (2012)
15. Chang, J.S.K. et al.: Note on deriving weights from pairwise comparison matrices in AHP. Inform. Manag. Sci. **19**(3), 507–517 (2008)
16. Chen, Q., Triantaphillou, E.: Estimating data for multi-criteria decision making problems: optimization techniques. In: Pardalos, P.M., Floudas, C. (eds.) Encyclopedia of Optimization, vol. 2. Kluwer Academic, Boston (2001)
17. Chiclana, F., Herrera, F., Herrera-Viedma, E.: Integrating multiplicative preference relations in a multipurpose decision making model based on fuzzy preference relations. Fuzzy Sets Syst. **112**, 277–291 (2001)
18. Chiclana, F. Herrera, F., Herrera-Viedma, E., Alonso, S.: Some induced ordered weighted averaging operators and their use for solving group decision-making problems based on fuzzy preference relations. Eur. J. Operat. Res. **182**, 383–399 (2007)
19. Chiclana, F., Herrera-Viedma, E., Alonso, S.: A note on two methods for estimating pairwise preference values. IEE Trans. Syst. Man Cybern. **39**(6), 1628–1633 (2009)
20. Choo, E., Wedley, W.: A common framework for deriving preference values from pairwise comparison matrices. Comput. Oper. Res. **31**(6), 893–908 (2004)
21. Cook, W., Kress, M.: Deriving weights from pairwise comparison ratio matrices: An axiomatic approach. Eur. J. Oper. Res. **37**(3), 355–362 (1988)
22. Crawford, G, Williams, C.: A note on the analysis of subjective judgment matrices. J. Math. Psychol. **29**, 25–40 (1985)
23. Dopazo, E., Gonzales-Pachon, J.: Consistency-driven approximation of a pairwise comparison matrix. Kybernetika **39**(5), 561–568 (2003)
24. Dubois, D., Prade, H.: Fuzzy Sets and Systems: Theory and Application. Academic Press, New York (1980)
25. Fan, Z.-P., Ma, J., Zhang, Q.: An approach to multiple attribute decision making based on fuzzy preference information on alternatives. Fuzzy Sets Syst. **131**, 101–106 (2002)
26. Fan, Z.-P., Hu, G.-F., Xiao, S.-H.: A method for multiple attribute decision-making with the fuzzy preference relation on alternatives. Comput. Ind. Eng. **46**, 321–327 (2004)
27. Fan, Z.-P., Ma, J., Jiang, Y.-P., Sun, Y.-H., Ma, L.: A goal programming approach to group decision making based on multiplicative preference relations and fuzzy preference relations. Eur. J. Oper. Res. **174**, 311–321 (2006)
28. Fedrizzi, M., Giove, S.: Incomplete pairwise comparison and consistency optimization. Eur. J. Oper. Res. **183**(1), 303–313 (2007)

29. Fedrizzi, M., Brunelli, M.: On the priority vector associated with a reciprocal relation and a pairwise comparison matrix. Soft Comput. **14**(2), 639–645 (2010)
30. Fiedler, M., Nedoma, J., Ramík, J., Rohn, J., Zimmermann, K.: Linear Optimization Problems with Inexact Data. Springer, Berlin/Heidelberg/New York/Hong Kong/London/Milan/Tokyo (2006)
31. Figueira, J., Greco, S., Ehrgott, M.: Multiple Criteria Decision Analysis: State of the Art Surveys. Springer, New York (2005)
32. Fodor, J., Roubens, M.: Fuzzy Preference Modeling and Multicriteria Decision Support. Kluwer, Dordrecht (1994)
33. Gantmacher, F.R.: The Theory of Matrices, vol.1. Chelsea Publ. Comp. (1977)
34. Gong, Z.-W.: Least-square method to priority of the fuzzy preference relations with incomplete information. Int. J. Approx. Reason. **47** (2008), 258–264.
35. Gong, Z., Li, L., Cao, J., Zhou, F.: On additive consistent properties of the intuitionistic fuzzy preference relation. Int. J. Inform. Tech. Dec. Making **9**(6), 1009–1025 (2010)
36. Harker, P.T.: Alternative modes of questioning in the analytic hierarchy process. Math. Modell. **9**(3–5), 353–360 (1987)
37. Harker, P.T.: Incomplete pairwise comparisons in the analytic hierarcy process. Math. Modell. **9**(11), 837–848 (1987)
38. Herrera-Viedma, E., Herrera, F., Chiclana, F., Luque, M.: Some issues on consistency of fuzzy preference relations. Eur. J. Oper. Res. **154**, 98–109 (2004)
39. Herrera, F., Herrera-Viedma, E., Chiclana, F.: Multiperson decision making based on multiplicative preference relations. Eur. J. Oper. Res. **129**, 372–385 (2001)
40. Herrera, F., Herrera-Viedma, E.: Choice functions and mechanisms for linguistic preference relations. Eur. J. Oper. Res. **120**, 144–161 (2000)
41. Herrera-Viedma, E., Chiclana, F., Herrera, F., Alonso, S.: A group decision-making model with incomplete fuzzy preference relations based on additive consistency. IEEE Trans. Syst. Man Cybern. B **37**, 176–189 (2007)
42. Herrera-Viedma, E., Alonso, S., Chiclana, F., Herrera, F.: A consensus model for group decision-making with incomplete fuzzy preference relations. IEEE Trans. Fuzzy Syst. **15**(5), 863–877 (2007)
43. Hovanov, N.V., Kolari, J.W., Sokolov, M.V.: Deriving weights from general pairwise comparison matrices. Math. Soc. Sci. **55**, 205–220 (2008)
44. International MCDM Society, website: http://www.mcdmsociety.org/
45. Kim, S.H, Choi, S.H, Kim, J.K.: An interactive procedure for multiple attribute group decision making with incomplete information: Range-based approach. Eur. J. Oper. Res. 118, 139–152 (1999)
46. Lee, H.-S.: A fuzzy method for evaluating suppliers. In: Wang, L. et al. (eds.) FSKD, LNAI 4223, pp. 1035–1043. Springer, Berlin (2006)
47. Lee, H.-S., Tseng, W.-K.: Goal programming methods for constructing additive consistency fuzzy preference relations. In: Gabrys, B. et al. (eds.) KES 2006, Part II, LNAI 4252, pp. 910–916. Springer, Berlin (2006)
48. Lee, H.-S., Yeh, C.-H.: Fuzzy multi-criteria decision making based on fuzzy preference relation. In: Lovrek, I. et al. (eds.) KES 2008, Part II, LNAI 5178, pp. 980–985. Springer, Berlin (2008)
49. Lee, H.-S., Shen, P.-D., Chyr, W.-L.: Prioritization of incomplete fuzzy preference relation. In: Lovrek, I. et al. (eds.) KES 2008, Part II, LNAI 5178, pp. 974–979. Springer, Berlin (2008)
50. Lin, C.C: A revised framework for deriving preference values from pairwise comparison matrices. Eur. J. Oper. Res. **176**, 1145–1150 (2007)
51. Lipovetsky, S., Conklin, M.W.: Robust estimation of priorities in AHP. Eur. J. Oper. Res. **137**, 110–122 (2002)
52. Ma, J. et al.: A method for repairing the inconsistency of fuzzy preference relations. Fuzzy Sets Syst. **157**, 20–33 (2006)

53. Mikhailov, L.: A fuzzy programming method for deriving priorities in the analytic hierarchy process. J. Oper. Res. Soc. **5**, 342–349 (2000)
54. Nishizawa, K.: Estimation of unknown comparisons in incomplete AHP and it's compensation. In: Report of the Research Institute of Industrial Technology, Nihon University, vol 77 (2004)
55. Ramík, J., Vlach, M.: Measuring consistency and inconsistency of pair comparison systems. Kybernetika **49**(3), 465–486 (2013)
56. Ramík, J.: Fuzzy preference matrix with missing elements and its application to ranking of alternatives. In: Proc. of the 31th International conference Mathematical Methods in Economics 2013, pp. 767–772, September 11–13, College of Polytechnics Jihlava, Jihlava (2013)
57. Ramík, J., Korviny, P.: Inconsistency of pairwise comparison matrix with fuzzy elements based on geometric mean. Fuzzy Sets Syst. **161**, 1604–1613 (2010)
58. Ramík, J., Vlach, M.: Generalized Concavity in Optimization and Decision Making. Kluwer Publ., Boston-Dordrecht-London (2001)
59. Saaty, T.L.: The Analytic Hierarchy Process. McGraw-Hill, New York (1980)
60. Saaty, T.L.: Fundamentals of Decision Making and Priority Theory with the AHP. RWS Publications, Pittsburgh (1994)
61. Saaty, T.L., Vargas, L.: Models, Methods, Concepts and Applications of the Analytic Hierarchy Process. Kluwer, Boston (2000)
62. Saaty, T.L.: Decision-making with the AHP: Why is the principal eigenvector necessary. Eur. J. Oper. Res. **145**, 85–91 (2003)
63. Shiraishi, S., Obata, T., Daigo, M.: Properties of a positive reciprocal matrix and their application to AHP. J. Oper. Res. Soc. Jpn **41**(3), 404–414 (1998)
64. Srdjevic, B.: Combining different prioritization methods in the analytic hierarchy process synthesis. Comput. Oper. Res. **32**, 1897–1919 (2005)
65. Switalski, Z.: General transitivity conditions for fuzzy reciprocal preference matrices. Fuzzy Sets Syst. **137**, 85–100 (2003)
66. Tanino, T.: Fuzzy preference orderings in group decision making. Fuzzy Sets Syst. **12**, 117–131 (1984)
67. Tanino, T.: Fuzzy preference relations in group decision making. In: Kacprzyk, J., Roubens, M. (eds.), Non-Conventional Preference Relations in Decision Making, pp. 54–71. Springer, Berlin (1988)
68. Thurstone, L.L.: A law of comparative judgment. Psychol. Rev. **34**, 273–286 (1927)
69. Xu, Z.S.: On compatibility of interval fuzzy preference relations. Fuzzy Optim. Decis. Making **3**, 217–225 (2004)
70. Xu, Z.S.: Goal programming models for obtaining the priority vector of incomplete fuzzy preference relation. Int. J. Approx. Reas. **36**, 261–270 (2004)
71. Xu, Z.S., Da, Q.L.: A least deviation method to obtain a priority vector of a fuzzy preference relation. Eur. J. Oper. Res. 164, 206–216 (2005)
72. Xu, Z.S., Chen, J.: Some models for deriving the priority weights from interval fuzzy preference relations. Eur. J. Oper. Res. **184**, 266–280 (2008)
73. Xu, Z.S.: A procedure for decision making based on incomplete fuzzy preference relation. Fuzzy Optim. Decis. Making **4**, 175–189 (2005)
74. Xu, Z.S.: Multiple-attribute group decision making with different formats of preference information on attributes. IEEE Trans. Syst. Man Cybern. B **37**, 1500–1511 (2007)
75. Xu, Z.S., Chen, J.: A subjective and objective integrated method for MAGDM problems with multiple types of exact preference formats. IDEAL, 145–154 (2007)
76. Xu, Z.S., Chen, J.: Group decision-making procedure based on incomplete reciprocal relations. Soft Comput. **12**, 515–521 (2008)
77. Xu, Z.S., Chen, J.: MAGDM linear-programming models with distinct uncertain preference structures. IEEE Trans. Syst. Man Cybern. B **38**, 1356–1370 (2008)
78. Yuen, K.K.F.: Pairwise opposite matrix and its cognitive prioritization operators: comparisons with pairwise reciprocal matrix and analytic prioritization operators. J. Oper. Res. Soc. **63**, 322–338 (2012)

79. Wang, Y.-M., Fan, Z.-P.: Group decision analysis based on fuzzy preference relations: logarithmic and geometric least squares methods. Appl. Math. Comput. **194**, 108–119 (2007)
80. Wang, Y.-M., Parkan, C.: Multiple attribute decision making based on fuzzy preference information on alternatives: ranking and weighting. Fuzzy Sets Syst. **153**, 331–346 (2005)
81. Wang, Y.-M., Fan, Z.-P., Hua, Z.: A chi-square method for obtaining a priority vector from multiplicative and fuzzy preference relations. Eur. J. Oper. Res. **182**, 356–366 (2007)

Chapter 3
Preference Matrices with Fuzzy Elements in Decision Making

Abstract This chapter is aimed on pairwise comparison matrices with fuzzy elements. Fuzzy elements of the pairwise comparison matrix are applied whenever the decision maker is not sure about the value of his/her evaluation of the relative importance of elements in question. We particularly deal with pairwise comparison matrices with fuzzy number elements and investigate some properties of such matrices. In comparison with pairwise comparison matrices with crisp elements investigated in the previous chapter, here we investigate pairwise comparison matrices with elements from alo-group over a real interval. In some sense, this chapter is a continuation of the second part of the previous chapter. Such an approach allows for a generalization dealing with additive, multiplicative and fuzzy pairwise comparison matrices with fuzzy elements. Moreover, we deal with the problem of measuring the inconsistency of fuzzy pairwise comparison matrices by defining corresponding inconsistency indexes. Numerical examples are presented to illustrate the concepts and derived properties.

3.1 Introduction

In this chapter we again deal with pairwise comparison matrices that we have already investigated in Chap. 2. However, when comparing this chapter to Chap. 2, we deal with *pairwise comparison matrices with fuzzy elements*, or, shortly, *PCF matrices*. Fuzzy elements of the pairwise comparison matrix could be applied whenever the decision maker is not sure about the preference degree of his/her evaluation of elements in question.

A *decision making problem (DM problem)* is formulated as follows, see also Sect. 2.2 in Chap. 2: Let $X = \{x_1, x_2, \dots, x_n\}$ be a finite set of alternatives ($n > 2$). The aim is to rank the alternatives from the best to the worst (or, vice versa), using the information given by a DM in the form of $n \times n$ PCF matrix $\tilde{A} = \{\tilde{a}_{ij}\}$.

An ordinal *ranking* of alternatives is required to obtain the best alternative(s), however, it often occurs that the DM is not satisfied with the ordinal ranking among alternatives and therefore a cardinal ranking i.e. rating is required.

The chapter was written by "Jaroslav Ramík".

M. Gavalec et al., *Decision Making and Optimization*, Lecture Notes in Economics and Mathematical Systems 677, DOI 10.1007/978-3-319-08323-0_3

In the recent literature we can find papers dealing with applications of pairwise comparison method where evaluations require fuzzy quantities, for instance when evaluating regional projects, web pages, e-commerce proposals etc., see e.g. [1, 5, 7, 8, 11, 19]. In the paper [4] the author proposed a method for measuring inconsistency of fuzzy pair-wise comparison matrix based on Saaty's principal eigenvector method. However, this method is rather cumbersome and numerically difficult. The earliest work in AHP using fuzzy quantities as data was published by van Laarhoven and Pedrycz [18]. They compared fuzzy ratios described by triangular membership functions. The method of logarithmic least squares was used to derive local fuzzy priorities. Later on, using a geometric mean, Buckley et al. [3] determined fuzzy priorities of comparison ratios whose membership functions were assumed trapezoidal. The issue of consistency in AHP using fuzzy sets as elements of the matrix was first tackled by Salo in [17]. Departing from the fuzzy arithmetic approach, fuzzy weights using an auxiliary mathematical programming formulation describing relative fuzzy ratios as constraints on the membership values of local priorities were derived. Leung and Cao, see [9], proposed a notion of tolerance deviation of fuzzy relative importance that is strongly related to Saaty's consistency ratio.

The former works that solved the problem of finding a rank of the given alternatives based on some PCF matrix are [2, 6, 9–13, 16, 20]. In [20] some simple linear programming models for deriving the priority vectors from various interval fuzzy preference relations are proposed. Leung and Cao [9] proposed a new definition of the PCF reciprocal matrix by setting deviation tolerances based on an idea of allowing inconsistent information. Mahmoudzadeh and Bafandeh [10] further discussed Leung and Cao's work and proposed a new method of fuzzy consistency test by direct fuzzification of QR (Quick Response) algorithm which is one of the numerical methods for calculating eigenvalues of an arbitrary matrix. Ramik and Korviny in [15] investigated inconsistency of pairwise comparison matrix with fuzzy elements based on geometric mean. They proposed an inconsistency index which, however, does not measure inconsistency as well as uncertainty ideally.

In what follows we shall investigate pairwise comparison matrices with elements being fuzzy quantities of the alo-group over an interval of the real line **R**. Such an approach allows for unifying the theory dealing with additive, multiplicative and fuzzy pairwise comparison matrices, see also Chap. 2, Examples 2.11, 2.12. Particularly, we shall deal with PC matrices where the elements are *fuzzy numbers* of the alo-group over a real interval, we shall call them shortly *PCFN matrices*. Moreover, for PCFN matrices we assume that all diagonal elements are crisp, particularly they are equal to the identity element of $\mathcal{G}$, i.e. $\tilde{a}_{ii} = e$ for all $i \in \{1, 2, \ldots, n\}$:

$$\tilde{A} = \begin{bmatrix} e & \tilde{a}_{12} & \cdots & \tilde{a}_{1n} \\ \tilde{a}_{21} & e & \cdots & \tilde{a}_{2n} \\ \vdots & \vdots & \ddots & \vdots \\ \tilde{a}_{n1} & \tilde{a}_{n2} & \cdots & e \end{bmatrix} . \tag{3.1}$$

We shall derive some properties of such matrices, see also Chap. 1, Sects. 1.11 and 1.12. To avoid misunderstanding we shall strictly distinguish: On one side, a *fuzzy PC matrix* where the matrix elements are crisp numbers from the unit interval [0, 1] we have already investigated in Chap. 2. On the other side, we consider a *PC matrix with fuzzy elements*, where the elements are fuzzy numbers on **R**, see also Chap. 1, Sect. 1.10. In the former case, a fuzzy PC matrix is a valued relation on X in the sense of Definition 1.14, in other words, a fuzzy relation on X in the sense of Definition 1.17. The latter case will be investigated here in this chapter.

In Sect. 3.2 we present some basic concepts and ideas of PC matrices with elements being fuzzy numbers of the alo-group over a real interval. In Sect. 3.3 some methods for deriving priorities from PCFN matrices are proposed. The material and the results presented in these two sections are completely new and have not been published yet. Moreover, Sect. 3.3 consists of four subsections dealing with the particular character of the alo-groups of the real line: additive, multiplicative, fuzzy additive and fuzzy multiplicative one. In Sect. 3.4, illustrative numerical examples are presented and discussed.

3.2 PC Matrices with Elements Being Fuzzy Sets of Alo-Group over a Real Interval

Let G be a proper interval of the real line **R** and $\leq$ is the total order on G inherited from the usual order on **R**, $\mathcal{G} = (G, \odot, \leq)$ be a real alo-group. We also assume that $\mathcal{G}$ is a divisible and continuous alo-group. Then G is an open interval, see e.g. [11].

A *pairwise comparison system over* G is a pair $(X, \tilde{A})$ constituted by a set $X = \{x_1, x_2, \ldots, x_n\}$ and a fuzzy relation $\tilde{A} : X \times X \to \mathcal{F}_{LRN}(G)$, where $\tilde{A}(x_i, x_j) = \tilde{a}_{ij} \in \mathcal{F}_{LRN}(G)$, $\mathcal{F}_{LRN}(G)$ is the set of all triangular (L, R)-fuzzy numbers on G and $\tilde{A} = \{\tilde{a}_{ij}\}$ is an $n \times n$ PCF matrix. In the context of a DM problem, the element $\tilde{a}_{ij}$ can be interpreted as an uncertain measure on G of the preference of x_i over x_j.

In most of the literature on fuzzy sets a triangular fuzzy number is defined by means of linear (L, R) functions, here we consider a more general case: strictly monotone functions or constant functions.

Let $\tilde{A} = \{\tilde{a}_{ij}\}$ be an $n \times n$ PCF matrix where each element $\tilde{a}_{ij}$ of $\tilde{A}$ is a triangular (L, R)-fuzzy number with the membership function $\mu_{\tilde{a}_{ij}}, i, j \in \{1, 2, \ldots, n\}, i \neq j$, given as follows:

$$\mu_{\tilde{a}_{ij}}(x) = \begin{cases} L_{ij}^{-1}(x) & if x \in [L_{ij}(0), L_{ij}(1)], \\ R_{ij}^{-1}(x) & if x \in [R_{ij}(1), R_{ij}(0)], \\ 0 & otherwise, \end{cases} \tag{3.2}$$

where L_{ij} is either an increasing continuous function, $L_{ij} : [0, 1] \to \mathbf{R}$, or L_{ij} is a constant function mapping interval [0, 1] into a given point $y \in \mathbf{R}$. Similarly, R_{ij}

is either a decreasing continuous function, $R_{ij} : [0,1] \to \mathbf{R}$, or R_{ij} is a constant function mapping interval $[0,1]$ into a given point $y \in \mathbf{R}$. We assume that $L_{ij}(1) = R_{ij}(1)$. By L_{ij}^{-1}, or R_{ij}^{-1}, we denote the inverse functions of the increasing function L_{ij}, or, decreasing function R_{ij}, respectively. If L_{ij}, or R_{ij} is a constant function, then for $y \in \mathbf{R}$ we define $L_{ij}^{-1}(y) = R_{ij}^{-1}(y) = 1$. The functions L_{ij} and R_{ij} are called the *left and right membership generating functions of* $\tilde{a}_{ij}$, respectively.

Moreover, for all $i \in \{1,\ldots,n\}$ we assume that

$$\mu_{\tilde{a}_{ii}}(x) = \begin{cases} 1 & if x = e, \\ 0 & otherwise, \end{cases} \tag{3.3}$$

where e is the *identity element of* $\mathcal{G}$. Then the matrix $\tilde{A} = \{\tilde{a}_{ij}\}$ is called the *PCFN matrix on* $\mathcal{G}$, shortly, *PCFN matrix*.

It is clear that each entry $\tilde{a}_{ij}$ of the PCFN matrix $\tilde{A} = \{\tilde{a}_{ij}\}$ can be identified with a triple $(a_{ij}^L; a_{ij}^M; a_{ij}^R)_{L_{ij},R_{ij}}$, where L_{ij} and R_{ij} are the membership generating functions with the properties (3.2), (3.3), and $a_{ij}^L = L_{ij}(0), a_{ij}^M = L_{ij}(1) = R_{ij}(1), a_{ij}^R = R_{ij}(0)$. For the sake of simplicity, if there is no danger of misunderstanding, we shall omit the subscripts referring to functions L_{ij}, R_{ij}, i.e. we simply write $\tilde{a}_{ij} = (a_{ij}^L; a_{ij}^M; a_{ij}^R)$. The elements of the PCFN matrix with the above mentioned properties will be called simply the *triangular fuzzy numbers*.

Remark 3.1. Notice that the crisp numbers (non-fuzzy numbers) are special cases of triangular fuzzy numbers. If for some $i, j \in \{1,2,\ldots,n\}$, $\tilde{a}_{ij}$ is a crisp number, then the corresponding functions L_{ij} and R_{ij} are constant and $\tilde{a}_{ij} = (a_{ij}^L; a_{ij}^M; a_{ij}^R)$ with $a_{ij}^L = a_{ij}^M = a_{ij}^R$. Moreover, the elements on the main diagonal of the matrix are crisp. On the other hand, if $\tilde{a}_{ij}$ is a proper fuzzy number (non-crisp one), then the corresponding L_{ij} and/or R_{ij} are strictly monotone functions and $\tilde{a}_{ij} = (a_{ij}^L; a_{ij}^M; a_{ij}^R)$ with $a_{ij}^L < a_{ij}^M = a_{ij}^R$, or $a_{ij}^L = a_{ij}^M < a_{ij}^R$, eventually $a_{ij}^L < a_{ij}^M < a_{ij}^R$. Notice that here the superscripts L, M and R mean "Left value", "Middle value" and "Right value", respectively.

Remark 3.2. The triangular fuzzy numbers $\tilde{a} = (a^L; a^M; a^R)$ are also appropriate in group decision making (GDM), where a^L can be interpreted as the minimum possible value of DMs judgments, a^R is interpreted as the maximum possible value of DMs judgments, and a^M—the mean value of the DMs judgments—is interpreted as the mean value, or, the most possible value of DMs judgments, see e.g. [5].

Let $\tilde{A} = \{\tilde{a}_{ij}\}$ be a PCFN matrix, $\alpha \in]0,1], i, j \in \{1,2,\ldots,n\}$. We define the α-cut, $[\tilde{a}_{ij}]_\alpha$, of $\tilde{a}_{ij}$ as:

$$[\tilde{a}_{ij}]_\alpha = \{x \in \mathbf{R} | \mu_{\tilde{a}_{ij}}(x) \geq \alpha\}. \tag{3.4}$$

Then we obtain

$$[\tilde{a}_{ij}]_\alpha = [L_{ij}(\alpha), R_{ij}(\alpha)], \tag{3.5}$$

where L_{ij}, R_{ij} are the membership generating functions of $\tilde{a}_{ij}$.

For $\alpha = 0$, we define the zero-cut, $[\tilde{a}_{ij}]_0$, as:

$$[\tilde{a}_{ij}]_0 = [L_{ij}(0), R_{ij}(0)]. \tag{3.6}$$

By the well known representation theorem, each fuzzy number can be represented by a family of its α-cuts, hence, each entry $\tilde{a}_{ij}$ of the PCFN matrix $\tilde{A} = \{\tilde{a}_{ij}\}$ can be equivalently represented as a family of α-cuts of $\tilde{a}_{ij}$, particularly, a family of the closed intervals

$$[L_{ij}(\alpha), R_{ij}(\alpha)], \alpha \in [0, 1]. \tag{3.7}$$

With respect to the notation introduced in Remark 3.1, we denote for $\alpha \in [0, 1]$:

$$a_{ij}^L(\alpha) = L_{ij}(\alpha), \tag{3.8}$$

$$a_{ij}^R(\alpha) = R_{ij}(\alpha). \tag{3.9}$$

Now, we introduce a reciprocity property for PCFN matrices. Our definition is based on α-cuts and extends the definitions of reciprocity presented in the literature, e.g. [14].

Let $\tilde{A} = \{\tilde{a}_{ij}\}$ be an $n \times n$ PCFN matrix on $\mathcal{G} = (G, \odot, \leq)$, $\alpha \in [0, 1]$. $\tilde{A}$ is said to be *α-$\odot$-reciprocal*, if the following condition (C1) holds:

(C1)

For every $i, j \in \{1, 2, \ldots, n\}$ and every $a_{ij} \in [\tilde{a}_{ij}]_\alpha$ there exists $a_{ji} \in [\tilde{a}_{ji}]_\alpha$ such that

$$a_{ij} \odot a_{ji} = e. \tag{3.10}$$

$\tilde{A}$ is said to be *$\odot$-reciprocal*, if condition (C1) holds for all $\alpha \in [0, 1]$.

Remark 3.3. If $\tilde{A} = \{\tilde{a}_{ij}\}$ is a PCFN matrix with crisp elements then condition (C1) coincides with the classical definitions of reciprocity, see e.g. [11]:

- $A = \{a_{ij}\}$ is additive-reciprocal if $a_{ij} = -a_{ji}$ for all i and j; here $a_{ij} \in]-\infty, +\infty[$.
- $A = \{a_{ij}\}$ is multiplicative-reciprocal if $a_{ij} = \frac{1}{a_{ji}}$ for all i and j; here $a_{ij} \in]0, +\infty[$.
- $A = \{a_{ij}\}$ is fuzzy-reciprocal if $a_{ij} = 1 - a_{ji}$ for all i and j; here $a_{ij} \in]0, 1[$.

Proposition 3.1. *Let $\tilde{A} = \{\tilde{a}_{ij}\}$ be a PCFN matrix, $\alpha \in [0, 1]$. $\tilde{A}$ is α-$\odot$-reciprocal, if and only if*

(i)

$$a_{ij}^L(\alpha) \odot a_{ji}^R(\alpha) = e \text{ and } a_{ij}^R(\alpha) \odot a_{ji}^L(\alpha) = e. \tag{3.11}$$

(ii)

$$[a_{ji}^{L}(\alpha), a_{ji}^{R}(\alpha)] = [(R_{ij}(\alpha))^{(-1)}, (L_{ij}(\alpha))^{(-1)}]. \tag{3.12}$$

(iii)

$$[a_{ji}^{L}(\alpha), a_{ji}^{R}(\alpha)] = [(a_{ij}^{R}(\alpha))^{(-1)}, (a_{ij}^{L}(\alpha))^{(-1)}]. \tag{3.13}$$

Proof. Let (iii) be satisfied, then $a_{ij}^{L}(\alpha) \le a_{ij} \le a_{ij}^{R}(\alpha)$ if and only if

$$(a_{ij}^{R}(\alpha))^{(-1)} \le (a_{ij})^{(-1)} \le (a_{ij}^{L}(\alpha))^{(-1)}.$$

Setting $a_{ji} = (a_{ij})^{(-1)}$ we obtain the required (i).

(ii) follows directly from (i) and (3.7).

(iii) follows directly from (ii), (3.8) and (3.9). The equivalence of the three conditions is proven.

Proposition 3.1 extends the known result on interval reciprocal comparison matrices, see [14, 20]. Particularly, if L_{ij} and R_{ij} are the membership generating functions of elements $\tilde{a}_{ij}$, where $\tilde{A} = \{\tilde{a}_{ij}\}$ is an $\odot$-reciprocal PCFN matrix, then $R_{ij}^{(-1)}$ and $L_{ij}^{(-1)}$ are the membership generating functions of the symmetric elements $\tilde{a}_{ji}$.

Example 3.1. Consider $\odot = +$, let $\tilde{A} = \{(a_{ij}^{L}; a_{ij}^{M}; a_{ij}^{R})_{L_{ij}R_{ij}}\}$ be a PCFN matrix with the non-linear membership generating functions for elements $\tilde{a}_{ij}$: $L_{ij}(t) = (a_{ij}^{M} - a_{ij}^{L})\sqrt{t} + a_{ij}^{L}$, $R_{ij}(t) = a_{ij}^{R} - (a_{ij}^{R} - a_{ij}^{M})t^2$, where $t \in [0, 1]$ and $1 \le i < j \le 3$.

Particularly, let $\tilde{A}$ be as follows:

$$\tilde{A} = \begin{pmatrix} 0 & (1;2;4)_{L_{12}R_{12}} & (4;6;7)_{L_{13}R_{13}} \\ (-4;-2;-1)_{L_{21}R_{21}} & 0 & (3;4;4)_{L_{23}L_{23}} \\ (-7;-6;-4)_{L_{31}R_{31}} & (-4;-4;3)_{L_{32}R_{32}} & 0 \end{pmatrix},$$

Here, for the symmetric elements, we have $L_{ji}(t) = (R_{ij}(t))^{(-1)} = -a_{ij}^{R} + (a_{ij}^{R} - a_{ij}^{M})t^2$, $R_{ji}(t) = (L_{ij}(t))^{(-1)} = -(a_{ij}^{M} - a_{ij}^{L})\sqrt{t} - a_{ij}^{L}$, where $t \in [0, 1]$ and $1 \le i < j \le 3$. By Proposition 3.1, $\tilde{A}$ is $+$-reciprocal, i.e. α-$+$-reciprocal for all $\alpha \in [0, 1[$.

Now, consider a similar PCFN matrix $\tilde{B}$ given as follows:

$$\tilde{B} = \begin{pmatrix} 0 & (1;2;4) & (4;6;7) \\ (-3;-2;-1) & 0 & (3;4;4) \\ (-7;-6;-4) & (-4;-4;3) & 0 \end{pmatrix}.$$

Here, all membership generating functions L_{ij} and R_{ij} are linear, i.e. $L_{ij}(t) = (a_{ij}^M - a_{ij}^L)t + a_{ij}^L$, $R_{ij}(t) = a_{ij}^R - (a_{ij}^R - a_{ij}^M)t$, where $t \in [0,1]$ and $1 \leq i, j \leq 3, i \neq j$.

Clearly, $\tilde{B}$ is not +-reciprocal as (3.11) is not satisfied for $\tilde{a}_{12} = (1;2;4)$ and $\tilde{a}_{21} = (-3;-2;-1)$ for all $\alpha \in [0,1[$.

Considering an $\odot$-reciprocal matrix, we shall be interested in a consistency property of the PCFN matrix. Our definition shall satisfy definitions of consistency presented in the literature, e.g. [14]. We start with the crisp case.

Let the PCFN matrix $A = \{a_{ij}\}$ be crisp. Then we have he following definition, see e.g. [12, 13].

$A = \{a_{ij}\}$ is *$\odot$-consistent* if for all $i, j, k \in \{1, 2, \ldots, n\}$

$$a_{ij} = a_{ik} \odot a_{kj}. \tag{3.14}$$

We obtain the following result, the proof of which is easy, see e.g. [12].

Proposition 3.2. *$A = \{a_{ij}\}$ is $\odot$-consistent if and only if there exists a vector $w = (w_1, w_2, \ldots, w_n)$, $w_i \in G$ such that*

$$w_i \div w_j = a_{ij} \text{ for all } i, j \in \{1, 2, \ldots, n\}. \tag{3.15}$$

Let $\alpha \in [0,1]$, a PCFN matrix $\tilde{A} = \{\tilde{a}_{ij}\}$ is said to be *α-$\odot$-consistent*, if the following condition (C2) holds:

(C2)

For every $i, j, k \in \{1, 2, \ldots, n\}$, there exist $a_{ij} \in [\tilde{a}_{ij}]_\alpha$, $a_{ik} \in [\tilde{a}_{ik}]_\alpha$ and $a_{kj} \in [\tilde{a}_{kj}]_\alpha$ such that

$$a_{ij} = a_{ik} \odot a_{kj}.$$

$\tilde{A}$ is said to be *$\odot$-consistent*, if condition (C2) holds for all $\alpha \in [0,1]$.

Remark 3.4. Let $\alpha, \beta \in [0,1], \alpha \geq \beta$. If $\tilde{A} = \{\tilde{a}_{ij}\}$ is α-$\odot$-consistent, then $\tilde{A}$ is β-$\odot$-consistent.

Let $\alpha \in [0,1]$, a vector $w = (w_1, w_2, \ldots, w_n)$ with $w_i \in G$ for all $i \in \{1, 2, \ldots, n\}$, is an *$\alpha$-$\odot$-consistent vector with respect to* $\tilde{A} = \{\tilde{a}_{ij}\}$ if for every $i, j \in \{1, 2, \ldots, n\}$ there exist $a_{ij} \in [\tilde{a}_{ij}]_\alpha$ such that

$$w_i \div w_j = a_{ij}. \tag{3.16}$$

Example 3.2. Consider $\odot = \cdot$, let $\tilde{C} = \{\tilde{c}_{ij}\}$ be a PCFN matrix with the membership generating functions for elements $\tilde{c}_{ij}$: L_{ij}, R_{ij}, where $1 \leq i < j \leq 3$. Here, for the symmetric elements, we assume that $L_{ji}(t) = (R_{ij}(t))^{-1}$ and $R_{ji}(t) =$

$(L_{ij}(t))^{-1}$, where $t \in [0,1]$ and $1 \le i < j \le 3$. Matrix $\tilde{C}$ is given as follows:

$$\tilde{C} = \begin{pmatrix} 1 & (1;2;2) & (2;6;8) \\ (\frac{1}{2};\frac{1}{2};1) & 1 & (2;3;4) \\ (\frac{1}{8};\frac{1}{6};\frac{1}{2}) & (\frac{1}{4};\frac{1}{3};\frac{1}{2}) & 1 \end{pmatrix},$$

By Proposition 3.1, $\tilde{C}$ is $\cdot$-reciprocal, i.e. α-$\cdot$-reciprocal for all $\alpha \in [0,1]$. Moreover, $\tilde{C}$ is $\cdot$-consistent, i.e. α-$\cdot$-consistent for all $\alpha \in [0,1]$.

Moreover, let $\tilde{D} = \{\tilde{d}_{ij}\}$ be a PCFN matrix with the membership generating functions for elements $\tilde{d}_{ij}$: L_{ij}, R_{ij}, where $1 \le i < j \le 3$. Again, for the symmetric elements, we assume that $L_{ji}(t) = (R_{ij}(t))^{-1}$ and $R_{ji}(t) = (L_{ij}(t))^{-1}$, where $t \in [0,1]$ and $1 \le i < j \le 3$. Matrix $\tilde{D}$ is given as follows:

$$\tilde{D} = \begin{pmatrix} 1 & (1;2;2) & (7;8;9) \\ (\frac{1}{2};\frac{1}{2};1) & 1 & (2;3;3) \\ (\frac{1}{9};\frac{1}{8};\frac{1}{7}) & (\frac{1}{3};\frac{1}{3};\frac{1}{2}) & 1 \end{pmatrix}.$$

Here, $\tilde{D}$ is a 3×3 PCFN matrix with triangular fuzzy numbers as elements. Again, $\tilde{D}$ is $\cdot$-reciprocal, however, $\tilde{D}$ is not $\alpha - \cdot$-consistent as evidently (C2) is not satisfied for any $\alpha \in [0,1]$.

Remark 3.5. If $\tilde{A}$ is a PCFN matrix with crisp elements, i.e. $\tilde{A} = A$ and $\tilde{A}$ is $\odot$-consistent, then A is $\odot$-consistent in the sense of definition (3.14).

Proposition 3.3. *Let $\tilde{A} = \{\tilde{a}_{ij}\}$ be an $\odot$-reciprocal PCFN matrix, $\alpha \in [0,1]$. $\tilde{A}$ is α-$\odot$-consistent if and only if there exists a vector $w^\alpha = (w_1^\alpha, w_2^\alpha, \ldots, w_n^\alpha)$ with $w_i^\alpha \in G$ for all $i \in \{1,2,\ldots,n\}$ such that*

$$w_i^\alpha \div w_j^\alpha \in [\tilde{a}_{ij}]_\alpha \text{ for all } i,j \in \{1,2,\ldots,n\}, \tag{3.17}$$

or, equivalently

$$a_{ij}^L(\alpha) \le w_i^\alpha \div w_j^\alpha \le a_{ij}^R(\alpha) \text{ for all } i,j \in \{1,2,\ldots,n\}. \tag{3.18}$$

Proof. The proof follows directly from (C2), Proposition 3.2 and from (3.5).

Clearly, if $\tilde{A}$ is α-$\odot$-consistent, $\alpha \in [0,1]$, then $\tilde{A}$ is β-$\odot$-consistent for all β, $0 \le \beta \le \alpha$. Hence, if $\tilde{A}$ is 1-$\odot$-consistent, then $\tilde{A}$ is $\odot$-consistent.

If, at least for one triple of elements and some $\alpha \in [0,1]$, the condition (C2) is not satisfied, then the matrix $\tilde{A}$ is $\odot$-*inconsistent*. It is important to measure an intensity of $\odot$-inconsistency as in some cases the PCFN matrix can be "close" to an $\odot$-consistent matrix, or, in the other cases $\odot$-inconsistency can be strong, meaning that the PCFN matrix can be "far" from any $\odot$-consistent matrix.

3.3 Methods for Deriving Priorities from PCFN Matrices

In this section, we assume that all matrices are $n \times n\odot$-reciprocal PCFN matrices on $\mathcal{G} = (G, \odot, \leq)$, where G is an open interval in $]-\infty, +\infty[$. We define the priority vector for ranking the alternatives which will be based on Proposition 3.3, particularly on the optimal solution of the following optimization problem:

(P1$\odot$)

$$\alpha \longrightarrow \max; \tag{3.19}$$

subject to

$$a_{ij}^{L}(\alpha) \leq w_i \div w_j \leq a_{ij}^{R}(\alpha) \text{ for all } i, j \in \{1, 2, \ldots, n\}, i < j, \tag{3.20}$$

$$\bigodot_{k=1}^{n} w_k = e, \tag{3.21}$$

$$0 \leq \alpha \leq 1, w_k \in G, \text{ for all } k \in \{1, 2, \ldots, n\}. \tag{3.22}$$

If optimization problem (P1$\odot$) has a feasible solution, i.e. system of constraints (3.20)–(3.22) has a solution, then (P1$\odot$) has also an optimal solution. Let α^* and $w^* = (w_1^*, \ldots, w_n^*)$ be an optimal solution of problem (P1$\odot$). Then $\alpha^* \in [0, 1]$ and α^* is called the *$\odot$-consistency grade of $\tilde{A}$*, denoted by $G_{\odot}(\tilde{A})$, i.e.

$$G_{\odot}(\tilde{A}) = \alpha^*. \tag{3.23}$$

Here, $w^* = (w_1^*, \ldots, w_n^*)$ is an α^*-$\odot$-consistent vector with respect to $\tilde{A}$ called the *$\odot$-priority vector of $\tilde{A}$*.

If optimization problem (P1$\odot$) has no feasible solution, then we define

$$G_{\odot}(\tilde{A}) = 0. \tag{3.24}$$

Generally, problem (P1$\odot$) is a nonlinear optimization problem that can be efficiently solved e.g. by the dichotomy method, which is a sequence of optimization problems, with optimal solutions converging to the optimal solution of problem (P1$\odot$), see e.g. [14]. For instance, given $\alpha \in [0, 1]$, $\odot = +$, then problem of finding a feasible solution of (P1$\odot$) can be solved as an LP problem (with variables $w_1, \ldots, w_n$).

Notice that the definition of $\odot$-consistency grade of a PCFN matrix is an extension of the definition given in [14], where the operation $\odot$ is multiplication.

Proposition 3.4. *If $w^* = (w_1^*, \ldots, w_n^*)$ is an $\odot$-priority vector of a PCFN matrix $\tilde{A} = \{\tilde{a}_{ij}\}$, then w^* is unique.*

Proof. We have to prove that the optimal solution α^* and $w^* = (w_1^*, \dots, w_n^*)$ of problem (P1⊙) is unique. Let $w^+ = (w_1^+, \dots, w_n^+)$ be another optimal solution of problem (P1⊙) such that $w^* \neq w^+$. Then $w^0 = (w_1^0, \dots, w_n^0)$ with $w_i^0 = (w_i^* \odot w_i^+)^{(\frac{1}{2})}$ for all $i \in \{1, \dots, n\}$ is also an optimal solution of problem (P1⊙) with the value α^0 of the objective function (3.19). By strict monotonicity of L_{ij} and R_{ij} we obtain $\alpha^0 > \alpha^*$, which is a contradiction. Therefore, $w^* = w^+$.

Remark 3.6. If $\tilde{A}$ is an ⊙-consistent PCFN matrix, then $G_\odot(\tilde{A}) = 1$ and, particularly, the crisp matrix $A^M = \{a_{ij}^M\}$ is ⊙-consistent. By Proposition 3.2 there exists an ⊙-consistent vector $w^M = (w_1^M, \dots, w_n^M)$ with respect to A^M. Therefore, w^M is also a priority vector of $\tilde{A}$. If $\tilde{A}$ is crisp, then the concept of ⊙-priority vector of $\tilde{A}$ coincides with the same concept for crisp matrices, see [12].

Remark 3.7. The optimal solution α^* and $w^* = (w_1^*, \dots, w_n^*)$ of problem (P1⊙) should be unique as decision makers ask for unique decision, i.e. unique ranking of the alternatives in X. The sufficient condition for uniqueness of the priority vector $w^* = (w_1^*, \dots, w_n^*)$ is that all elements $\tilde{a}_{ij}$ of the PCFN matrix $\tilde{A}$ are triangular (L, R)-fuzzy numbers and, particularly, that the core of each $\tilde{a}_{ij}$,

$$Core(\tilde{a}_{ij}) = \{t \in G | \mu_{a_{ij}}(t) = 1\},$$

is a singleton. This is the reason why it has no sense to extend the fuzzy elements of PC matrix to (L, R)-fuzzy intervals (i.e. to trapezoidal fuzzy numbers).

If optimization problem (P1⊙) has no feasible solution, then $\tilde{A}$ is inconsistent and the ⊙-consistency grade of $\tilde{A}$ is zero, i.e. $G_\odot(\tilde{A}) = 0$. The inconsistency of $\tilde{A}$ will be measured by the minimum of the ⊙-mean distance of the *ratio matrix* $W = \{w_i \div w_j\}$ to matrix $\{[\tilde{a}_{ij}]_0\}$ where the elements are zero-cuts of $\tilde{a}_{ij}$, i.e. intervals $[\tilde{a}_{ij}]_0$.

Let $w = (w_1, \dots, w_n), w_i \in G$ for all $i \in \{1, \dots, n\}$. Denote

$$I_\odot(\tilde{A}, w) = \min\Big\{\Big(\bigodot_{i<j} \|a_{ij} \div (w_i \div w_j)\|\Big)^{\left(\frac{2}{n(n-1)}\right)} \;\Big|\; a_{ij} \in [\tilde{a}_{ij}]_0 \text{ for all } 1 \leq i < j \leq n\Big\}. \tag{3.25}$$

where $\| \dots \|$ is the $\mathcal{G}$-norm.

Proposition 3.5. (i) *If* $w = (w_1, \dots, w_n) \in G^n$, *then*

$$I_\odot(\tilde{A}, w) \geq e, \tag{3.26}$$

where e is the identity element of $\mathcal{G}$.

(ii) $\alpha \in [0, 1], w = (w_1, \dots, w_n)$ *is a feasible solution of (P1⊙), if and only if*

$$I_\odot(\tilde{A}, w) = e. \tag{3.27}$$

Proof. (i) Inequality (3.26) follows directly from definition (1.57) and from the fact that $\|a\| = \max\{a, a^{(-1)}\} \geq e$ for each $a \in G$.

(ii) 1. As conditions (3.20)–(3.22) are satisfied, setting $a_{ij} = w_i \div w_j$, by (3.20) we obtain $a_{ij} \in [\tilde{a}_{ij}]_0$, hence $a_{ij} \div (w_i \div w_j) = e$ for all $i, j \in \{1, 2, \ldots, n\}$, consequently, (3.27) is true.

2. Suppose that

$$I_{\odot}(\tilde{A}, w^*) = e,$$

for some $w^* = (w_1^*, \ldots, w_n^*) \in G^n$. Without loss of generality we may assume that $\bigodot_{k=1}^{n} w_k^* = e$, i.e. (3.21) holds. Then there exist $a_{ij}^* \in [\tilde{a}_{ij}]_0$ for all $i, j \in \{1, 2, \ldots, n\}, i < j$, such that

$$I_{\odot}(\tilde{A}, w^*) = \Big(\bigodot_{i<j} \|a_{ij}^* \div (w_i^* \div w_j^*)\|\Big)^{\left(\frac{2}{n(n-1)}\right)} = e.$$

Now, for all $i, j \in \{1, 2, \ldots, n\}, i < j$ we have $\|a_{ij}^* \div (w_i^* \div w_j^*)\| = e$, hence $a_{ij}^* \div (w_i^* \div w_j^*) = e$, which is equivalent to $a_{ij}^* = w_i^* \div w_j^*$. Therefore, $a_{ij}^L(0) \leq w_i^* \div w_j^* \leq a_{ij}^R(0)$, i.e. (3.20) is satisfied. We conclude that $\alpha^* = 0$ and $w^* = (w_1^*, \ldots, w_n^*)$ is a feasible solution of (P1$\odot$).

Now, we define a priority vector in case there is no feasible solution of (P1$\odot$). Clearly, such priority vector cannot become an α-$\odot$-consistency vector of $\tilde{A}$ for some $\alpha > 0$.

Consider the following optimization problem (P2$\odot$).

(P2$\odot$)

$$I_{\odot}(\tilde{A}, w) \longrightarrow \min; \tag{3.28}$$

subject to

$$\bigodot_{k=1}^{n} w_k = e, \tag{3.29}$$

$$w_k \in G, \text{ for all } k \in \{1, 2, \ldots, n\}. \tag{3.30}$$

We define the *$\odot$-consistency index of* $\tilde{A}$, $I_{\odot}(\tilde{A})$, as

$$I_{\odot}(\tilde{A}) = I_{\odot}(\tilde{A}, w^*), \tag{3.31}$$

where $w^* = (w_1^*, \ldots, w_n^*)$ is the optimal solution of (P2$\odot$).

Notice that optimization problem (P2$\odot$) has always an optimal solution as constraints (3.29), (3.30) are always solvable and (3.25) is continuous on G and bounded from below by the value e. If (P1$\odot$) has no feasible solution, then the optimal solution $w^* = (w_1^*, \dots, w_n^*)$ of (P2$\odot$) is called the $\odot$*-priority vector of* $\tilde{A}$.

Remark 3.8. In particular, assume that $\tilde{A}$ is crisp, i.e. $\tilde{A} = A$. If A is $\odot$-consistent, then by (3.23) we obtain $G_\odot(A) = 1$ and $I_\odot(A) = e$. However, if A is $\odot$-inconsistent, then $G_\odot(A) = 0$ and by the properties of the distance function (3.25), we obtain $I_\odot(A) > e$.

Remark 3.9. If $\odot$-consistency grade $G_\odot(\tilde{A}) > 0$, then by Proposition 3.4 the $\odot$-priority vector of $\tilde{A}$ is unique. However, if $G_\odot(\tilde{A}) = 0$, then the uniqueness of optimal solution of (P2$\odot$) is not sure. Depending on the particular operation $\odot$, problem (P2$\odot$) may have multiple optimal solutions—priority vectors. This is an unfavorable situation from the point of view of the DM. An appropriate modification of (3.25) so that the unique optimal solution of (P2$\odot$) is achieved is an open problem.

In the following subsections we investigate four particular cases of the alo-group: additive, multiplicative, fuzzy additive, and fuzzy multiplicative.

3.3.1 PCFN Matrix on Additive Alo-Group

Additive alo-group $\mathcal{R} = (]-\infty, +\infty[, +, \leq)$ is a continuous alo-group with: $e = 0$, $x^{(-1)} = -x$, $x \div y = x - y$, $\|a\| = max\{a, a^{-1}\} = |a|$; and $d_{\mathcal{R}}(a,b) = \|a \div b\| = |a - b| \in [0, +\infty[$. The alo-group $\mathcal{R}$ is divisible and the (n)-root of a is denoted as $x = a/n$. The mean $m_+(a_1, a_2, \dots, a_n)$ is the arithmetic mean denoted as $\frac{1}{n}\sum_{i=1}^{n} a_i$.

Let $\tilde{A} = \{\tilde{a}_{ij}\}$ be an $n \times n$ PCFN matrix on $\mathcal{G}$, $G =]-\infty, +\infty[$. An appropriate priority vector based on the PCFN matrix $\tilde{A}$ for ranking the given alternatives can be defined by the optimal solution of the following optimization problem (P1+).

(P1+)

$$\alpha \longrightarrow \max; \tag{3.32}$$

subject to

$$a_{ij}^L(\alpha) \leq w_i - w_j \leq a_{ij}^R(\alpha) \text{ for all } i, j \in \{1, 2, \dots, n\}, i < j, \tag{3.33}$$

$$\sum_{k=1}^{n} w_k = 0, \tag{3.34}$$

$$0 \leq \alpha \leq 1. \tag{3.35}$$

If optimization problem (P1+) has a feasible solution, i.e. system of constraints (3.33)–(3.35) is solvable, then (P1+) has also an optimal solution. Let α^* and $w^* = (w_1^*, \dots, w_n^*)$ be an optimal solution of problem (P1+). Then $\alpha^* \geq 0$ and

α^* is called the *+-consistency grade of* $\tilde{A}$, denoted by $G_+(\tilde{A})$, i.e.

$$G_+(\tilde{A}) = \alpha^*. \tag{3.36}$$

If optimization problem (P1+) has no feasible solution, then we define

$$G_+(\tilde{A}) = 0. \tag{3.37}$$

This nonlinear optimization problem can be solved using the dichotomy method mentioned before, for more details, see [14]. The optimal solution of problem (P1+), $w^* = (w_1^*, \ldots, w_n^*)$, is an α^*-+-consistent vector with respect to $\tilde{A}$ called the *+-priority vector of* $\tilde{A}$. In case $G_+(\tilde{A}) = 0$, and no feasible solution of (P1+) exists, then we define the +-priority vector of $\tilde{A}$ as an optimal solution of the following optimization problem.

(P2+)

$$I_+(\tilde{A}, w) \longrightarrow \min; \tag{3.38}$$

subject to

$$\sum_{k=1}^{n} w_k = 0, \tag{3.39}$$

where

$$I_+(\tilde{A}, w) = \min\{\frac{2}{n(n-1)} \sum_{i<j} |a_{ij} - w_i + w_j| \mid a_{ij} \in [\tilde{a}_{ij}]_0 \text{ for all } 1 \leq i < j \leq n\}. \tag{3.40}$$

It is clear that the optimization problem (P2+) has always an optimal solution. We obtain the *+-consistency index of* $\tilde{A}$, $I_+(\tilde{A})$, by

$$I_+(\tilde{A}) = I_+(\tilde{A}, w^*), \tag{3.41}$$

where $w^* = (w_1^*, \ldots, w_n^*)$ is the optimal solution of (P2+). If there is no feasible solution of (P1+), then $w^* = (w_1^*, \ldots, w_n^*)$ is called the *+-priority vector of* $\tilde{A}$.

Remark 3.10. In particular, assume that $\tilde{A}$ is crisp, i.e. $\tilde{A} = A$. If A is +-consistent, then by (3.23) we obtain $G_+(A) = 1$ and $I_+(A) = 0$. However, if A is +-inconsistent, then $G_+(A) = 0$ and $I_+(A) > 0$.

Some numerical examples can be found in Sect. 3.4.

3.3.2 PCFN Matrix on Multiplicative Alo-Group

Multiplicative alo-group $\mathcal{R}^+ = (]0, +\infty[, \bullet, \leq)$ is a continuous alo-group with: $e = 1$, $x^{(-1)} = x^{-1} = 1/x$, $x^{(n)} = x^n$, $x \div y = \frac{x}{y}$, $\|a\| = max\{a, a^{-1}\}$; and $d_{\mathcal{R}^+}(a, b)$ is given by $\|a \div b\| = \|\frac{a}{b}\| = max\{\frac{a}{b}, \frac{b}{a}\} \in [1, +\infty[$. The alo-group $\mathcal{R}^+$ is divisible and the (n)-root of a is $x = a^{1/n}$. The mean $m_\bullet(a_1, a_2, \dots, a_n)$ is the geometric mean $\left(\prod_{i=1}^{n} a_i\right)^{1/n}$. Here, for the sake of clarity, instead of usual multiplication symbol $\cdot$ we use the symbol $\bullet$.

An appropriate priority vector based on the PCFN matrix $\tilde{A} = \{\tilde{a}_{ij}\}$ for ranking the given alternatives is defined by the optimal solution of the following optimization problem (P1$\bullet$).

(P1$\bullet$)

$$\alpha \longrightarrow \max; \tag{3.42}$$

subject to

$$a_{ij}^L(\alpha) \leq \frac{w_i}{w_j} \leq a_{ij}^R(\alpha) \text{ for all } i, j \in \{1, 2, \dots, n\}, i < j, \tag{3.43}$$

$$\prod_{k=1}^{n} w_k = 1, \tag{3.44}$$

$$0 \leq \alpha \leq 1, w_k > 0, \text{ for all } k \in \{1, 2, \dots, n\}. \tag{3.45}$$

If optimization problem (P1$\bullet$) has a feasible solution, i.e. system of constraints (3.43)–(3.45) is solvable, then (P1$\bullet$) has also an optimal solution.

Let α^* and $w^* = (w_1^*, \dots, w_n^*)$ be an optimal solution of problem (P1$\bullet$). Then $\alpha^* \geq 0$ and α^* is called the *$\bullet$-consistency grade of* $\tilde{A}$, denoted by $G_\bullet(\tilde{A})$, i.e.

$$G_\bullet(\tilde{A}) = \alpha^*. \tag{3.46}$$

Here, $w^* = (w_1^*, \dots, w_n^*)$ is an α^*-$\bullet$-consistent vector with respect to $\tilde{A}$ called the *$\bullet$-priority vector of* $\tilde{A}$.

If optimization problem (P1$\bullet$) has no feasible solution, then we define

$$G_\bullet(\tilde{A}) = 0. \tag{3.47}$$

Problem (P1$\bullet$) is an optimization problem that can be efficiently solved by the dichotomy method described in the previous subsection, see also [14]. Then, for given $\alpha \in [0, 1]$, problem (P1$\bullet$) can be solved as an LP problem (with variables $w_1, \dots, w_n$).

Let $w = (w_1, \dots, w_n), w_i \in]0, +\infty[$ for all $i \in \{1, \dots, n\}$. Similarly to (3.25) and (3.40) denote

$$I_{\bullet}(\tilde{A}, w) = \min\{(\prod_{i<j} \| \frac{a_{ij}}{\frac{w_i}{w_j}} \|)^{\frac{2}{n(n-1)}} \mid a_{ij} \in [\tilde{a}_{ij}]_0 \text{ for all } 1 \le i < j \le n\}. \quad (3.48)$$

Now, we define a suitable priority vector also in case if no feasible solution of (P1•) exists. In contrast to the case of $G_{\bullet}(\tilde{A}) > 0$, this priority vector cannot be an α-•-consistency vector of $\tilde{A}$ for some $\alpha > 0$. However, it can be defined as an optimal solution of the following optimization problem.

(P2•)

$$I_{\bullet}(\tilde{A}, w) \longrightarrow \min; \quad (3.49)$$

subject to

$$\prod_{k=1}^{n} w_k = 1, \quad (3.50)$$

$$w_k > 0, \text{ for all } k \in \{1, 2, \dots, n\}. \quad (3.51)$$

It is clear that the optimization problem (P2•) has always an optimal solution. We obtain the *•-consistency index of* $\tilde{A}$, $I_{\bullet}(\tilde{A})$, by

$$I_{\bullet}(\tilde{A}) = I_{\bullet}(\tilde{A}, w^*), \quad (3.52)$$

where $w^* = (w_1^*, \dots, w_n^*)$ is the optimal solution of (P2•). If there is no feasible solution of (P1•), then $w^* = (w_1^*, \dots, w_n^*)$ is called the *•-priority vector of* $\tilde{A}$.

Remark 3.11. In particular, assume that $\tilde{A}$ is crisp, i.e. $\tilde{A} = A$. If A is •-consistent, then by (3.23) we obtain $G_{\bullet}(A) = 1$ and also $I_{\bullet}(A) = 1$. However, if A is •-inconsistent, then $G_{\bullet}(A) = 0$ and $I_{\bullet}(A) > 1$.

Remark 3.12. The same consistency index $I_{\bullet}(\tilde{A})$, can be achieved when solving problem (P2•), where instead of constraint (3.50) we use

$$\sum_{k=1}^{n} w_k = 1. \quad (3.53)$$

Then the •-priority vector of $\tilde{A}$ is calculated as $(\frac{w_1}{c}, \dots, \frac{w_n}{c})$, where $c = (\prod_{k=1}^{n} w_k)^{\frac{1}{n}}$.

Some illustrative examples can be found in Sect. 3.4.

3.3.3 PCFN Matrix on Fuzzy Additive Alo-Group

Fuzzy additive alo-group $\mathscr{R}_a = (]-\infty, +\infty[, +_f, \leq)$ is a continuous alo-group with:

$$a +_f b = a + b - 0.5,\ e = 0.5,\ a^{(-1)} = 1 - a,$$
$$a^{(n)} = n.a - \frac{n-1}{2},\ a \div_f b = a - b + 0.5,\ \|a\| = max\{a, 1-a\}, \tag{3.54}$$

and

$$d_{\mathscr{R}_a}(a,b) = \|a -_f b\| = max\{a - b + 0.5, b - a + 0.5\}.$$

The fuzzy additive alo-group $\mathscr{R}_a$ is divisible and the (n)-root of a is $x = \frac{a}{n} + \frac{n-1}{2n}$.

The product $a_1 \odot a_2 \odot \ldots \odot a_n = \sum_{k=1}^{n} a_k - \frac{n-1}{2}$ and mean $m_{+_f}(a_1, a_2, \ldots, a_n) = \frac{1}{n} \sum\limits_{i=1}^{n} a_i$.

The entry a_{ij} of a crisp pairwise comparison matrix $A = \{a_{ij}\}$ assumes different meanings: In additive PC matrices, it is a preference difference; in multiplicative PC matrices, it represents a preference ratio. PCFN matrices can be defined on additive or multiplicative alo-groups investigated already in the previous subsections. In fuzzy PC matrices, each entry a_{ij} quantifies in [0, 1] the distance from the value 0.5 expressing the indifference. Here, the fuzzy entries of PCFN matrices can be defined (by nonzero membership grades) on the unit interval [0, 1] and their membership functions can be defined as zero outside the unit interval. An appropriate group operation $+_f$ is defined in (3.54). By formula (3.14), the matrix $A = \{a_{ij}\}$ is $+_f$*-consistent* if

$$a_{ij} = a_{ik} +_f a_{kj} = a_{ik} + a_{kj} - 0.5. \tag{3.55}$$

In the literature, this type of consistency is called *fuzzy additive consistency*, see e.g. [13], or, *additive transitivity*, see e.g. [20]. Here, we call the appropriate alo-group $\mathscr{R}_a$, the *fuzzy additive alo-group*. As the group operation $+_f$ is not closed on the interval [0, 1], the basic set where this operation is closed is $]-\infty, +\infty[$.

Let $\tilde{A} = \{\tilde{a}_{ij}\}$ be an $n \times n$ PCFN matrix with the entries $\tilde{a}_{ij}$ being triangular (L, R)-fuzzy numbers on [0, 1]. An appropriate priority vector based on the PCFN matrix $\tilde{A}$ for ranking the given alternatives can be defined by the optimal solution of the following optimization problem (P1$+_f$).

(P1$+_f$)

$$\alpha \longrightarrow \max; \tag{3.56}$$

subject to

$$a_{ij}^{L}(\alpha) \leq w_i - w_j + 0.5 \leq a_{ij}^{R}(\alpha) \text{ for all } i, j \in \{1, 2, \dots, n\}, i < j, \tag{3.57}$$

$$\sum_{k=1}^{n} w_k = \frac{n}{2}, \tag{3.58}$$

$$0 \leq \alpha \leq 1. \tag{3.59}$$

If optimization problem (P1$+_f$) has a feasible solution, i.e. system of constraints (3.57)–(3.59) is solvable, then (P1$+_f$) has also an optimal solution. Let α^* and $w^* = (w_1^*, \dots, w_n^*)$ be an optimal solution of problem (P1$+_f$). Then $\alpha^* \geq 0$ and α^* is called the *$+_f$-consistency grade of* $\tilde{A}$, denoted by $G_{+_f}(\tilde{A})$, i.e.

$$G_{+_f}(\tilde{A}) = \alpha^*. \tag{3.60}$$

If optimization problem (P1$+_f$) has no feasible solution, then we define

$$G_{+_f}(\tilde{A}) = 0. \tag{3.61}$$

This optimization problem can be solved using the dichotomy method mentioned before, see [14]. The optimal solution of problem (P1$+_f$), $w^* = (w_1^*, \dots, w_n^*)$, is an α^*-$+f$-consistent vector with respect to $\tilde{A}$ called the *$+_f$-priority vector of* $\tilde{A}$. If no feasible solution of (P1$+_f$) exists, then we define the $+_f$-priority vector of $\tilde{A}$ as an optimal solution of the following optimization problem.

(P2$+_f$)

$$I_{+_f}(\tilde{A}, w) \longrightarrow \min; \tag{3.62}$$

subject to

$$\sum_{k=1}^{n} w_k = \frac{n}{2}. \tag{3.63}$$

Here,

$$I_{+_f}(\tilde{A}, w) = \min\Big\{\frac{2}{n(n-1)} \sum_{i<j} \|a_{ij} - w_i + w_j\| \;\Big|\; a_{ij} \in [\tilde{a}_{ij}]_0 \text{ for all } 1 \leq i < j \leq n\Big\}. \tag{3.64}$$

It is clear that the optimization problem (P2$+_f$) has always an optimal solution. We define the *$+_f$-consistency index of* $\tilde{A}$, $I_{+_f}(\tilde{A})$, as

$$I_{+_f}(\tilde{A}) = I_{+_f}(\tilde{A}, w^*), \tag{3.65}$$

where $w^* = (w_1^*, \dots, w_n^*)$ is the optimal solution of (P2$+_f$). If there is no feasible solution of (P1$+_f$), then $w^* = (w_1^*, \dots, w_n^*)$ is called the *$+_f$-priority vector of $\tilde{A}$.*

Some illustrative examples can be found in Sect. 3.4.

3.3.4 PCFN Matrix on Fuzzy Multiplicative Alo-Group

Fuzzy multiplicative alo-group $]\mathbf{0},\mathbf{1}[_{\mathbf{m}}=(]0,1[,\bullet_f,\leq)$, see also [11–13], is a continuous alo-group with:

$$a \bullet_f b = \frac{ab}{ab + (1-a)(1-b)}, e = 0,5, a^{(-1)} = 1 - a, \qquad (3.66)$$
$$\|a\| = max\{a, 1-a\},$$

hence

$$\|a \div_f b\| = max\{\frac{a(1-b)}{a(1-b) + (1-a)b}, \frac{(1-a)b}{a(1-b) + (1-a)b}\} \in]0,1[. \qquad (3.67)$$

We have also the product

$$a_1 \bullet_f \dots \bullet_f a_n = \frac{\prod_{i=1}^{n} a_i}{\prod_{i=1}^{n} a_i + \prod_{i=1}^{n}(1-a_i)}, \qquad (3.68)$$

and mean

$$m_{\bullet_f}(a_1, \dots, a_n) = \frac{(\prod_{i=1}^{n} a_i)^{1/n}}{(\prod_{i=1}^{n} a_i)^{1/n} + (\prod_{i=1}^{n}(1-a_i))^{1/n}}. \qquad (3.69)$$

Fuzzy multiplicative alo-group $]\mathbf{0},\mathbf{1}[_{\mathbf{m}}$ is divisible and continuous. For more details and properties, see [14].

In fuzzy PC matrices, each entry a_{ij} quantifies in [0, 1] the distance from the value 0.5 expressing the indifference. Here, PCFN matrices is defined on the unit interval]0, 1[. An appropriate group operation $\bullet_f$ is defined in (3.66). By formula (3.14), the matrix $A = \{a_{ij}\}$ is *$\bullet_f$-consistent* if

$$a_{ij} = a_{ik} \bullet_f a_{kj} = \frac{a_{ik}a_{kj}}{a_{ik}a_{kj} + (1-a_{ik})(1-a_{kj})}. \qquad (3.70)$$

In the literature, this type of consistency is called *fuzzy multiplicative consistency*, see e.g. [13], or, *multiplicative transitivity*, see e.g. [20]. Here, we call the appropriate alo-group $\mathbf{]0,1[_m}$, the *fuzzy multiplicative alo-group*. Notice that the group operation $\bullet_f$ is closed on the interval $]0,1[$.

Let $\tilde{A} = \{\tilde{a}_{ij}\}$ be a PCFN matrix on $]0,1[$. By problem (P1$\odot$) an appropriate priority vector for ranking the alternatives can be defined by the optimal solution of the following optimization problem (P1$\bullet_f$).

(P1$\bullet_f$)

$$\alpha \longrightarrow \max; \tag{3.71}$$

subject to

$$a_{ij}^L(\alpha) \le \frac{w_i(1-w_j)}{w_i(1-w_j)+(1-w_i)w_j} \le a_{ij}^R(\alpha) \text{ for all } i,j \in \{1,2,\ldots,n\}, i<j, \tag{3.72}$$

$$\frac{\prod_{i=1}^{n} w_i}{\prod_{i=1}^{n} w_i + \prod_{i=1}^{n}(1-w_i)} = 0.5, \tag{3.73}$$

$$0 \le \alpha \le 1,\ 0 \le w_k \le 1, \text{ for all } k \in \{1,2,\ldots,n\}. \tag{3.74}$$

If optimization problem (P1$\bullet_f$) has a feasible solution, i.e. system of constraints (3.72)–(3.74) has a solution, then (P1$\bullet_f$) has also an optimal solution.

Let α^* and $w^* = (w_1^*,\ldots,w_n^*)$ be an optimal solution of problem (P1$\bullet_f$). Then $\alpha^* \ge 0$ and α^* is called the *$\bullet_f$-consistency grade of* $\tilde{A}$, denoted by $G_{\bullet_f}(\tilde{A})$, i.e.

$$G_{\bullet_f}(\tilde{A}) = \alpha^*. \tag{3.75}$$

Here, $w^* = (w_1^*,\ldots,w_n^*)$ is an α^*-$\bullet_f$-consistent vector with respect to $\tilde{A}$ called the *$\bullet_f$-priority vector of* $\tilde{A}$.

If optimization problem (P1$\bullet_f$) has no feasible solution, then we define

$$G_{\bullet_f}(\tilde{A}) = 0. \tag{3.76}$$

Problem (P1$\bullet_f$) is an optimization problem that can be solved by the dichotomy method mentioned in the previous subsection, see also [14], or other nonlinear optimization SW, e.g. Solver in Excel.

Let $w = (w_1,\ldots,w_n), w_i \in]0,1[$ for all $i \in \{1,\ldots,n\}$. Similarly to (3.64) we denote

$$I_{\bullet_f}(\tilde{A}, w) = \min\{\frac{(\prod_{i<j} b_{ij})^{\frac{2}{n(n-1)}}}{(\prod_{i<j} b_{ij})^{\frac{2}{n(n-1)}} + (\prod_{i<j}(1-b_{ij}))^{\frac{2}{n(n-1)}}} \mid a_{ij} \in [\tilde{a}_{ij}]_0$$

$$\text{for all } 1 \le i < j \le n\}, \tag{3.77}$$

where

$$b_{ij} = \|\frac{a_{ij}(1-w_i)w_j}{a_{ij}(1-w_i)w_j + (1-a_{ij})w_i(1-w_j)}\|. \tag{3.78}$$

Here $\|\ldots\|$ is the $\mathcal{G}$-norm.

Now, we shall derive a suitable priority vector also in case there is no feasible solution of (P1$\bullet_f$) exists. Consider the following optimization problem.

(P2$\bullet_f$)

$$I_{\bullet_f}(\tilde{A}, w) \longrightarrow \min; \tag{3.79}$$

subject to

$$\frac{\prod_{i=1}^{n} w_i}{\prod_{i=1}^{n} w_i + \prod_{i=1}^{n}(1-w_i)} = 0.5, \tag{3.80}$$

$$0 \le \alpha \le 1,\ 0 \le w_k \le 1,\ \text{for all } k \in \{1, 2, \ldots, n\}. \tag{3.81}$$

We define the *$\bullet_f$-consistency index of $\tilde{A}$*, $I_{\bullet_f}(\tilde{A})$, as

$$I_{\bullet_f}(\tilde{A}) = I_{\bullet_f}(\tilde{A}, w^*) \tag{3.82}$$

where $w^* = (w_1^*, \ldots, w_n^*)$ is the optimal solution of (P2$\bullet_f$).

If there is no feasible solution of (P1$\bullet_f$), then $w^* = (w_1^*, \ldots, w_n^*)$ is called the *$\bullet_f$-priority vector of $\tilde{A}$*.

Remark 3.13. In particular, assume that $\tilde{A}$ is crisp, i.e. $\tilde{A} = A$. If A is $\bullet_f$-consistent, then we obtain $G_{\bullet_f}(A) = 1$ and also $I_{\bullet_f}(A) = 0.5$. On the other hand, if A is $\bullet_f$-inconsistent, then $G_{\bullet_f}(A) = 0$ and $I_{\bullet_f}(A) > 0.5$.

Remark 3.14. Instead of problem (P1$\bullet_f$) we can solve the equivalent problem (P1$\bullet_f$*) in a simpler form:

(P1$\bullet_f$*)

$$\alpha \longrightarrow \max; \tag{3.83}$$

subject to

$$a_{ij}^L(\alpha) \leq \frac{v_i}{v_i + v_j} \leq a_{ij}^R(\alpha) \text{ for all } i, j \in \{1, 2, \ldots, n\}, i < j, \tag{3.84}$$

$$\prod_{i=1}^{n} v_i = 1, \tag{3.85}$$

$$0 \leq \alpha \leq 1, \; 0 \leq v_k, \text{ for all } k \in \{1, 2, \ldots, n\}. \tag{3.86}$$

The equivalence of (P1$\bullet_f$) and (P1$\bullet_f$*) can be shown by substitution $v_k = \frac{w_k}{1-w_k}$, or, eventually, $w_k = \frac{v_k}{1+v_k}$ for all $k \in \{1, 2, \ldots, n\}$.

Remark 3.15. Instead of problem (P1$\bullet_f$*) we can solve the equivalent problem (P1$\bullet_f$**) in a more convenient form:

(P1$\bullet_f$**)

$$\alpha \longrightarrow \max; \tag{3.87}$$

subject to

$$a_{ij}^L(\alpha) \leq \frac{u_i}{u_i + u_j} \leq a_{ij}^R(\alpha) \text{ for all } i, j \in \{1, 2, \ldots, n\}, i < j, \tag{3.88}$$

$$\sum_{i=1}^{n} u_i = 1, \tag{3.89}$$

$$0 \leq \alpha \leq 1, \; 0 \leq u_k, \text{ for all } k \in \{1, 2, \ldots, n\}. \tag{3.90}$$

In order to show the equivalence, we apply substitution $u_k = \frac{1}{c} v_k$, where $c = \sum_{i=1}^{n} v_i$, or, eventually, $v_k = \frac{1}{d} u_k$, and $d = \prod_{i=1}^{n} u_i$ for all $k \in \{1, 2, \ldots, n\}$.

All three problems (P1$\bullet_f$), (P1$\bullet_f$*) and (P1$\bullet_f$**) are equivalent.

Remark 3.16. The same substitutions as those applied in problem (P1$\bullet_f$) can be used also in problem (P2$\bullet_f$), to obtain the equivalent problems (P2$\bullet_f$*) and (P2$\bullet_f$**) in a simpler form. Here, in formula (3.77) for calculating $I_{\bullet_f}(\tilde{A}, v)$, formula (3.78) is simplified to

$$b_{ij} = \| \frac{a_{ij} v_i}{a_{ij} v_i + (1 - a_{ij}) v_j} \|.$$

Some numerical examples can be found in the following section.

3.4 Illustrative Numerical Examples

In this section we present 8 numerical examples of 3×3 PCFN matrices illustrating the particular cases of the above presented concepts. The first matrix in each example is the crisp (i.e. nonfuzzy) matrix, the second matrix is the fuzzy one. All PCFN matrices are reciprocal which means that the membership generating functions (either linear, or nonlinear) for matrix elements: L_{ij}, R_{ij}, satisfy condition (3.12), i.e. $L_{ji}(t) = (R_{ij}(t))^{(-1)}$ and $R_{ji}(t) = (L_{ij}(t))^{(-1)}$, where $t \in [0,1]$ and $1 \leq i \neq j \leq 3$. In both cases the corresponding consistency grade and consistency index as well as the priority vector is calculated.

Example 3.3. Let $\tilde{A}_1$ and $\tilde{B}_1$ be PCFN matrices on the additive alo-group $\mathcal{R}$:

$$\tilde{A}_1 = \begin{pmatrix} (0;0;0) & (2;2;2) & (8;8;8) \\ (-2;-2;-2) & (0;0;0) & (6;6;6) \\ (-8;-8;-8) & (-6;-6;-6) & (0;0;0) \end{pmatrix} = \begin{pmatrix} 0 & 2 & 8 \\ -2 & 0 & 6 \\ -8 & -6 & 0 \end{pmatrix},$$

$$\tilde{B}_1 = \begin{pmatrix} 0 & (1;2;2) & (7;8;9) \\ (-2;-2;-1) & 0 & (3;4;4) \\ (-9;-8;-7) & (-4;-4;-3) & 0 \end{pmatrix}.$$

Here, $\tilde{A}_1$, $\tilde{B}_1$ are 3×3 PCFN matrices, particularly, PCFN matrices with triangular fuzzy number elements. $\tilde{A}_1$ is a crisp matrix (nonfuzzy), +-reciprocal and +-consistent, hence the consistency grade $G_+(\tilde{A}_1) = 1$ and the +-priority vector of $\tilde{A}_1$ is $w^* = (3.333, 1.333, -4.666)$. The consistency index $I_+(\tilde{A}_1) = 0$.

$\tilde{B}_1$ is a +-reciprocal PCFN matrix (noncrisp). As there is no feasible solution of the corresponding problem (P1+), the consistency grade $G_+(\tilde{B}_1) = 0$. The consistency index $I_+(\tilde{B}_1) = 0.333$, hence, $\tilde{B}_1$ is +-inconsistent and the +-priority vector of $\tilde{B}_1$ is $w^* = (2.964, 0.922, -3.886)$.

Example 3.4. Let $\tilde{A}_2$ and $\tilde{B}_2$ be PCFN matrices on the multiplicative alo-group $\mathcal{R}^+$:

$$\tilde{A}_2 = \begin{pmatrix} (1;1;1) & (2;2;2) & (8;8;8) \\ (\frac{1}{2};\frac{1}{2};\frac{1}{2}) & (1;1;1) & (4;4;4) \\ (\frac{1}{8};\frac{1}{8};\frac{1}{8}) & (\frac{1}{4};\frac{1}{4};\frac{1}{4}) & (1;1;1) \end{pmatrix} = \begin{pmatrix} 1 & 2 & 8 \\ \frac{1}{2} & 1 & 4 \\ \frac{1}{8} & \frac{1}{4} & 1 \end{pmatrix},$$

$$\tilde{B}_2 = \begin{pmatrix} 1 & (1;2;2) & (7;8;9) \\ (\frac{1}{2};\frac{1}{2};1) & 1 & (2;3;3) \\ (\frac{1}{9};\frac{1}{8};\frac{1}{7}) & (\frac{1}{3};\frac{1}{3};\frac{1}{2}) & 1 \end{pmatrix}.$$

Here, $\tilde{A}_2$ is a crisp matrix (nonfuzzy), $\bullet$-reciprocal and $\bullet$-consistent. Hence, the consistency grade $G_\bullet(\tilde{A}_2) = 1$ and the $\bullet$-priority vector of $\tilde{A}_2$ is $w^* = (2.520, 1.260, 0.315)$. The consistency index $I_\bullet(\tilde{A}_2) = 1$. Therefore, PCFN matrix $\tilde{A}_2$ is $\bullet$-reciprocal as well as $\bullet$-consistent in the classical sense.

$\tilde{B}_2$ is a $\bullet$-reciprocal PCFN matrix (noncrisp). As there is no feasible solution of the corresponding problem (P1$\bullet$), the consistency grade $G_{\bullet}(\tilde{B}_2) = 0$. The consistency index $I_{\bullet}(\tilde{B}_2) = 1.053$, hence, $\tilde{B}_2$ is $\bullet$-inconsistent. The $\bullet$-priority vector of $\tilde{B}_2$ is $w^* = (2.377, 1.182, 0.356)$.

Example 3.5. Let $\tilde{A}_3$ and $\tilde{B}_3$ be PCFN matrices on the fuzzy additive alo-group $\mathscr{R}_a$, i.e. we set $\odot = +_f$. The entries of matrix $\tilde{B}_3$ are triangular fuzzy numbers defined on interval $]0, 1[$:

$$\tilde{A}_3 = \begin{pmatrix} \frac{1}{2} & \frac{2}{3} & \frac{11}{12} \\ \frac{1}{3} & \frac{1}{2} & \frac{3}{4} \\ \frac{1}{12} & \frac{1}{4} & \frac{1}{2} \end{pmatrix},$$

$$\tilde{B}_3 = \begin{pmatrix} 0.5 & (0.5; 0.55; 0.6) & (0.8; 0.85; 0.9) \\ (0.4; 0.45; 0.5) & 0.5 & (0.4; 0.5; 0.6) \\ (0.1; 0.15; 0.2) & (0.4; 0.5; 0.6) & 0.5 \end{pmatrix}.$$

Here, $\tilde{A}_3$, $\tilde{B}_3$ are 3×3 PCFN matrices, particularly, PCFN matrices with triangular fuzzy number elements on $]0, 1[$. $\tilde{A}_3$ is a crisp matrix (nonfuzzy), $+_f$-reciprocal and $+_f$-consistent. Hence, the consistency grade $G_{+_f}(\tilde{A}_3) = 1$ and the $+_f$-priority vector of $\tilde{A}_3$ is $w^* = (0.694, 0.528, 0.278)$. The consistency index $I_{+_f}(\tilde{A}_3) = 0.5$. We can conclude that $\tilde{A}_3$ is $+_f$-reciprocal as well as $+_f$-consistent in a "classical sense".

$\tilde{B}_3$ is a $+_f$-reciprocal PCFN matrix (noncrisp). The elements of $\tilde{B}_3$ are triangular fuzzy numbers. The consistency grade $G_{+_f}(\tilde{B}_3) = 0$. The consistency index $I_{+_f}(\tilde{B}_3) = 0.533 > 0.5$, hence, $\tilde{B}_3$ is $+_f$-inconsistent. The $+_f$-priority vector of $\tilde{B}_3$ is $w^* = (0.654, 0.479, 0.367)$.

Example 3.6. Let $\tilde{A}_4$ and $\tilde{B}_4$ be PCFN matrices on the fuzzy multiplicative alo-group $]\mathbf{0}, \mathbf{1}[_{\mathrm{m}}$, i.e. $\odot = \bullet_f$. The following PCFNs are defined on interval $]0, 1[$:

$$\tilde{A}_4 = \begin{pmatrix} \frac{1}{2} & \frac{2}{3} & \frac{6}{7} \\ \frac{1}{3} & \frac{1}{2} & \frac{3}{4} \\ \frac{1}{7} & \frac{1}{4} & \frac{1}{2} \end{pmatrix},$$

$$\tilde{B}_4 = \begin{pmatrix} 0.5 & (0.6; 0.7; 0.8) & (0.75; 0.8; 0.9) \\ (0.2; 0.3; 0.4)) & 0.5 & (0.7; 0.75; 0.8) \\ (0.1; 0.2; 0.25) & (0.2; 0.25; 0.3) & 0.5 \end{pmatrix}.$$

Here, $\tilde{A}_4$, $\tilde{B}_4$ are 3×3 PCFN matrices, particularly, PCFN matrices with triangular fuzzy number elements on $]0, 1[$. $\tilde{A}_4$ is a crisp, $\bullet_f$-reciprocal and $\bullet_f$-consistent. Hence, the consistency grade $G_{\bullet_f}(\tilde{A}_4) = 1$ and the $\bullet_f$-priority vector of $\tilde{A}_4$ is $w^* = (2.289, 1.145, 0.382)$. The consistency index $I_{\bullet_f}(\tilde{A}_4) = 0.5$. We can conclude that $\tilde{A}_4$ is $\bullet_f$-reciprocal as well as $\bullet_f$-consistent in a "classical sense".

$\tilde{B}_4$ is a $\bullet_f$-reciprocal PCFN matrix (noncrisp). The elements of $\tilde{B}_4$ are triangular fuzzy numbers. As there is no feasible solution of the corresponding problem (P1$\bullet_f$), the consistency grade $G_{\bullet_f}(\tilde{B}_4) = 0$. The consistency index $I_{\bullet_f}(\tilde{B}_4) = 0.548 > 0.5$, hence, $\tilde{B}_4$ is $\bullet_f$-inconsistent. The $\bullet_f$-priority vector of $\tilde{B}_4$ is $w^* = (0.570, 0.881, 1.990)$.

3.5 Conclusion

This chapter investigated pairwise comparison matrices with fuzzy elements. Fuzzy elements of the pairwise comparison matrix are applied whenever the decision maker is not sure about the value of his/her evaluation of the relative importance of elements in question. In comparison with pairwise comparison matrices with crisp elements investigated in the literature, here we investigated pairwise comparison matrices with elements from abelian linearly ordered group (alo-group) over a real interval. We generalized the concept of reciprocity and consistency of pairwise comparison matrices with triangular fuzzy numbers (PCFN matrices). Moreover, we introduced the concept of priority vector which is a generalization of the crisp concept. Such an approach allows for a generalization dealing both with the PCFN matrices on the additive, multiplicative and also fuzzy alo-groups. It also unifies several approaches known from the literature, see e.g. [3, 11, 14, 15, 17, 20]. We also dealt with the problem of measuring the inconsistency of PCFN matrices by defining corresponding indexes. The first index called the consistency grade G is the maximal α of the α-cut, such that the corresponding PCFN matrix is α-consistent. On the other hand, the consistency index I of the PCFN matrix measures the distance of the zero-cut of the PCFN matrix to the closest priority vector. If the PCFN matrix is crisp, then G is equal to 1 and the consistency index I is equal to the identity element e. Eight numerical examples were presented to illustrate the concepts and derived properties.

References

1. Bozbura, F.T., Beskese, A.: Prioritization of organizational capital measurement indicators using fuzzy AHP. Int. J. Approx. Reason. **44**(2), 124–147 (2007)
2. Bozoki, S., Rapcsak, T.: On Saaty's and Koczkodaj's inconsistencies of pairwise comparison matrices. WP 2007-1. http://www.oplab.sztaki.hu/WP_2007_1_Bozoki_Rapcsak.pdf (June 2007)
3. Buckley, J.J.: Fuzzy hierarchical analysis. Fuzzy Sets Syst. **17**(1), 233–247 (1985) [ISSN 0165-0114]
4. Buckley, J.J., et al.: Fuzzy hierarchical analysis revisited. Eur. J. Oper. Res. **129**, 48–64 (2001)
5. Buyukozkan, G.: Multi-criteria decision making for e-marketplace selection. Internet Res. **14**(2), 139–154 (2004)

6. Dagdeviren, M., Yüksel, I.: Developing a fuzzy analytic hierarchy process (AHP) model for behavior-based safety management. Inf. Sci. **178**, 1717–1723 (2008)
7. Kulak, O., Kahraman, C.: Fuzzy multi-attribute selection among transportation companies using axiomatic design and analytic hierarchy process. Inf. Sci. **170**, 191–210 (2005)
8. Lee, K.L., Lin, S.C.: A fuzzy quantified SWOT procedure for environmental evaluation of an international distribution center. Inf. Sci. **178**(2), 531–549 (2008)
9. Leung, L.C., Cao, D.: On consistency and ranking of alternatives in fuzzy AHP. Eur. J. Oper. Res. **124**, 102–113 (2000) [ISSN 0377-2217]
10. Mahmoudzadeh, M., Bafandeh, A.R.: A new method for consistency test in fuzzy AHP. J. Intell. Fuzzy Syst. **25**(2), 457–461 (2013)
11. Mikhailov, L.: Deriving priorities from fuzzy pairwise comparison judgments. Fuzzy Sets Syst. **134**, 365–385 (2003)
12. Mikhailov, L.: A fuzzy approach to deriving priorities from interval pairwise comparison judgements. Eur. J. Oper. Res. **159**, 687–704 (2004)
13. Mikhailov, L., Tsvetinov, P.: Evaluation of services using a fuzzy analytic hierarchy process. Appl. Soft Comput. **5**, 23–33 (2004)
14. Ohnishi, S., Dubois, D., et al.: A fuzzy constraint based approach to the AHP. In: Uncertainty and Intelligent Information Systems, pp. 217–228. World Scientific, Singapore (2008)
15. Ramik, J., Korviny, P.: Inconsistency of pair-wise comparison matrix with fuzzy elements based on geometric mean. Fuzzy Sets Syst. **161**, 1604–1613 (2010) [ISSN 0165-0114]
16. Saaty, T.L.: Multicriteria Decision Making - The Analytical Hierarchy Process, vol. I. RWS Publications, Pittsburgh (1991)
17. Salo, A.A.: On fuzzy ratio comparison in hierarchical decision models. Fuzzy Sets Syst. **84**, 21–32 (1996) [ISSN 0165-0114]
18. Van Laarhoven, P.J.M., Pedrycz, W.: A fuzzy extension of Saaty's priority theory. Fuzzy Sets Syst. **11**(4), 229–241 (1983) [ISSN: 0165-0114]
19. Vaidya, O.S., Kumar, S.: Analytic hierarchy process: An overview of applications. Eur. J. Oper. Res. **169**(1), 1–29 (2006)
20. Xu, Z.S., Chen, J.: Some models for deriving the priority weights from interval fuzzy preference relations. Eur. J. Oper. Res. **184**, 266–280 (2008)

Part II
Special Matrices in Max-Min Algebra

Chapter 4
Optimization Problems Under Max-Min Separable Equation and Inequality Constraints

Abstract Equation and inequality systems, in which functions of the form

$$\max_{j \in J}(\min(a_j, r_j(x_j)))$$

occur are studied (J is a finite index set, a_j are real numbers, $r_j(x_j)$ are strictly increasing functions). The functions occur either on one side of the relations or on both sides of them. In the former case we call the relations one-sided, in the latter case two-sided. Properties of equation and inequality systems with max, min-separable functions on one or both sides of the relations, as well as optimization problems under (max, min)-separable equation and inequality constraints are studied. For the optimization problems with one-sided (max, min)-separable constraints an explicit solution formula is derived, a duality theory is developed and some optimization problems on the set of points attainable by the functions occurring in the constraints are solved. Solution methods for some classes of optimization problems with two-sided equation and inequality constraints are proposed in the last part of this chapter.

4.1 Introduction

Under the concept of extremally linear functions we understand functions $f : \mathbf{R}^n \to \mathbf{R}$ of the form

$$f(x) = f(x_1, \ldots, x_n) = c_1 \otimes x_1 \oplus \ \ldots \ \oplus c_n \otimes x_n,$$

where $\oplus$ is one of the extremal operations "max" or "min" and $\otimes$ is an appropriate group or semigroup operation chosen in such a way that the pair $(\oplus, \otimes)$ behaves like the pair of addition and multiplication in classic linear algebra. We will consider pairs $(\max, +)$ and $(\max, \min)$ and will speak accordingly about $(\max, +)$-*linear* or $(\max, \min)$-*linear functions*. Let us note that the operations max, min, + are special cases of the T-norms introduced in Chap. 1. Algebraic structures with such

The chapter was written by "Karel Zimmermann"

M. Gavalec et al., *Decision Making and Optimization*, Lecture Notes in Economics and Mathematical Systems 677, DOI 10.1007/978-3-319-08323-0_4

pairs of operations appeared in the literature under different names as e.g. "extremal algebra" ([18, 19]), "path algebra" ([8]), "minimax algebra" ([7]), "max-algebra" ([2, 14]), "incline algebra" ([3]), "idempotent algebra" ([9, 10, 12, 13]), "tropical algebra" ([11,12]), "fuzzy algebra" ([5]). They are applied to solving some problems concerning synchronization of events, network capacity problems, problems in the fuzzy set theory and others ([1,2, 15, 16]). The optimization problems we are going to investigate consist in minimizing objective functions having the form

$$f(x) = f(x_1, \ldots, x_n) = \max(f_1(x_1), f_2(x_2), \ \ldots f_n(x_n)),$$

where $f_j(x_j),\ j = 1, \ \ldots, \ n$ are real continuous functions. Such functions f will be called *max-separable functions*. In the optimization problems considered in this text functions f_j are as a rule monotone or unimodal (e.g. strictly convex or concave) functions, which allow to find effectively maximum or minimum on a closed finite interval. The sets of feasible solutions of the optimization problems are described by finite systems of equations and inequalities, in which either *max-plus linear functions* or *max-min linear functions* occur. If such functions occur only on one side of the equations or inequalities with a constant on the other side, we speak about *"one-sided" equation or inequality constraints*, otherwise we speak about *"two-sided" equation or inequality constraints*. Let us remark that since the operation $\oplus = \max$ is a semigroup operation, it is not possible to transfer variables from one side of the relations to the other side so easily like in the usual linear algebra and methods for solving problems with one-sided constraints and two-sided constraints are as a rule substantially different.

We will bring some examples showing the motivation for the research mentioned above and possible areas of its applications.

Example 4.1. Let us assume that passengers are to be transported from n given places $P_j,\ j \in J = \{1, \ \ldots, \ n\}$ to m destinations $D_i,\ i \in I = \{1, \ \ldots, \ m\}$ by n transportation units (e.g. trains) $T_j,\ j \in J$, the departure time of which is equal to x_j. The transportation time from place P_j to destination D_i is equal to a positive number t_{ij}. The passengers travelling from P_j to destination D_i reach therefore their destination at time $x_j + t_{ij}$. It follows that last passengers from all places P_j, which have to be transported to destination D_i will arrive at D_i at time $\max_{j \in J}(t_{ij} + x_j)$. Let us assume that we require that the arrival time of the last passengers travelling to D_i must be equal to $\hat{b}_i$. We require further that the departure times $x_j,\ j \in J$ must lie within given finite bounds i.e. $x_j \in [\underline{x}_j, \overline{x}_j],\ j \in J$. To satisfy these requirements, we have to find $x_j,\ j \in J$, which satisfy the system of equations

$$\max_{j \in J}(t_{ij} + x_j) = \hat{b}_i,\ i \in I,\ \underline{x}_j \leq x_j \leq \overline{x}_j,\ j \in J.$$

This system is a system of one-sided (max, +)-linear equations and inequalities. Let us denote the set of solution of the system by $M(\hat{b})$. Let $\hat{x}_j,\ j \in J$ be recommended departure times from P_j, and $f_j(x_j) = |x_j - \hat{x}_j|,\ j \in J$, $f(x) = \max_{j \in J} f_j(x_j)$. Let us consider the optimization problem

$$f(x) \longrightarrow \min$$

subject to

$$x \in M(\hat{b}).$$

The optimal solution x^{opt} of this optimization problem is a feasible vector of departure times, which is as close as possible to the recommended vector of departure times $\hat{x}$ in the sense of the Tshebyshev norm $\|x - \hat{x}\| = f(x)$.

If $M(\hat{b}) = \emptyset$, then we will try to find a vector b^{opt}, which solves another optimization problem:

$$\left\| b - \hat{b} \right\| = \max_{i \in I} \left| b_i - \hat{b}_i \right| \longrightarrow \min$$

subject to

$$M(b) \neq \emptyset.$$

In other words the optimal solution b^{opt} of this problem is the nearest vector of arrival times to destinations D_i, $i \in I$, for which there exist feasible departure times from P_j, $j \in J$.

Example 4.2. Let us assume that m places P_i, $i \in I = \{1, 2, \ldots, m\}$ are connected with n places R_j, $j \in J = \{1, 2 \ldots, n\}$ by roads of given capacities $a_{ij} \geq 0$, $i \in I$, $j \in J$. We have to connect places $R_j, j \in J$ with place T and choose the capacities x_j of the road connecting the two places R_j and T in such a way that $x_j \in [\underline{x}_j, \overline{x}_j]$. The capacity of the road connecting P_i with T via place R_j is therefore equal to $\min(a_{ij}, x_j)$. We will require further to find such capacities x_j, $j \in J$ that for each $i \in I$ the maximum capacity of the roads $P_i R_j T$ over all $j \in J$ is equal to a given positive value $\hat{b}_i$, i.e.

$$\max_{j \in J}(\min(a_{ij}, x_j)) = \hat{b}_i, \; i \in I \qquad (*)$$

Let us set

$$M(b) = \{x \mid \max_{j \in J}(\min(a_{ij}, x_j)) = b_i, i \in I, \; x_j \in [\underline{x}_j, \overline{x}_j], \; j \in J\}.$$

Let us assume that $M(\hat{b}) = \emptyset$. In such a case we want to find $b \in \mathbf{R}^m$, which is the nearest to $\hat{b}$ and $M(b) \neq \emptyset$ holds. In other words, we will solve the following minimization problem:

$$\left\| b - \hat{b} \right\| = \max_{i \in I} \left| b_i - \hat{b}_i \right| \longrightarrow \min$$

subject to

$$M(b) \neq \emptyset.$$

Let b^{opt} denote the optimal solution of this problem. Then $M(b^{opt}) \neq \emptyset$. If set $M(b^{opt})$ contains more than one elements x and we may continue the calculations and find in some sense the most appropriate feasible capacity vector in $M(b^{opt})$. For instance, we can be given some required capacities $\hat{x} = (\hat{x}_1, \ \ldots, \ \hat{x}_n)$. Let us set in this case $f_j(x_j) = \alpha_j \left| x_j - \hat{x}_j \right|$, for all $j \in J$, $f(x) = \max_{j \in J} f_j(x_j)$, where coefficients $\alpha_j > 0$ express the importance of the recommendation that x_j should be equal to $\hat{x}_j$. Functions $f_j(x_j)$ can be interpreted as penalties for the violation of the recommendation "x_j should be equal to $\hat{x}_j$". To find a feasible capacity vector, for which the maximum penalty $f_j(x_j)$ is minimal means to solve the following optimization problem:

$$f(x) \longrightarrow \min$$

subject to

$$x \in M(b^{opt})$$

Example 4.3. Let us assume that we have m pairs of fuzzy sets A_i, B_i, $i \in I = \{1, \ldots, m\}$ with a finite support $J = \{1, \ldots, n\}$ and membership functions $\mu_i : J \rightarrow [0, 1]$, $\nu_i : J \rightarrow [0, 1]$ respectively. We have to find fuzzy set X with membership function $\mu_X : J \rightarrow [0, 1]$. Let functions $\mu_{iX} : J \rightarrow [0, 1]$, $\nu_{iX} : J \rightarrow [0, 1]$ be defined as follows:

$$\mu_{iX}(j) = \mu_i(j) \wedge \mu_X(j), \ \nu_{iX}(j) = \nu_i(j) \wedge \mu_X(j), \ j \in J,$$

where symbol $\wedge$ is used to denote the minimum of two numbers, i.e. $\alpha \wedge \beta = \min(\alpha, \beta)$ for any real numbers α, β. Then for each $i \in I$ function μ_{iX} is the membership function of the intersection of fuzzy sets A_i and X, and function ν_{iX} is the membership function of the intersection of fuzzy sets B_i and X. The expressions

$$H_i^{(1)}(X) = \max_{j \in J}(\mu_{iX}(j)), \ H_i^{(2)}(X) = \max_{j \in J}(\nu_{iX}(j))$$

are the heights of the intersections of fuzzy sets A_i, X and B_i, X respectively. The heights $H_i^{(1)}(x)$, $H_i^{(2)}(x)$ express the maximal achievable membership value of intersections of sets A_i, X and B_i, X respectively. We will interpret the sets A_i, B_i as fuzzy goals and will find set X, or in other words values $\mu_X(j)$, $j \in J$, which are bounded by given bounds $\underline{x}_j \in [0, 1]$, $\overline{x}_j \in [0, 1]$, $j \in J$ and satisfy the system of equalities

$$H_i^{(1)}(X) = H_i^{(2)}(X), \ i \in I.$$

This requirement means that for each $i \in I$ the maximal achievable membership of the intersections of set X with A_i and B_i should be the same (i.e. in some sense goals $A_i,\ B_i$ are equally important). The required relations can be interpreted as special fuzzy constraints. Let us introduce the following notations for all $i \in I,\ j \in J$:

$$a_{ij} = \mu_i(j),\ b_{ij} = \nu_i(j), x_j = \mu_(j).$$

We will consider now the following set of feasible solutions $\mu_X(j), j \in J$ (i.e. feasible values of membership function μ_X):

$$\tilde{M}(\underline{x}, \overline{x}) = \{x \in [0,1]^n \mid H_i^{(1)}(X) = H_i^{(2)}(X),\ i \in I,\ \underline{x} \le x \le \overline{x}\}.$$

By making use of the introduced notations we obtain:

$$\tilde{M}(\underline{x}, \overline{x}) = \{x \in [0,1]^n \mid \max_{j \in J}(a_{ij} \wedge x_j) = \max_{j \in J}(b_{ij} \wedge x_j),\ i \in I,\ \underline{x} \le x \le \overline{x}\}.$$

Set $\tilde{M}(\underline{x}, \overline{x})$ is therefore described by a system of (max, min)-linear equations and the given upper and lower bounds of the variables.

4.2 Systems of One-Sided Max-Separable Equations and Inequalities: Unified Approach

The aim of this section is to present a unified approach to solving optimization problems under the constraints, which are described by systems of one-sided max −separable equations and inequalities the special case of which are the systems of one-sided (max, +)− and (max, min)−linear equations and inequalities. For this purpose we will introduce the following notations:
$I = \{1,\ \ldots,\ m\},\ J = \{1, \ldots,\ n\}$, let $\mathbf{R}$ denote the set of real numbers, $A = \{a_{ij}\}$ real $m \times n$ matrix $i \in I,\ j \in J, \alpha \wedge \beta = \min(\alpha, \beta)$ for any $\alpha,\ \beta \in \mathbf{R}$. We assume that $r_{ij} : \mathbf{R} \longrightarrow \mathbf{R}$ are for all $i \in I,\ j \in J$ continuous strictly increasing functions, $R(x) = \{r_{ij}(x_j)\}$ is the matrix of functions $r_{ij}(x_j)$. We will consider in the sequel systems having the form

$$\max_{j \in J}(a_{ij} \wedge r_{ij}(x_j) \ge b_i,\ i \in I_1, \tag{4.1}$$

$$\max_{j \in J}(a_{ij} \wedge r_{ij}(x_j) \le b_i,\ i \in I_2, \tag{4.2}$$

where $I = I_1 \cup I_2$ and $b = (b_1,\ \ldots,\ b_m) \in \mathbf{R}^m$ is given.

The set of all solutions of system (4.1), (4.2) will be denoted $M(b)$.

Lemma 4.1. *System of inequalities (4.2) can be replaced by system*

$$x_j \leq \overline{x}_j(b), \; j \in J, \tag{4.3}$$

where

$$\overline{x}_j(b) = \min_{k \in I_j(b)} r_{kj}^{-1}(b_k), \; j \in J, \tag{4.4}$$

where $I_j(b) = \{k \in I_2 \mid a_{kj} > b_k\}$ *and we set* $\overline{x}_j(b) = \infty$ *if* $I_j(b) = \emptyset$.

Proof. Let x satisfy (4.2). Then

$$a_{ij} \wedge r_{ij}(x_j) \leq b_i, \; i \in I_2, j \in J.$$

It must be therefore $r_{kj}(x_j) \leq b_k$ for those $k \in I_2$, for which $a_{kj} > b_k$. Since the inverse function r_{kj}^{-1} is according to our assumptions strictly increasing, it follows that (4.4) must hold for all $j \in J$. □

Lemma 4.2. *Let for all* $i \in I_1, \; j \in J$

$$T_{ij}(b_i) = \{x_j \mid a_{ij} \wedge r_{ij}(x_j) \geq b_i, \; x_j \leq \overline{x}_j(b)\}.$$

Then

$$M(b) \neq \emptyset \iff \textit{for all } i \in I_1 \textit{ there exists } j(i) \in J \textit{ such that } T_{ij(i)}(b_i) \neq \emptyset.$$

Proof. Let there exists an index $i \in I_1$ such that for all $j \in J$

$$T_{ij}(b_i) = \emptyset.$$

Let us consider a fixed $j \in J$ and find out under which conditions set $T_{ij}(b_i)$ can be empty. Let us first assume that $a_{ij} \geq b_i$. Then $T_{ij}(b_i) = \emptyset$ if $r_{ij}(x_j) \geq b_i$ implies that $x_j > \overline{x}_j(b)$, i.e. if $x_j > r_{ij}^{-1}(\overline{x}(b)) > \overline{x}_j(b)$. Let us assume now that $a_{ij} < b_i$. Then we have evidently $a_{ij} \wedge r_{ij}(x_j) < b_i$ for an arbitrary x_j and therefore $T_{ij}(b_i) = \emptyset$.

Summarizing we obtain that $T_{ij}(b_i) = \emptyset$ if and only if either

$$a_{ij} \geq b_i \text{ and at the same time } r_{ij}(x_j) \geq b_i \implies x_j > \overline{x}_j(b)$$

or

$$a_{ij} < b_i.$$

If now the former possibility takes place and for some index $j \in J$ inequality $x_j > \overline{x}_j(b)$ holds, then $x \notin M(b)$ according to Lemma 4.1. If only the latter case occurs, then for any x_j inequalities $a_{ij} \wedge r_{ij}(x_j) < b_i$ for all $j \in J$ hold and therefore

$\max_{j\in J}(a_{ij} \wedge r_{ij}(x_j)) < b_i$ so that (4.1) is not fulfilled and $x \notin M(b)$. This proves the implication $\Longrightarrow$.

Let us assume now that $\forall i \in I_1$ there exists an index $j(i) \in J$ such that $T_{ij(i)}(b_i) \neq \emptyset$. We have to prove that $M(b) \neq \emptyset$. Let $K_j = \{i \in I_1 \mid T_{ij}(b_i) \neq \emptyset\}$ for all $j \in J$, $T_j(b) = \bigcap_{i\in K_j} T_{ij}(b_i)$ if $K_j \neq \emptyset$, $T_j(b) = \{x_j \in \mathbf{R} \mid x_j \leq \overline{x}_j(b)\}$ otherwise. Let us note that if $K_j \neq \emptyset$, then $T_j(b) \neq \emptyset$ and $T_j(b) = [\max_{i\in K_j} r_{ij}^{-1}(b_i), \overline{x}_j(b)]$. Let us choose $\tilde{x}_j \in T_j(b)$ for all $j \in J$. Then $\tilde{x}_j \leq \overline{x}_j(b)$ for all $j \in J$. Let $i \in I_1$ be arbitrarily chosen so that $T_{ij(i)}(b_i) \neq \emptyset$ and $i \in K_{j(i)}$. We have $\tilde{x}_{j(i)} \in T_{j(i)}(b) \subseteq T_{ij(i)}(b_i)$. Therefore $a_{ij(i)} \wedge r_{ij(i)}(\tilde{x}_{j(i)}) \geq b_i$ so that $\max_{j\in J}(a_{ij} \wedge r_{ij}(\tilde{x}_j)) \geq b_i$. Since $i \in I_1$ was arbitrarily chosen, we conclude that $\tilde{x}$ satisfies relations (4.1), (4.2) so that $\tilde{x} \in M(b)$ and $M(b) \neq \emptyset$. Therefore we proved also the implication $\Longleftarrow$ and the lemma is proved. □

Remark 4.1. Let us note that since $\overline{x}_j(b) \in T_j(b)$, $j \in J$, we obtain as a consequence of Lemma 4.2 that $M(b) \neq \emptyset$ if and only if $\overline{x}(b) \in M(b)$.

4.3 Optimization Problems with Feasible Set $M(b)$

In this section we will consider optimization problems of the form

$$f(x) = \max_{j\in J} f_j(x_j) \tag{4.5}$$

subject to

$$x \in M(b), \tag{4.6}$$

where $f_j : \mathbf{R} \longrightarrow \mathbf{R}$ are continuous functions for all $j \in J$. Let us introduce the following notations:

$$\min_{x_j\in T_{ij}(b_i)} f_j(x_j) = f_j(x_j^{(i)}) \tag{4.7}$$

for all $i \in I_1$, $j \in J$ such that $T_{ij}(b_i) \neq \emptyset$,

$$\min_{j\in J,\ T_{ij}(b_i)\neq\emptyset} f_j(x_j^{(i)})) = f_{k(i)}(x_{k(i)}^{(i)})),\ i \in I_1, \tag{4.8}$$

$$P_s = \{i \in I_1 \mid k(i) = s\},\ s \in J, \tag{4.9}$$

$$T_s^*(b) = \bigcap_{i\in P_s} T_{is}(b_i)\ \textit{for } s \in J \textit{ such that } P_s \neq \emptyset, \tag{4.10}$$

$$T_s^*(b) = \{x_s \mid x_s \leq \overline{x}_s(b)\}\ \textit{for } s \in J \textit{ such that } P_s = \emptyset. \tag{4.11}$$

Theorem 4.1. *Let $M(b) \neq \emptyset$ and let*

$$f_s(x_s^*) = \min_{x_s \in T_s^*(b)} f_s(x_s),\ s \in J. \tag{4.12}$$

Then $x^* = (x_1^*, \dots, x_n^*)$ is the optimal solution of optimization problem (4.5), (4.6).

Proof. First we will prove that x^* is a feasible solution. It follows immediately from the definition of x^* that $x^* \leq \overline{x}(b)$ so that x^* satisfies relations (4.2). To prove that x^* is a feasible solution, it remains to prove that x^* satisfies (4.1). Let $i \in I$ be arbitrary. Let $k(i) = s$ be the index from formula (4.8). Then we have $x_s^* \in T_s^*(b) \subseteq T_{is}^*(b_i) = T_{ik(i)}(b_i)$ and thus $a_{is} \wedge r_{is}(x_s^*) \geq b_i$ and we obtain $\max_{j \in J}(a_{ij} \wedge r_{ij}(x_j^*)) \geq a_{is} \wedge r_{is}(x_s^*) \geq b_i$. Since $i \in I_1$ was arbitrary, we obtain that x^* satisfies (4.1) and therefore $x^* \in M(b)$.

It remains to prove that x^* is the optimal solution, i.e. that $f(x) \geq f(x^*)$ for any $x \in M(b)$. Let us assume that $\tilde{x}$ is an element of $M(b)$ such that $f(\tilde{x}) < f(x^*)$. Let $f(x^*) = f_s(x_s^*),\ s \in J$. Then $f_s(\tilde{x}_s) < f(x^*) = f_s(x_s^*)$ and it must hold that $\tilde{x}_s \notin T_s^*(b)$.

Since we assumed that r_{ij}'s are increasing functions, it holds for any nonempty sets $T_{i_1 j}$, $T_{i_2 j}$ either $T_{i_1 j} \subseteq T_{i_2 j}$ or $T_{i_2 j} \subseteq T_{i_1 j}$. It follows that if P_s is nonempty, then there exists index $h \in P_s$ such that $T_s^*(b) = T_{hs}(b_h)$ and it holds $\min_{j \in J} f_j(x_j^{(h)}) = f_s(x_s^{(h)}) = f_s(x_s^*) = f(x^*)$. Since $f_s(\tilde{x}_s) < f_s(x_s^*) = f_s(x_s^{(h)})$, $\tilde{x}_s \notin T_{hs}(b_h)$. Since $\tilde{x} \in M(b)$, there must exist an index $v \in J,\ v \neq s$ such that $\tilde{x}_v \in T_{hv}(b_h)$. We have then

$$f(\tilde{x}) \geq f_v(\tilde{x}_v) \geq \min_{x_v \in T_{hv}(b_h)} f_v(x_v) = f_v(x_v^{(h)}) \geq f_s(x_s^{(h)}) = f_s(x_s^*) = f(x^*),$$

which is in contradiction with the assumption that $f(\tilde{x}) < f(x^*)$. This contradiction completes the proof. □

Remark 4.2. Let us note that it is possible to prove that Theorem 4.1 follows as a special case immediately from the results in [18] but we preferred to present a separate direct proof here for completeness of this publication.

Remark 4.3. The optimization problem

$$f(x) \longrightarrow \min$$

subject to

$$x \in M(\hat{b})$$

considered in Example 4.1 is a special case of problem (4.5), (4.6) above if we set $r_{ij}(x_j) = t_{ij} + x_j$ and choose a_{ij} sufficiently large or formally equal to ∞ assuming that we set $\alpha \wedge \infty = \alpha$ for any real α. Such systems of equations and inequalities were called (max, +)-linear systems and their properties were extensively studied e.g. in [2, 6, 17].

Similarly the problem considered in Example 4.2 is a special case of problem (4.5), (4.6) if we set $r_{ij}(x_j) = x_j$. Such problems were mentioned probably for the first time in [17].

4.4 Duality Theory

Explicit solution of the optimization problems derived in the preceding section makes possible to develop a duality theory for max-separable optimization problems. We will formulate two max-separable optimization problems, which are interconnected in a similar way as the dual optimization problems in the classical linear or convex optimization.

Let us consider the optimization problem

$$f(x) = \max_{j \in J} f_j(x_j) \longrightarrow \max \tag{4.13}$$

subject to

$$\max_{j \in J}(a_{ij} \wedge r_{ij}(x_j)) \leq 0, \ i \in I, \tag{4.14}$$

where we assume that $I = \{1, \ldots, m\}$, $J = \{1, \ldots, n\}$, $f_j : \mathbf{R} \longrightarrow \mathbf{R}$ are continuous strictly increasing functions. We assume further that a_{ij} are given real numbers and similarly as above $r_{ij} : \mathbf{R} \longrightarrow \mathbf{R}$ are continuous strictly increasing functions for all $i \in I, \ j \in J$. The optimal solution of this problem will be denoted x^{opt}.

Let us define sets I_j for all $j \in J$ as follows: $I_j = \{i \in I \mid a_{ij} > 0\}$ and set

$$\overline{x}_j = \min_{i \in I_j} r_{ij}^{-1}(0), \ \ j \in J \tag{4.15}$$

We will assume that for all $j \in J$ set I_j is non-empty so that $\overline{x}_j < \infty, \ j \in J$. Taking into account that $f_j, \ j \in J$ are strictly increasing functions, we obtain that $x^{opt} = \overline{x}$. The optimal value of the objective function is

$$f(x^{opt}) = \max_{j \in J} f_j(\overline{x}_j) = \max_{j \in J} f_j(\min_{i \in I_j} r_{ij}^{-1}(0)). \tag{4.16}$$

Let us consider further the optimization problem

$$g(u) = \max_{i \in I} u_i \longrightarrow \min \tag{4.17}$$

subject to

$$\max_{i \in I}(a_{ij} \wedge r_{ij}(f_j^{-1}(u_i)) \geq 0, \ j \in J. \tag{4.18}$$

We will find the optimal solution of this problem using the explicit method described in the preceding section.

Let sets T_{ij} for $i \in I, \ j \in J$ be defined as follows:

$$T_{ij} = \{u_i \mid a_{ij} \wedge r_{ij}(f_j^{-1}(u_i)) \geq 0\}. \tag{4.19}$$

Let us note that $T_{ij} \neq \emptyset$ if and only if $a_{ij} \geq 0$ so that for each $j \in J$ we have $V_j = \{i \in I \mid T_{ij} \neq \emptyset\} = \{i \in I \mid a_{ij} \geq 0\}$. If $T_{ij} \neq \emptyset$, then $T_{ij} = \{u_i \mid u_i \geq f_j(r_{ij}^{-1}(0))\}$. Therefore if T_{ij} is nonempty, then $T_{ij} = [u_i^{(j)}, \infty)$, where $u_i^{(j)} = \min\{u_i \mid u_i \in T_{ij}\} = f_j(r_{ij}^{-1}(0))$.

We will find the optimal solution of optimization problem (4.17)–(4.18) using Theorem 4.1. Since we assumed that $V_j \neq \emptyset$, the set of feasible solutions of the optimization problem is nonempty and the objective function $g(u)$ is on the set of feasible solutions bounded from below and continuous. Therefore there exists a finite optimal value of the objective function. Let us set

$$\min_{i \in V_j} u_j^{(i)} = u_{i(j)}^{(j)}, \ \ j \in J.$$

If there are more indices $i(j) \in I$ satisfying this equation, we will choose an arbitrary one. Let us note that since $V_j \neq \emptyset$ for all $j \in J$, we obtain for each $j \in J$ a finite value $u_{i(j)}^{(j)} = f_j(r_{i(j)j}^{-1}(0))$.

Let

$$P_s = \{j \in J \mid i(j) = s\}, \ \ s \in I,$$

$$T_s = \bigcap_{j \in P_s} T_{sj} \ \ if \ \ P_s \neq \emptyset,$$

$$T_s = \mathbf{R} \ \ if \ \ P_s = \emptyset.$$

Let us note that if $P_s \neq \emptyset$, we have :

$$T_s = [\max_{j \in P_s} u_s^{(j)}, \infty) = [u_s^{j(s)}, \infty) = T_{sj(s)}$$

for some $j(s) \in P_s \subseteq J$.

Let us define $u^* \in \mathbf{R}^m$ as follows:

$$u_s^* = \max_{j \in P_s} u_s^{(j)} \ \ \texttt{if} \ \ P_s \neq \emptyset,$$

$$u_s^* = -\infty \ \ \texttt{if} \ \ P_s = \emptyset.$$

Then according to Theorem 4.1 u^* is the optimal solution of optimization problem (4.17)–(4.18). The corresponding optimal value of the objective function is

$$g(u^*) = \max_{s \in I} u_s^* = \max_{s \in I, P_s \neq \emptyset} u_s^*.$$

We will prove now the following strong duality theorem:

Theorem 4.2. *Let $I_j \neq \emptyset$ for all $j \in J$ and let $a_{ij} \neq 0$ $\forall i \in I,\ j \in J$. Then the optimal values of optimization problems (4.13)–(4.14) and (4.17)–(4.18) are equal, i.e. it holds that*

$$\max_{j \in J} f_j(\overline{x}_j) = \max_{s \in I} u_s^*. \tag{4.20}$$

Proof. Note that according to the assumption that $a_{ij} \neq 0$ $\forall i \in I,\ j \in J$ it is $I_j = V_j$ and therefore

$$f(\overline{x}) = \max_{j \in J} f_j(\overline{x}_j) = \max_{j \in J} f_j(\min_{i \in I_j} r_{ij}^{-1}(0)) = \max_{j \in J} f_j(\min_{i \in V_j} r_{ij}^{-1}(0)). \tag{4.21}$$

Let us assume that

$$g(u^*) = \max_{s \in I} u_i^* = u_h^*$$

Then we have:

$$g(u^*) = u_h^* = \max_{j \in P_h} \min_{i \in V_j} f_j(r_{ij}^{-1}(0)) = \max_{j \in P_h} f_j(\min_{i \in V_j} r_{ij}^{-1}(0)). \tag{4.22}$$

Note that the last equality in (4.22) follows from the assumption that $f_j,\ j \in J$ are increasing functions.

If we compare formulas (4.21) and (4.22), we see that it remains to prove that

$$\max_{j \in P_h} f_j(\min_{i \in V_j} r_{ij}^{-1}(0)) = \max_{j \in J} f_j(\min_{i \in V_j} r_{ij}^{-1}(0)).$$

In other words we have to prove that the terms $f_j(\min_{i \in V_j} r_{ij}^{-1}(0))$ for $j \in J \setminus P_h$ do not influence the value u_h^*, i.e. that

$$f_j(\min_{i \in V_j} r_{ij}^{-1}(0)) \leq u_h^*,\ j \in (J \setminus P_h). \tag{4.23}$$

To prove (4.23), let us assume that $q \in J \setminus P_h$ is arbitrarily chosen and consider the value

$$f_q(\min_{i \in V_q} r_{iq}^{-1}(0)) = f_q(r_{i(q)q}^{-1}(0)).$$

It follows that $q \in P_{i(q)}$ and therefore

$$f_q(\min_{i \in V_q} r_{iq}^{-1}(0)) = f_q(r_{i(q)q}^{-1}(0)) \leq \max_{j \in P_{i(q)}} f_j(r_{i(q)j}^{-1}(0)) = u_{i(q)}^* \leq \max_{i \in I} u_i^* = u_h^*. \tag{4.24}$$

Since $q \in (J \setminus P_h)$ was arbitrarily chosen, the inequality (4.24) completes the proof. □

Remark 4.4. (a) Note that system of inequality constraints with arbitrary right hand sides

$$\max_{j \in J}(\tilde{a}_{ij} \wedge \tilde{r}_{ij}(x_j)) \leq b_i, \ i \in I,$$

can be easily transformed to form (4.14) if we set

$$a_{ij} = \tilde{a}_{ij} - b_i, \ r_{ij}(x_j) = \tilde{r}_{ij}(x_j) - b_i.$$

(b) If we want to avoid infinite components in u^*, we can set $u_i^* = \alpha$ if $P_i = \emptyset$, where α is any number such that $\alpha \leq \max_{s \in I, P_s \neq \emptyset} u_s^*$.
(c) Let us remark that for any feasible solution x of problem (4.13)–(4.14) and any feasible solution u of problem (4.17)–(4.18) the following inequalities hold:

$$f(x) \leq f(\overline{x}) = g(u^*) \leq g(u).$$

These inequalities are an analogue of the weak duality theorem from the classic optimization theory.

In what follows we will illustrate the theoretical results by small numerical examples.

Example 4.4. Let $m = 2, \ n = 3, I = \{1, 2\}, \ j \in J = \{1, 2, 3\}$. Let

$$R(t) = \{r_{ij}(t)\} = \begin{pmatrix} 7t & 4t - 1 & 1t \\ 5t - 1 & 6t - 1 & 8t - 1 \end{pmatrix}$$

Let

$$A = \{a_{ij}\} = \begin{pmatrix} -1, & 3, & -1 \\ 4, & 7, & 2 \end{pmatrix}$$

Let us consider the maximization problem

$$f(x) = \max(x_1, \ x_2, \ x_3) \longrightarrow \max$$

subject to

$$\max(-1 \wedge 7x_1,\ 3 \wedge 4x_2 - 1,\ -1 \wedge x_3) \le 0$$

$$\max(4 \wedge 5x_1 - 1,\ 7 \wedge 6x_2 - 1,\ 2 \wedge 8x_3 - 1) \le 0$$

The optimal solution is $\overline{x} = (1/5, 1/6, 1/8)$,

the optimal value of the objective function is $f(\overline{x}) = \max(1/5, 1/6, 1/8) = 1/5$.

The corresponding dual minimization problem has the following form:

$$g(u) = \max(u_1, u_2) \longrightarrow \min$$

subject to

$$\max(-1 \wedge 7u_1, 4 \wedge 5u_2 - 1) \ge 0$$

$$\max(3 \wedge 4u_1 - 1, 7 \wedge 6u_2 - 1) \ge 0$$

$$\max(-1 \wedge u_1, 2 \wedge 8u_2 - 1) \ge 0$$

The optimal solution is $u^* = (-\infty, 1/5)$.

The optimal value of the objective function is $g(u^*) \max(-\infty, 1/5) = 1/5$.

Other optimal solutions with finite components are $u^*(\alpha) = (\alpha, 1/5)$, where $\alpha \le 1/5$.

The next example illustrates the importance of the assumption that $a_{ij} \ne 0$ in Theorem 4.2.

Example 4.5. Let $m = 2$, $n = 1$ and let us consider the maximization problem

$$f(x) = x_1 \longrightarrow \max$$

subject to

$$0 \wedge (8x_1 - 1) \le 0$$

$$1 \wedge (6x_1 - 1) \le 0$$

The optimal solution is $\overline{x} = (1/6)$, the optimal value of the objective function is $f(\overline{x}) = 1/6$.

The corresponding dual minimization problem has the following form:

$$g(u) = \max(u_1, u_2) \longrightarrow \min$$

subject to

$$\max(0 \wedge (8u_1 - 1), 1 \wedge (6u_2 - 1)) \ge 0.$$

The optimal solution is $u^* = (1/8, -\infty)$, the optimal value of the objective function is $g(u^*) = 1/8$. We see that since the condition $a_{ij} \neq 0$ is not fulfilled, it may arise a duality gap. We have in this case $a_{11} = 0$ and it is $f(\overline{x}) = 1/6 > g(u^*) = 1/8$.

The next example shows what can happen if the condition $I_j \neq \emptyset,\ j \in J$ is not fulfilled.

Example 4.6. Let us consider the minimization problem

$$f(x) = \max(x_1,\ x_2,\ x_3) \quad \longrightarrow \quad \max$$

subject to

$$\max(-1 \wedge 7x_1,\ 3 \wedge 4x_2 - 1,\ -1 \wedge x_3) \leq 0,$$
$$\max(-4 \wedge 5x_1 - 1,\ 7 \wedge 6x_2 - 1,\ 2 \wedge 8x_3 - 1) \leq 0.$$

In this case we have $I_1 = \emptyset$ and therefore $\overline{x} = (\infty, 1/6, 1/8)$ so that $f(\overline{x}) = \infty$.

The corresponding dual minimization problem has the following form:

$$g(u) = \max(u_1, u_2) \longrightarrow \min$$

subject to

$$\max(-1 \wedge 7u_1, -4 \wedge 5u_2 - 1) \geq 0,$$
$$\max(3 \wedge 4u_1 - 1, 7 \wedge 6u_2 - 1) \geq 0,$$
$$\max(-1 \wedge u_1, 2 \wedge 8u_2 - 1) \geq 0.$$

Since it is in the first line $-1 < 0$ and $-4 < 0$, it is for any $u_1,\ u_2$ always $\max(-1 \wedge 7u_1, -4 \wedge 5u_2) < 0$, so that the set of the feasible solutions of the dual optimization problem is empty.

Let us note that the result of the last example can be generalized as follows.

If the condition $a_{ij} \neq 0,\ i \in I, j \in J$ is fulfilled and there exists an index $j_0 \in J$ such that $I_{j_0} = \emptyset$, then $f(\overline{x}) = f_{j_0}(\overline{x}_{j_0}) = \infty$ and the j_0-th inequality of the dual minimization problem is not fulfilled, i.e. $\max_{i \in I}(a_{ij_0} \wedge r_{ij_0}(f_{j_0}^{-1}(u_i))) < 0$ and therefore the set of feasible solutions of the dual minimization problem is empty. This result is similar like in the classic duality theory.

We will present further examples of dual pairs for some special subclasses of max-separable optimization problems.

Example 4.7. Let us consider maximization problem of the form

$$f(x) = \max_{j \in J} f_j(x_j) = \max(c_j + x_j) \quad \longrightarrow \quad \max$$

subject to

$$\max_{j \in J}(d_{ij} + x_j) \leq 0,\ i \in I.$$

In this case we have in the general formulation (4.13) - (4.14): $f_j(x_j) = c_j + x_j$, $r_{ij}(x_j) = d_{ij} + x_j$ and $a_{ij} = \infty$. Such problems are called in the literature (max, +)-linear.

Taking into account that in this case $f_j^{-1}(t) = t - c_j$ for all $j \in J$, we obtain the following form of the left hand sides of the constraints in the dual minimization problem (4.18):

$$\max_{i \in I}(a_{ij} \wedge r_{ij}(f_j^{-1}(u_i))) = \max_{i \in I}(r_{ij}(f_j^{-1}(u_i))) = \max_{i \in I}(d_{ij} + u_i - c_j)$$

Therefore the corresponding dual minimization problem has in this case the following form:

$$g(u) = \max_{i \in I} u_i \quad \longrightarrow \quad \min$$

subject to

$$\max_{i \in I}(d_{ij} + u_i - c_j) \geq 0, \ j \in J.$$

The components of the optimal solutions of the maximization problem are

$$\overline{x}_j = \min_{i \in I}(-d_{ij}), \ j \in J,$$

the components of the optimal solution of the dual minimization problem are

$$u_i^* = \max_{j \in P_i} \min_{i \in I}(f_j(r_{ij}^{-1}(0))) = \max_{j \in P_i} \min_{i \in I}(c_j - d_{ij}), \ i \in I.$$

If conditions $I_j \neq \emptyset \ \forall j \in J$ and $a_{ij} \neq 0 \ \forall i \in I, j \in J$ are fulfilled, the equality of the optimal values of the objective functions $\max_{j \in J}(c_j + \overline{x}_j)) = \max_{i \in I} u_i^*$ holds.

Example 4.8. Let us consider maximization problem of the form

$$f(x) = \max_{j \in J} f_j(x_j) = \max(x_j) \quad \longrightarrow \quad \max$$

subject to

$$\max_{j \in J}(d_{ij} \wedge x_j) \leq b_i, \ i \in I.$$

Such inequality systems are referred in the literature as (max, min)-linear. Subtracting b_i from the both sides of the inequalities we obtain:

$$\max_{j \in J}(d_{ij} - b_i) \wedge (x_j - b_i) \leq 0, \ i \in I.$$

In this way we transformed the constraints to the standard form (4.14), where $f_j(x_j) = x_j, a_{ij} = d_{ij} - b_i$ and $r_{ij}(x_j) = x_j - b_i$.

The dual minimization problem has the form:

$$g(u) = \max_{i \in I} u_i \quad \longrightarrow \quad \min$$

subject to

$$\max_{i \in I}((d_{ij} - b_i) \wedge (u_i - b_i)) \geq 0, \; i \in I.$$

If $I_j \neq \emptyset$ and $d_{ij} - b_i \neq 0$ for all $i \in I, j \in J$, the optimal values of the objective functions are equal.

We will bring two more numerical problems illustrating the theoretical results of the two last examples.

Example 4.9. Let us consider the (max, +)-linear maximization problem

$$f(x) = \max(x_1, x_2, x_3) \quad \longrightarrow \quad \max$$

subject to

$$\max(5 + x_1, 3 + x_2, 5 + x_3) \leq 0,$$

$$\max(6 + x_1, 7 + x_2, 1 + x_3) \leq 0,$$

$$\max(8 + x_1, 4 + x_2, 6 + x_3) \leq 0.$$

We have in this case $m = n = 3, a_{ij} = \infty$ for all $i \in I, j \in J$, the optimal solution of the problem is $\overline{x} = (-8, -7, -6)$, the optimal value of the objective function is

$$f(\overline{x}) = \max(-8, -7, -6) = -6.$$

The dual minimization problem is

$$g(u) = \max_{i \in I} u_i \quad \longrightarrow \quad \min$$

subject to

$$\max(5 + u_1, 6 + u_2, 8 + u_3) \geq 0,$$

$$\max(3 + u_1, 7 + u_2, 4 + u_3) \geq 0,$$

$$\max(5 + u_1, 1 + u_2, 6 + u_3) \geq 0.$$

The optimal solution is $u^* = (-\infty, -6, -6)$. The optimal value of the objective function is

$$g(u^*) = \max(-\infty, -6, -6) = -6.$$

Example 4.10. Let us consider the following (max, +)-linear maximization problem, in which $m = n = 3$:

$$f(x) = \max(x_1, x_2, x_3) \quad \longrightarrow \quad \max$$

subject to

$$\max(5 \wedge x_1, 3 \wedge x_2, 5 \wedge x_3) \leq 4,$$
$$\max(6 \wedge x_1, 7 \wedge x_2, 1 \wedge x_3) \leq 2,$$
$$\max(8 \wedge x_1, 4 \wedge x_2, 6 \wedge x_3) \leq 5.$$

The constraints can be transformed to the standard form (4.14) by subtracting b_i from the left hand side of the i-th inequality:

$$\max(1 \wedge x_1 - 4, -1 \wedge x_2 - 4, 1 \wedge x_3 - 4) \leq 0,$$
$$\max(4 \wedge x_1 - 2, 5 \wedge x_2 - 2, -1 \wedge x_3 - 2) \leq 0,$$
$$\max(3 \wedge x_1 - 5, -1 \wedge x_2 - 5, 1 \wedge x_3 - 5) \leq 0.$$

The optimal solution is $\overline{x} = (2, 2, 4)$, the optimal value of the objective function is

$$f(\overline{x}) = \max(2, 2, 4) = 4.$$

The dual minimization problem has the form

$$g(u) = \max(u_1, u_2, u_3) \quad \longrightarrow \quad \min$$

subject to

$$\max(1 \wedge u_1 - 4, 4 \wedge u_2 - 2, 3 \wedge u_3 - 5) \geq 0,$$
$$\max(-1 \wedge u_1 - 4, 5 \wedge u_2 - 2, -1 \wedge u_3 - 5) \geq 0,$$
$$\max(1 \wedge u_1 - 4, -1 \wedge u_2 - 2, 1 \wedge u_3 - 5) \geq 0.$$

The optimal solution is $u^* = (2, 2, 4)$, the optimal value of the objective function is

$$g(u^*) = \max(2, 2, 4) = 4.$$

Remark 4.5. Let us note that we did not consider purely (max, min)-linear objective function $f(x) = \max_{j \in J}(c_j \wedge x_j)$ in Example 4.8 and Example 4.10. In this case functions $f_j(x_j) = c_j \wedge x_j$ would not satisfy the assumption of Theorem 4.2 that $f_j,\ j \in J$ must be strictly increasing functions of x_j.

4.5 Optimization Problems on Attainable Sets of Equation Systems

Let us consider now systems of equations

$$\max_{j\in J}(a_{ij} \wedge r_{ij}(x_j)) = \hat{b}_i, \quad i \in I, \tag{4.25}$$

where $I = \{1, \ldots, m\}$, $J = \{1, \ldots, n\}$, $r_{ij} : \mathbf{R} \longrightarrow \mathbf{R}$ are for all i, j again continuous and strictly increasing functions. The set of all solutions of such systems will be denoted by $M^{=}(\hat{b})$. If $M^{=}(\hat{b}) = \emptyset$, it arises a question which is the next to $\hat{b}$ right hand side of system (4.25), for which the system of equations is solvable.

Definition 4.1. Let $\mathbf{A} = \{a_{ij}\}$. Let us define set $\mathbf{R(A)}$ as follows:

$$\mathbf{R(A)} = \{b \in \mathbf{R}^m \mid \text{there exists } x \in \mathbf{R}^n \text{ such that } \max_{j\in J}(a_{ij} \wedge r_{ij}(x_j)) = b_i\ ,\ \forall i \in I\} \tag{4.26}$$

Set $\mathbf{R(A)}$ will be called *attainable set* of equation system (4.25) and points $b \in \mathbf{R(A)}$ will be called attainable points or *attainable right hand sides* of equation system with the left hand sides like in (4.25).

We will solve the following optimization problem on the attainable set $\mathbf{R(A)}$:

$$\left\|b - \hat{b}\right\| = \max_{i\in I}\left|b_i - \hat{b}_i\right| \quad \longrightarrow \quad \min \tag{4.27}$$

subject to

$$b \in \mathbf{R(A)}.$$

The optimal solution of this problem will be denoted b^{opt}. Let us note that if $\hat{b} \in \mathbf{R(A)}$, then the optimal value of the objective function of problem (4.27) is equal to zero. Otherwise this optimal value is always positive.

Note that optimization problem (4.27) can be reformulated as follows:

$$\left\|b - \hat{b}\right\| \quad \longrightarrow \min \tag{4.28}$$

subject to

$$M(b) \neq \emptyset,$$

or

$$t \quad \longrightarrow \quad \min \left\|b - \hat{b}\right\| \leq t \tag{4.29}$$

subject to

$$b \in \mathbf{R}(\mathbf{A}).$$

Taking into account the definition of the norm $\left\| b - \hat{b} \right\|$, we can conclude that problem (4.29) is equivalent to

$$t \longrightarrow \min \tag{4.30}$$

subject to

$$\left| b_i - \hat{b}_i \right| \le t,\ i \in I,\ b \in \mathbf{R}(\mathbf{A}),$$

or

$$t \longrightarrow \min \tag{4.31}$$

$$\hat{b}_i - t \le b_i \le \hat{b}_i + t,\ i \in I,\ \ b \in \mathbf{R}(\mathbf{A}).$$

and since we assume that $b \in \mathbf{R}(\mathbf{A})$, we can reformulate (4.31) as

$$t \longrightarrow \min \tag{4.32}$$

subject to

$$\hat{b}_i - t \le \max_{j \in J}(a_{ij} \wedge r_{ij}(x_j)) \le \hat{b}_i + t,\ i \in I,\ x \in \mathbf{R}^n.$$

Let us note that the requirement

$$\max_{j \in J}(a_{ij} \wedge r_{ij}(x_j)) \le \hat{b}_i + t,\ \ i \in I \tag{4.33}$$

following from (4.32) means that it must be $a_{ij} \wedge r_{ij}(x_j) \le \hat{b}_i + t,\ i \in I,\ j \in J$, which is equivalent to $x_j \le r_{ij}^{-1}(\hat{b}_i + t),\ i \in I_j(t)$, where for all $j \in J$ we set $I_j(t) = \{i \mid a_{ij} > \hat{b}_i + t\}$. It means that it must be

$$x_j \le \hat{x}_j(\hat{b} + t) = \min_{i \in I_j(t)} (r_{ij}^{-1}(\hat{b}_i + t)), \tag{4.34}$$

where $\hat{b} + t = (\hat{b}_1 + t,\ \ldots,\ \hat{b}_m + t)$.

Taking into account these considerations we can reformulate optimization problem (4.32) as follows:

$$t \longrightarrow \min \tag{4.35}$$

subject to

$$\max_{j \in J}(a_{ij} \wedge r_{ij}(x_j)) \geq \hat{b}_i - t,\ x_j \leq \hat{x}_j(\hat{b} + t),\ i \in I,\ j \in J.$$

We will denote the set of feasible solutions of optimization problem (4.35) by $M(t)$ and the corresponding optimal value of t will be denoted by t^{opt}. Let us note that if $\hat{b} \in \mathbf{R}(\mathbf{A})$, we have $t^{opt} = 0$, otherwise it always $t^{opt} > 0$. Since $M(t) \neq \emptyset$ for a sufficiently large t, a nonnegative and finite t^{opt} of optimization problem (4.35) always exists. Our next aim is to propose a method for determining t^{opt} and b^{opt}.

First we will prove the following lemma

Lemma 4.3. *Let* $T_{ij}(t) = \{x_j \mid a_{ij} \wedge r_{ij}(x_j) \geq \hat{b}_i - t,\ x_j \leq \hat{x}_j(\hat{b}+t)\},\ i \in I, j \in J.$ *Then*

$$M(t) \neq \emptyset \text{ if and only if for all } i \in I \text{ exists } j(i) \in J \text{ such that } T_{ij(i)}(t) \neq \emptyset. \tag{4.36}$$

Proof. Let $T_{ij}(t) = \emptyset$ for all $j \in J$ for some $i \in I$. Then for any $j \in J$ either $x_j > \hat{x}_j(\hat{b}+t)$ or $a_{ij} \wedge r_{ij}(x_j) < \hat{b}_i - t$. We will show that in this situation $x \notin M(t)$ for any $x \in \mathbf{R}^n$. If $x_j > \hat{x}_j(\hat{b} + t)$, then it follows from (4.34) that there exists an index $k \in I_j(t)$ such that $a_{kj} \wedge r_{kj}(x_j) > \hat{b}_k + t$ so that $\max_{j \in J}(a_{kj} \wedge r_{kj}(x_j) > \hat{b}_k + t)$ and therefore $x \notin M(t)$. Let now $x \leq \hat{x}(\hat{b} + t)$ so that the assumption $T_{ij}(t) = \emptyset$ for all $j \in J$ for some $i \in I$ implies that $a_{ij} \wedge r_{ij}(x_j) < \hat{b}_i - t,\ j \in J$ and therefore $\max_{j \in J}(a_{ij} \wedge r_{ij}(x_j)) < \hat{b}_i - t$ and therefore $x \notin M(t)$. It follows that under our assumption that $T_{ij}(t) = \emptyset,\ j \in J$ no element of $M(t)$ can exist and $M(t) = \emptyset$.

Let now for any $i \in I$ there exists $j(i) \in J$ such that $T_{ij(i)}(t) \neq \emptyset$. We will show that $M(t) \neq \emptyset$. Let $P_j = \{i \mid j(i) = j\},\ j \in J$ and $T_j(t) = \bigcap_{i \in P_j} T_{ij(i)}\ \forall j \in J$ such that $P_j \neq \emptyset$. Let us set further $T_j(t) = [-\infty, \hat{x}_j(\hat{b}+t)]$ if $P_j = \emptyset$. Let us note that $T_j(t) = [\max_{i \in P_j}(\hat{b}_i - t), \hat{x}_j(\hat{b} + t)]$ if $P_j \neq \emptyset$. It follows that $\hat{x}_j(\hat{b} + t) \in T_j(t),\ j \in J$. Let $i \in I$ be arbitrarily chosen. Since $T_{ij(i)}(t) \neq \emptyset$ and we have $\hat{x}_{j(i)}(\hat{b} + t) \in T_{j(i)}(t) \subseteq T_{ij(i)}(t)$, we obtain that $a_{ij(i)} \wedge r_{ij(i)}(\hat{x}_{j(i)}) > \hat{b}_i - t$ and therefore

$$\max_{j \in J}(a_{ij} \wedge r_{ij}(\hat{x}_j(\hat{b} + t))) \geq a_{ij(i)} \wedge r_{ij(i)}(\hat{x}_{j(i)}) \geq \hat{b}_i - t.$$

Since $i \in I$ was arbitrarily chosen, we obtain eventually that $\hat{x}(\hat{b} + t) \in M(t)$ so that $M(t) \neq \emptyset$, which we wanted to prove. Therefore (4.36) is fulfilled. □

Let us note that $x_j(\hat{b} + t)$ for a $j \in J$ is in general not continuous function in t. Figure 4.1 below shows a graph of $x_j(\hat{b} + t)$ in case that $r_{ij}(x_j) = x_j$.

Lemma 4.4. *Let* $i \in I,\ j \in J$ *be arbitrarily chosen. There exists* τ_{ij} *such that* $T_{ij}(t) \neq \emptyset$ *if and only if* $t \geq \tau_{ij}$.

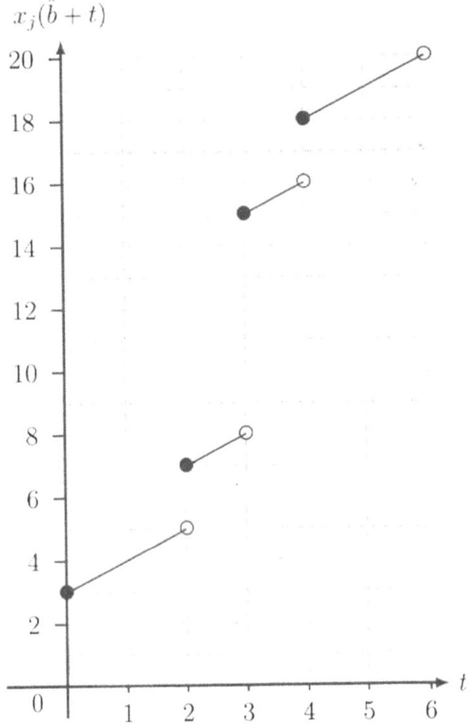

Fig. 4.1 A graph of $x_j(\hat{b}+t)$

Proof. We will show that for any of the sets $T_{ij}(t)$ there exists τ_{ij} such that $T_{ij}(t) \neq \emptyset$ if and only if $t \geq \tau_{ij}$.

Note that $T_{ij}(t) \neq \emptyset$ if and only if $\hat{b}_i - t \leq a_{ij} \wedge r_{ij}(\hat{x}_j(\hat{b}+t)) \leq \hat{b}_i + t$. It must be therefore $a_{ij} \geq \hat{b}_i - t$ and $r_{ij}(\hat{x}_j(\hat{b}+t)) \geq \hat{b}_i - t$. It follows that if $T_{ij}(t) \neq \emptyset$, it must be $t \geq \hat{b}_i - a_{ij}$ and at the same time $\hat{x}_j(\hat{b}+t) \geq r_{ij}^{-1}(\hat{b}_i - t)$. Let us note that $\hat{x}_j(\hat{b}+t)$ is in general a partially continuous (more exactly upper-semicontinuous) function in t (compare Fig. 4.1), $r_{ij}^{-1}(\hat{b}_i - t)$ is a strictly decreasing function of t and $a_{ij} \wedge r_{ij}(\hat{x}_j(\hat{b}+t))$ is equal to a constant a_{ij} if $r_{ij}(\hat{x}_j(\hat{b}+t) \geq a_{ij}$.

First, it follows if set $T_{ij}(t)$ is nonempty then $t \geq \tau_{ij}^{(1)}$, where $\tau_{ij}^{(1)}$ solves the equation $\hat{b}_i - t = a_{ij}$, i.e. $\tau_{ij}^{(1)} = \hat{b}_i - a_{ij}$.

Further, it remains to investigate the case when $a_{ij} \geq r_{ij}(\hat{x}_j(\hat{b}) + t) \geq \hat{b}_i - t$. In this case we have to find the minimum value of t, for which the inequality $\hat{x}_j(\hat{b} + t) \geq r_{ij}^{-1}(\hat{b} - t)$ holds. Since $\hat{x}_j(\hat{b}+t)$ is not in general continuous in t, the equation $\hat{x}_j(\hat{b}+t) = r_{ij}^{-1}(\hat{b}-t)$ may not be solvable with respect to t, but it is possible to find the minimum value of t, at which the inequality holds. If we denote this value

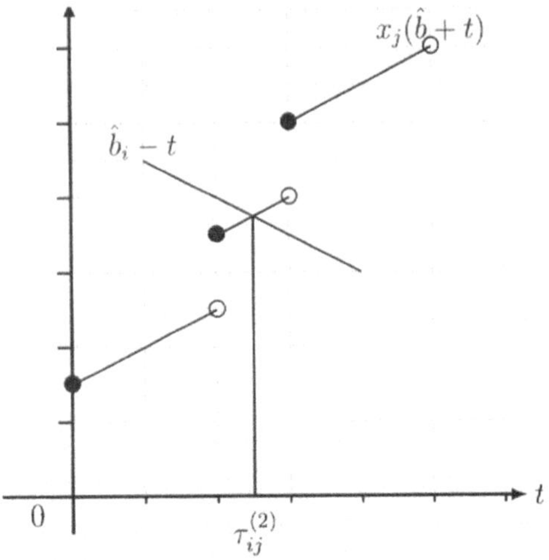

Fig. 4.2 $\tau_{ij}^{(2)}$ is an intersection of $x_j(\hat{b}+t)$ and $\hat{b}_i(t-1)$

$\tau_{ij}^{(2)}$, it will be $\hat{x}_j(\hat{b}+\tau_{ij}^{(2)}) \geq r_{ij}^{-1}(\hat{b}-\tau_{ij}^{(2)})$ and $\hat{x}_j(\hat{b}+t) < r_{ij}^{-1}(\hat{b}-t)$ for all $t < \tau_{ij}^{(2)}$. We will find $\tau_{ij}^{(2)}$ as follows.

Let $a_{k_1j} > a_{k_2j} > \cdots > a_{k_s} \geq a_{ij}$. Then we have $\hat{x}_j(\hat{b}+t) = \infty$ if $t \geq a_{k_1j}-b_{k_1}$ and $\hat{x}_j(\hat{b}+t) = r_{k_hj}^{-1}(\hat{b}_{k_h}+t)$ if $a_{k_hj}-b_{k_h} \leq t < a_{k_{h-1}j}-b_{k_{h-1}}$ for $h = 2, \ldots, s$. Let us set $t_h = a_{k_hj}-b_{k_h}$, $h = 1, \ldots, s$. Let

$$p = \max\{h \mid t_h \geq r_{ij}^{-1}(\hat{b}_i - t_h),\ 1 \leq h \leq s\}.$$

Then there exists the unique value $\tau_{ij}^{(2)} \in [t_p, t_{p+1}]$ such that $\hat{x}_j(\hat{b}+\tau_{ij}^{(2)}) \geq r_{ij}^{-1}(\hat{b}-\tau_{ij}^{(2)})$ and either $\hat{x}_j(\hat{b}+\tau_{ij}^{(2)}) = r_{ij}^{-1}(\hat{b}_{k_p}+\tau_{ij}^{(2)})$ or $\hat{x}_j(\hat{b}+\tau_{ij}^{(2)}) > r_{ij}^{-1}(\hat{b}_{k_p}+\tau_{ij}^{(2)})$ and $r_{ij}^{-1}(\hat{b}_{k_p}+t) < \hat{b}_i - t$ for all $t < \tau_{ij}^{(2)}$. Let us note that in the former case we can find an approximate value $\tau_{ij}^{(2)}$ numerically by making use of an appropriate search method and in the latter case we have $\tau^{(2)} = t_p$. The former case is shown in Fig. 4.2, the latter case is depicted in Fig. 4.3 below. Let us note that the existence and uniqueness of $\tau_{ij}^{(1)}$, $\tau_{ij}^{(2)}$ follows from the monotonicity and continuity of functions r_{ij}.

Let us set $\tau_{ij} = \max(\tau_{ij}^{(1)}, \tau_{ij}^{(2)})$. We have:

If $t < \tau_{ij} = \tau_{ij}^{(1)}$, then $a_{ij} \wedge r_{ij}(\hat{x}_j(\hat{b})) \leq a_{ij} < \hat{b}_i - t$ so that $T_{ij}(t) = \emptyset$;

If $t < \tau_{ij} = \tau_{ij}^{(2)}$, then $\hat{b}_i - t > a_{ij} \geq$ so that $a_{ij} \wedge r_{ij}(\hat{x}_j(\hat{b})) < \hat{b}_i - t$ and therefore $T_{ij}(t) = \emptyset$;

We obtain that $T_{ij}(t) \neq \emptyset$ if and only if $t \geq \tau_{ij}$, which completes the proof. □

Theorem 4.3. *There exists a value $\tau \geq 0$ such that $M(t) \neq \emptyset$ if and only if $t \geq \tau$.*

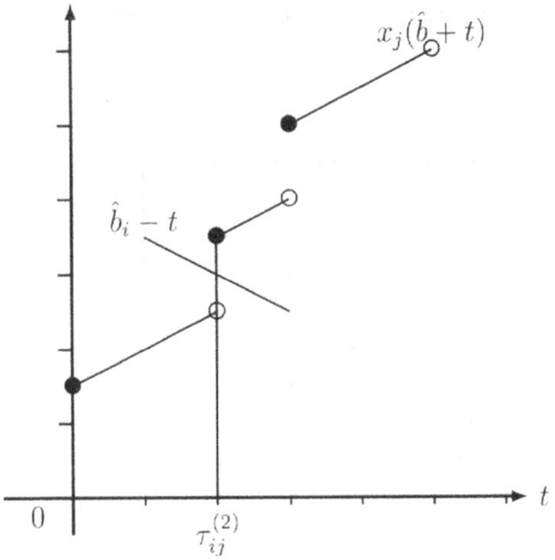

Fig. 4.3 The case $\tau_{ij}^{(2)} = t_p$

Proof. Let $\tau_{ij},\ i \in I,\ j \in J$ be defined as in Lemma 4.4. Let us set

$$\tau = \max_{i \in I} \min_{j \in J} \tau_{ij}.$$

Let us set $\tau_i = \min_{j \in J} \tau_{ij},\ i \in I$. Let index $i_0 \in I$ be defined as follows: $\tau_{i_0} = \max_{i \in I} \tau_i$. Then $\tau_{i_0} = \tau$ and if $t < \tau$, then $T_{i_0 j}(t) = \emptyset$ for all $j \in J$ and therefore according to Lemma 4.1 we obtain that $M(t) = \emptyset$. If on the other hand $t \geq \tau_{i_0} = \tau$, then $\hat{x}(\hat{b} + t) \in M(t)$ so that $M(t) \neq \emptyset$. This completes the proof. □

As a consequence of Theorem 4.3 we obtain that

$$t^{opt} = \tau\ ,\ b_i^{opt} = \max_{j \in J}(a_{ij} \wedge r_{ij}(\hat{x}_j(\hat{b} + \tau)))\ ,i \in I,$$

where t^{opt}, b^{opt} are the optimal solutions of problems (4.29), (4.27) respectively. The optimal value of the objective function of optimization problems (4.27), (4.28) is $\left\| b^{opt} - \hat{b} \right\| = \tau$ and $\hat{x}(\hat{b} + \tau)$ is the maximum element of set $M(\tau)$.

Remark 4.6. The computational complexity depends in general on functions r_{ij}. Values $\tau_{ij}^{(2)}$ can be in general obtained only by making use of an approximate search method, values $\tau_{ij}^{(1)}$ can be expressed explicitly. This fact leads to the idea to propose an approximate search method directly to find an approximate value of t^{opt}. We can first check whether $t^{opt} = 0$ by checking whether $M(0) \neq \emptyset$ (i.e. whether $\hat{x}(\hat{b}) \in M(0)$). If the answer is positive, the problem is solved. Otherwise we can easily find a positive value $\overline{t}$ such that $M(\overline{t}) \neq \emptyset$. It can be easily verified that we can choose as $\overline{t}$ e.g. any value greater than $\max_{i,j}(|a_{ij}|, |r_{ij}^{-1}(\hat{b}_i)|)$. In this case we have $t^{opt} \in (0, \overline{t}]$ and therefore we can make use of the usual search procedure on this interval.

Taking into account Remark 4.6 we will propose in the sequel an algorithm for finding an approximate ϵ-optimal value $t^{opt}(\epsilon)$ of optimisation problem (4.29) with the properties: $M(t^{opt}(\epsilon)) \neq \emptyset$ and $|t^{opt} - t^{opt}(\epsilon)| < \epsilon$ for a given positive ϵ. We will assume that $\hat{b} \notin \mathbf{R}(\mathbf{A})$ and $\overline{t} > 0$ has been chosen.

ALGORITHM I.

1 Input $m,\ n,\ a_{ij},\ r_{ij},\ \hat{b}_i,\ \overline{t},\ \epsilon,\ \underline{t} := 0$;
2 If $|\overline{t} - \underline{t}| < \epsilon$, go to 6 ;
3 Set $\overline{t}_1 := (\overline{t} - \underline{t})/2 + \underline{t}$;
4 If $M(\overline{t}_1) \neq \emptyset$, set $\overline{t} := \overline{t}_1$, go to 2;
5 Set $\underline{t} := (\overline{t}_1)/2 + \overline{t}_1$, go to 2;
6 $t^{opt}(\epsilon) := \overline{t}$, STOP.

Let us note that the verification of $M(\overline{t}_1) \neq \emptyset$ in step 4 can be done by verifying whether $\hat{x}(\hat{b} + \overline{t}_1) \in M(\overline{t}_1)$.

Remark 4.7. In what follows we will consider some special cases of systems (4.25) for which explicit expressions for τ_{ij}'s exist. In these cases explicit formulas for τ can be derived. Especially we will consider the case $r_{ij}(x_j) = p_{ij} + x_j$, where $p_{ij} \in \mathbf{R}$ and $a_{ij} = \infty\ \forall i \in I,\ j \in J$, which was used in Example 4.1 and the case $r_{ij}(x_j) = x_j,\ a_{ij} \in \mathbf{R},\ i \in I,\ j \in J$, which was applied in Example 4.2.

Let us first consider a special case from Example 4.1. We will assume therefore that for all $i \in I,\ j \in J$ we have $r_{ij}(x_j) = p_{ij} + x_j$ and $a_{ij} = \infty$. System (4.25) can be reformulated in this case as follows:

$$\max_{j \in J}(p_{ij} + x_j) = \hat{b}_i,\ i \in I. \tag{4.37}$$

The maximum element $\hat{x}(\hat{b} + t)$ can be expressed as follows:

$$\hat{x}_j(\hat{b} + t) = \min_{k \in I}(\hat{b}_k - p_{kj}) + t = \hat{b}_{k(j)} - p_{k(j)j} + t,\ j \in J. \tag{4.38}$$

We have further $r_{ij}^{-1}(\hat{b}_i - t) = \hat{b}_i - p_{ij} - t,\ i \in I,\ j \in J$. Therefore $\tau_{ij}^{(2)}$ solves in this case the equation

$$\hat{b}_i - p_{ij} - t = \hat{b}_{k(j)} - p_{k(j)j} + t,\ i \in I,\ j \in J. \tag{4.39}$$

and we obtain the following explicit formulas for $\tau_{ij}^{(1)}$:

$$\tau_{ij}^{(2)} = ((\hat{b}_i - p_{ij}) - (\hat{b}_{k(j)} - p_{k(j)j}))/2, \ i \in I, \ j \in J. \tag{4.40}$$

Since $a_{ij} = \infty$, $\tau_{ij}^{(1)} = -\infty$ does not take part in determining of τ and we obtain eventually the following explicit formula for τ:

$$\tau = \max_{i \in I} \min_{j \in J}((\hat{b}_i - p_{ij}) - (\hat{b}_{k(j)} - p_{k(j)j}))/2, i \in I, \ j \in J. \tag{4.41}$$

We will illustrate this case by a small numerical example.

Example 4.11. Let $m = n = 3$, $\hat{b} = (0, 0, 0)^T$,

$$P = \{p_{ij}\} = \begin{pmatrix} 3, \ 1, \ 5 \\ 4, \ 4, \ 6 \\ 7, \ 7, \ 3 \end{pmatrix}$$

We will consider the system

$$\max(3 + x_1, 1 + x_2, 5 + x_3) = 0,$$

$$\max(4 + x_1, 4 + x_2, 6 + x_3) = 0,$$

$$\max(7 + x_1, 7 + x_2, 3 + x_3) = 0.$$

In this case we have $r_{ij}(x_j) = a_{ij} + x_j$ for all $i \in I, \ j \in J$. It can be easily verified that $\hat{b} \notin \mathbf{R}(\hat{b})$, further we have

$$T_{ij}(t) = [0 - p_{ij} - t, \min_{k \in I}(-p_{kj} + t)], \ i \in I, \ j \in J.$$

It follows that

$$T = \{\tau_{ij}\} = 1/2\left(-p_{ij} - \min_{k \in I}(-p_{kj})\right),$$

and therefore

$$T = \{\tau_{ij}\} = \begin{pmatrix} 2, & 3, & 1/2 \\ 3/2, & 3/2, & 0 \\ 0, & 0, & 3/2 \end{pmatrix}.$$

Using the formula derived above we obtain $\tau = 1/2$, $\overline{x}(\hat{b} + \tau) = (-6.5, -6.5, -5.5)^T$, $b^{opt} = (-1/2, 1/2, 1/2)^T$. It can be easily verified that

$$\left\| b^{opt} - \hat{b} \right\| = \max(1/2, 1/2, 1/2) = 1/2 = t^{opt} = \tau.$$

Remark 4.8. Let us note that formula (4.41) for the special case of (max, +)- linear systems was derived in a different way in [10]. In [10] was derived for t^{opt} the formula

$$t^{opt} = 1/2(A(\hat{b}^{-}A)^{-})^{-}\hat{b}, \tag{4.42}$$

where we use the notation introduced in [10], i.e. we set $c^{-} = -c$ for any $c \in \mathbf{R}^m$, $Ax \in \mathbf{R}^m$ with $(Ax)_i = \max_{j \in J}(a_{ij} + x_j), i \in I$ for any (m,n)-matrix A with elements a_{ij} and $x \in \mathbf{R}^n$, $A^{-} = -A^T$ (A^T denotes the transposed matrix to A).

Using our notation we have

$$(\hat{b}^{-}A)_j^{-} = \left(\max_{k \in I}(-\hat{b}_k + a_{kj})\right)^{-} = -\max_{k \in I}(-\hat{b}_k + a_{kj}) = \min_{k \in I}(\hat{b}_k - a_{kj}),\ j \in J,$$

$$(A(\hat{b}^{-}A)^{-})_i^{-} = -\max_{j \in J}(a_{ij} + (\hat{b}^{-}A)_j^{-})) = \min_{j \in J}(-a_{ij} - (\hat{b}^{-}A)_j^{-}),\ \in I.$$

It follows that for all $i \in I$

$$(A(\hat{b}^{-}A)^{-})_i^{-} = \min_{j \in J}(-a_{ij} - (\hat{b}^{-}A)_j^{-}) = \min_{j \in J}(-a_{ij} - \min_{k \in I}(\hat{b}_k - a_{kj})),$$

and

$$A((\hat{b}^{-}A)^{-})^{-}\hat{b} = \max_{i \in I}(A((\hat{b}^{-}A)^{-})_i^{-} + \hat{b}_i) = \max_{i \in I}(\hat{b}_i - a_{ij} - \min_{k \in I}(\hat{b}_k - a_{kj})) = 2t^{opt},$$

If we replace a_{ij} by p_{ij}, we see that equality (4.42) is identical with (4.41).

Further we derive similarly a simplified formula for τ also in the second special case from Example 4.2, i.e. the case $r_{ij}(x_j) = x_j$ and $a_{ij} \in \mathbf{R}$. In this case we consider the system

$$\max_{j \in J}(a_{ij} \wedge x_j) = \hat{b}_i,\ i \in I. \tag{4.43}$$

for the maximum element $\hat{x}(\hat{b} + t)$ holds:

$$\hat{x}_j(\hat{b} + t) = \min_{k \in I_j(t)} \hat{b}_k + t = \hat{b}_{k_p} + t,\ j \in J, \tag{4.44}$$

where $I_j(t) = \{k \in I \mid a_{kj} > \hat{b}_k + t\}$ and indexes p and k_p are defined like in the proof of Lemma 4.4. We have further

$$\tau_{ij}^{(1)} = \hat{b}_i - a_{ij}.$$

$\tau_{ij}^{(2)}$ solves the equation $r_{ij}^{-1}(\hat{b}_i - t) = \hat{x}_j(\hat{b} + t)$ with respect to t. Since in our case $r_{ij}^{-1}(\hat{b}_i - t) = \hat{b}_i - t$, we obtain that either $\tau_{ij}^{(2)} = a_{k_p j} - \hat{b}_{k_p}$ or $\tau_{ij}^{(2)}$ solves the equation $\hat{b}_i - t = \hat{b}_{k_p} + t$, i.e. $\tau_{ij}^{(2)} = (\hat{b}_i - \hat{b}_{k_p})/2$. We obtain

$$\tau_{ij} = \max(\hat{b}_i - a_{ij}, \tau_{ij}^{(2)}, 0), \tag{4.45}$$

where $\tau_{ij}^{(2)}$ can be obtained explicitly by the procedure described above.

The formula for τ is then the following:

$$\tau = \max_{i \in I} \min_{j \in J}(\max(\hat{b}_i - a_{ij}, \tau_{ij}^{(2)}, 0). \tag{4.46}$$

We will illustrate determining values $\tau_{ij}^{(2)}$ for the case when $r_{ij}(x_j) = x_j$ by the following numerical example.

Example 4.12. Let us have $m = 5$, $I = \{1, 2, 3, 4, 5\}$, $j \in J$, let $a^{(j)} = (a_{1j}, \ \ldots, \ a_{5j})^T = (7, 3, 5, 10.5, 1)^T$ (i.e. $a^{(j)}$ is the $j - th$ column of a matrix with elements $a_{ij}, i \in I, j \in J$), $\hat{b} = (4, 1, 2, 4.5, 10)^T$. Then $I_j(t)$, $\hat{x}_j(\hat{b} + t)$ are defined as follows:

If $t \geq 6$, then $I_j(t) = \emptyset$, $\hat{x}_j(\hat{b}) + t = \infty$;

If $t \in [3, 6)$, then $I_j(t) = \{4\}$, $\hat{x}_j(\hat{b} + t) = \hat{b}_4 + t = 4.5 + t$;

If $t \in [2, 3)$, then $I_j(t) = \{1, 3, 4\}$, $\hat{x}_j(\hat{b} + t) = \hat{b}_3 + t = 2 + t$;

If $t \in [-9, 2)$, then $I_j(t) = \{1, 2, 3, 4\}$, $\hat{x}_j(\hat{b} + t) = \hat{b}_2 + t = 1 + t$;

If $t \in (-\infty, -9)$, then $I_j(t) = \{1, 2, 3, 4, 5\}$, $\hat{x}_j(\hat{b} + t) = \hat{b}_5 + t = 1 + t$.

Let us choose e.g. $i = 4$ so that we should solve the equation $4.5 - t = \hat{x}_j(\hat{b} + t)$. It can be easily verified that $4.5 - t < \hat{x}_j(\hat{b} + t)$ if $t \geq 2$ and $4.5 - t > \hat{x}_j(\hat{b} + t) = 1 + t$ for all $t < 2$ so that in this case no solution of the equation $4.5 - t = \hat{x}_j(\hat{b} + t)$ because of the discontinuity of function $\hat{x}_j(\hat{b} + t)$ and we set $\tau_{4j}^{(2)} = 2$. Further we have $\hat{b}_4 - a_{4j} = 4.5 - 10.5 = -6$ so that according to (4.45) we obtain

$$\tau_{4j} = \max(\hat{b}_4 - a_{4j}, \tau_{4j}^{(2)}) = \max(-6, 2) = 2.$$

Another situation will arise if we choose $i = 3$ so that $a_{3j} = 5$ and $\hat{b}_3 - t = 2 - t$. For determining value $\tau_{3j}^{(2)}$ consider the equation $2 - t = \hat{x}_j(\hat{b} + t)$. In this case $2 - t < \hat{x}_j(\hat{b} + t) = 1 + t$ if $t \geq 2$ and $2 - t > \hat{x}_j(\hat{b} + t) = 1 + t$ if t is sufficiently small (e.g. $t = 0$). Therefore there exists a solution of the equation $2 - t = \hat{x}_j(\hat{b} + t) = 1 + t$ on the interval $(0, 2)$ and we obtain $\tau_{3j}^{(2)} = 0.5$. Since $\hat{b}_3 - a_{3j} = 2 - 5 = -3$, we obtain in accordance with (4.45)

$$\tau_{3j} = \max(\hat{b}_3 - a_{3j}, \tau_{3j}^{(2)}) = \max(-3, 0.5) = 0.5.$$

Let us consider further index $i = 5$. Then $a_{5j} = 1, \hat{b}_5 - t = 10 - t, \tau_{5j}^{(1)} = \tau_{5j}^{(2)} = \hat{b}_5 - a_{5j} = 10 - 1 = 9$. Further equation $\hat{b}_5 - t = \hat{x}(\hat{b} + t)$ has no solution since for $t \geq 3$ it is $10 - t < \hat{x}(\hat{b} + t)$ and for $t < 3$ we have $10 - t > \hat{x}(\hat{b} + t)$ so that $\tau_{5j}^{(2)} = 3$ Therefore finally we obtain

$$\tau_{5j} = \max(\hat{b}_5 - a_{5j}, \tau_{5j}^{(2)}) = \max(9, 3) = 9.$$

We considered above optimization problems on attainable sets, in which we searched the nearest element of the attainable set $\mathbf{R(A)}$, to a given "target point" $\hat{b} \in \mathbf{R}^m$. In the sequel, we will consider a generalization of the optimization problems, in which instead of the "target point" a "target set" $\hat{B} \subset \mathbf{R}^m$ occurs. For this purpose we have to change appropriately the definition of the concept of attainable set in accordance with the chosen definition of a solution of such systems. The systems we will consider can be described as

$$\max_{j \in J}(a_{ij} \wedge r_{ij}(x_j) = b_i, \ i \in I, \ b = (b_1, \ \ldots, \ b_m) \in \hat{B}. \tag{4.47}$$

We will define further the corresponding attainable set as follows.

Definition 4.2. Set $\tilde{\mathbf{R}}(A, \hat{B}) = \{b \in \hat{B} \mid$ there exists $x \in \mathbf{R}^n$ such that $\max_{j \in J}(a_{ij} \wedge r_{ij}(x_j) = b_i, \ i \in I\}$ will be called attainable set of system (4.47).

We will consider optimization problems of the form

$$\rho(b, \hat{B}) \longrightarrow \min \tag{4.48}$$

subject to

$$b \in \tilde{\mathbf{R}}(A, \hat{B}),$$

where $\rho(b, \hat{B}) = \min_{c \in \hat{B}} \|b - c\|$ and $\hat{B}$ is a compact subset of $\mathbf{R}^m$. We will consider the simplest special case, in which $\hat{B} = [\underline{b}, \overline{b}] = \{b \,; \underline{b} \leq b \leq \overline{b}\}$, where $\underline{b} \leq \overline{b}$ are given elements of $\mathbf{R}^m$. In this case, problem (4.48) can be reformulated as follows:

$$t \longrightarrow \min \tag{4.49}$$

subject to

$$\underline{b}_i - t \leq \max_{j \in J}(a_{ij} \wedge r_{ij}(x_j)) \leq \overline{b}_i + t, \ i \in I.$$

Problem (4.49) can be reformulated as

$$t \longrightarrow \min \tag{4.50}$$

subject to

$$\underline{b}_i - t \le \max_{j\in J}(a_{ij} \wedge r_{ij}(x_j)),\ i \in I,\ x \le \hat{x}(\overline{b} + t),$$

where for all $j \in J$

$$\hat{x}_j(\overline{b} + t) = \min_{k\in I_j(t)} r_{kj}^{-1}(\overline{b}_k + t),\ I_j(t) = \{k \in I \mid a_{kj} \ge \overline{b}_k + t\}$$

Let us note that $t^{opt} = 0$ if and only if $\tilde{\mathbf{R}}(A, \hat{B}) \cap \hat{B} \neq \emptyset$.

Further we can proceed like in the case of problem (4.35) replacing for all $i \in I$ term $\hat{b}_i - t$ by $\underline{b}_i - t$ and $\hat{b}_i + t$ by $\overline{b}_i + t$. For the special cases (4.37) and (4.43) we can obtain explicit formulas for t^{opt}.

Remark 4.9. The set of feasible solutions of the optimization problem (4.27) was equal to $\mathbf{R(A)}$. Let us note that we could require that b satisfies simple additional constraints. For instance we can require that $b_i = \hat{b}_i,\ i \in \tilde{I}$, or that $b_i = \alpha,\ i \in \tilde{I}$, where $\tilde{I}$ is a nonempty proper subset of I and α is a given real number. The procedures proposed above can be easily adjusted for such simple modification problems of (4.27).

Remark 4.10. In real world applications described in Examples 4.1, 4.2 solving optimization problem (4.48) can be interpreted as a situation, in which no feasible solution with the right hand side $b \in \hat{B}$ exists, and we find the nearest right hand side generating a non-empty feasible set. We assume at the same time that the elements p_{ij} in Example 4.1 and a_{ij} in Example 4.2 remain unchanged, i.e. the travelling times in Example 4.1 and the capacities in Example 4.2 cannot be changed. Another situation which can be investigated arises if we can change the traveling times and capacities, but the right hand sides $\hat{b}$ (i.e. the prescribed arrival times in Example 4.1 and prescribed maximal capacities in Example 4.2) are fixed.

We will analyze now in more detail the situation mentioned in Remark 4.10 for system (4.37), which is a special case of (4.27) of the form

$$\max_{j\in J}(\hat{p}_{ij} + x_j) = b_i,\ i \in I, \tag{4.51}$$

where matrix $\hat{P}$ with elements $\hat{p}_{ij}$ is given and $b \in \mathbf{R}^m$.

Definition 4.3. Let for any fixed $b \in \mathbf{R}^m$ set $\hat{\mathbf{R}}(b)$ be defined as follows:

$$\hat{\mathbf{R}}(b) = \{P = \|p_{ij}\| \mid \max_{j\in J}(p_{ij} + x_j) = b_i,\ i \in I.\}$$

The set $\hat{\mathbf{R}}(b)$ will be called *attaining set* of system (4.51) at point b.

Set $\hat{\mathbf{R}}(b)$ can be characterized as a set of (m, n)-matrices P with elements p_{ij}, from which the given vector b can be attained (i.e. for which there exists a solution x satisfying (4.51)).

Let us consider the optimization problem

$$\left\| P - \hat{P} \right\| \longrightarrow \min \tag{4.52}$$

subject to

$$P \in \hat{\mathbf{R}}(b).$$

We can reformulate (4.52) as follows:

$$t \longrightarrow \min subject to \tag{4.53}$$

$$\left\| P - \hat{P} \right\| \leq t, \; P \in \hat{\mathbf{R}}(b),$$

or alternatively

$$t \longrightarrow \min \tag{4.54}$$

$$\hat{p}_{ij} - t \leq p_{ij} \leq \hat{p}_{ij} + t \leq t, \; i \in I, j \in J, P \in \hat{\mathbf{R}}(b).$$

Let us note that if $t, \; P$ satisfy (4.53), then there exists $x \in \mathbf{R}^n$ such that $\max_{j \in J}(p_{ij} + x_j) = b_i, \; i \in I$ so that inequalities $\max_{j \in J}(\hat{p}_{ij} + t + x_j) \geq b_i \geq \max_{j \in J}(\hat{p}_{ij} - t + x_j), \; i \in I$ hold for any $t \geq 0$. It follows that problem (4.51) is equivalent with

$$t \longrightarrow \min \tag{4.55}$$

subject to

$$\max_{j \in J}(\hat{p}_{ij} - t + x_j) \leq b_i \leq \max_{j \in J}(\hat{p}_{ij} + t + x_j), \; i \in I.$$

Transferring t to the opposite sides of inequalities we obtain eventually equivalent problem

$$t \longrightarrow \min \tag{4.56}$$

subject to

$$\max_{j \in J}(\hat{p}_{ij} + x_j) \leq b_i + t \; \text{ and } \; \max_{j \in J}(\hat{p}_{ij} + x_j) \geq b_i - t, \; i \in I.$$

Problem (4.56) can be solved by the same way as problem (4.32) if we assume that the original system has the special form (4.37) and replace $\hat{b}$ by b.

4.6 Two-Sided max-Separable Equation and Inequality Systems: Some Special Cases

Under two-sided max-separable equations and inequality systems we will understand systems of the form

$$\max_{j\in J}(a_{ij} \wedge r_{ij}(x_j)) \geq \max_{j\in J}(b_{ij} \wedge q_{ij}(x_j)),\ i \in I_1, \tag{4.57}$$

$$\max_{j\in J}(a_{ij} \wedge r_{ij}(x_j)) \leq \max_{j\in J}(b_{ij} \wedge q_{ij}(x_j)),\ i \in I_2, \tag{4.58}$$

where $I = I_1 \cup I_2$, $r_{ij} : \mathbf{R} \longrightarrow \mathbf{R}$, $q_{ij} : \mathbf{R} \longrightarrow \mathbf{R}$ are strictly increasing continuous functions, $a_{ij},\ b_{ij} \in \mathbf{R} \cup \{-\infty\}$, $A = \{a_{ij}\}$, $B = \{b_{ij}\}$, we set $\alpha \wedge -\infty = -\infty$ for all $\alpha \in \mathbf{R}$, and we assume that the equations have already been replaced by two inequalities. We will investigate in this section properties of some special cases of such systems.

Remark 4.11. Let us note that according to our knowledge, no effective method for solving two-sided systems in the general form mentioned above has been published. In this work we will present results concerning special cases of the systems, which are connected with polynomial or at least pseudopolynomial computation procedures and may have applications similar to those considered in the examples in Introduction. Let us remark that an appropriate choice of elements a_{ij}, b_{ij} of matrices $A,\ B$ makes possible to include in the inequalities also upper bounds on variables of the considered inequality system. In the sequel, we will sometimes write the lower and upper bound constraints of variables x_j separately in their explicit form $\underline{x}_j \leq x_j \leq \overline{x}_j$, where $\underline{x}_j,\ \overline{x}_j$ are given finite bounds. Let us note that the lower and upper bounds occur practically always in the applications, since the variables represent as a rule amounts of resources or capacities of certain objects, which are always bounded.

The first special case of system (4.57), (4.58) we will consider in this section is the case, in which the right hand side variables are different from the left hand side variables and finite lower and upper bounds on variables are given, i.e. the system has the form:

$$\max_{j\in J}(a_{ij} \wedge r_{ij}(x_j)) \geq \max_{j\in J}(b_{ij} \wedge q_{ij}(y_j)),\ i \in I_1, \tag{4.59}$$

$$\max_{j\in J}(a_{ij} \wedge r_{ij}(x_j)) \leq \max_{j\in J}(b_{ij} \wedge q_{ij}(y_j)),\ i \in I_2, \tag{4.60}$$

$$\underline{x} \leq x \leq \overline{x},\ \underline{y} \leq y \leq \overline{y}, \tag{4.61}$$

where $\underline{x} \leq \overline{x},\ \underline{y} \leq \overline{y}$ are given finite elements of $\mathbf{R}^n$.

To simplify the notation we will use where necessary for the left and right hand sides of the inequalities the notations

$$r_i(x) = \max_{j\in J}(a_{ij} \wedge r_{ij}(x_j)),\ q_i(y) = \max_{j\in J}(b_{ij} \wedge q_{ij}(y_j)),\ i \in I. \tag{4.62}$$

Note that the solutions x must satisfy the inequalities

$$x_j \le \overline{x}_j(\overline{y}) = \min_{k\in I_2^*}(r_{kj}^{-1}(q_k(\overline{y}))),\ y_j \le \overline{y}_j(\overline{x}) = \min_{k\in I_1^*}(q_{kj}^{-1}(r_k(\overline{x})))\ j \in J,$$

where $I_2^* = \{k \in I_2 \mid a_{kj} > q_k(\overline{y})\}$, $I_1^* = \{k \mid b_{kj} > r_k(\overline{x})\}$.

If either $\overline{x}_{j_0}(\overline{y}) < \underline{x}_{j_0}$ or $\overline{y}_{j_0}(\overline{x}) < \underline{y}_{j_0}$ for some $j_0 \in J$, then the set of solutions and of system (4.59)–(4.61) is empty.

We will assume further that the set of all solutions satisfying (4.59), (4.61) will be denoted M_1, set of all solutions satisfying (4.60), (4.61) will be denoted M_2 so that if M denotes the set of solutions satisfying (4.59), (4.60), (4.61), then $M = M_1 \cap M_2$.

Let us note that if $M \neq \emptyset$, then it must hold

$$r_i(\overline{x}(\overline{y})) \ge q_i(\underline{y}),\ i \in I_1,\ r_i(\underline{x}) \le q_i(\overline{y}(\overline{x})),\ i \in I_2 \tag{4.63}$$

and

$$\underline{x} \le \overline{x}(\overline{y}),\ \underline{y} \le \overline{y}(\overline{x}). \tag{4.64}$$

In what follows, we will investigate some special cases.

As the first special case we will consider two-sided systems (4.53)–(4.55), in which $I_2 = \emptyset$. We consider therefore a system having the form

$$\max_{j\in J}(r_{ij}(x_j) \ge q_i(y),\ i \in I,\ \underline{x} \le x \le \overline{x},\ \underline{y} \le y \le \overline{y}, \tag{4.65}$$

where $I = I_1$ and $q_i(y)$, $i \in I$ are defined as in (4.62). If y is fixed, the system can be considered as a one-sided system of the form (4.1)–(4.2), in which we set $b_i = q_i(y)$ for all $i \in I$ with a fixed y. Such system with a fixed y satisfies all assumptions of Lemma 4.1–4.2 and Theorem 4.1. Therefore we can verify for any fixed y whether the set of solutions of (4.65) is nonempty by making use of Lemma 4.2 and relations (4.63), (4.64). If system (4.65) has a nonempty set of solutions, we can solve for any fixed y optimization problem of the form (4.5)–(4.6) with $b_i = q_i(y)$, $i \in I$. Adjusting formulas (4.7)–(4.11), we can obtain using Theorem 4.1 an explicit formula of the optimal solution $x^*(y)$ of the optimization problem

$$f(x) \longrightarrow \min \tag{4.66}$$

subject to

$$\max_{j\in J}(r_{ij}(x_j) \ge q_i(y),\ i \in I,\ \underline{x} \le x \le \overline{x},\ \underline{y} \le y \le \overline{y}.$$

Note that this optimal solution of the problem (4.66) as well as the corresponding optimal value $f(x^*(y))$ of the objective function depends on the chosen fixed y.

Remark 4.12. Let us note that if system (4.65) has a nonempty set of solutions for $y = y^{(1)}$ and $y = y^{(2)}$ and inequality $y^{(1)} \le y^{(2)}$ holds, then it follows under the monotonicity assumptions about the functions r_{ij}, q_{ij} and f_j that $f(x^*(y^{(1)})) \le f(x^*(y^{(2)}))$. Therefore we have $f(x^*(\underline{y})) \le f(x^*(y))$ for any y, for which the set of solutions of (4.65) is nonempty. In other words if we interpret y as parameters in the right hand sides of (4.65), we can call $\underline{y}$ "the optimal choice of parameters y" in the sense of [19].

We will illustrate the theoretical considerations above by small numerical examples:

Example 4.13. Let us consider the special case, in which $a_{ij} = b_{ij} = \infty$, $r_{ij}(x_j) = c_{ij} + x_j$, $q_{ij}(y_j) = d_{ij} + y_j$, where c_{ij}, $d_{ij} \in \mathbf{R}$, i.e. we consider system of two-sided (max, +)-linear inequalities. Let us have further $m = n = 3$, so that $I = \{1, 2, 3\}$, $J = \{1, 2, 3\}$. Let us consider the optimization problem:

$$f(x) = \max(x_1, x_2, x_3) \longrightarrow \min$$

subject to

$$c_1(x) = \max(4 + x_1, 5 + x_2, 2 + x_3) \ge d_1(y) = \max(3 + y_1, 4 + y_2, 8 + y_3),$$
$$c_2(x) = \max(7 + x_1, 2 + x_2, 3 + x_3) \ge d_2(y) = \max(7 + y_1, 8 + y_2, 9 + y_3),$$
$$c_3(x) = \max(8 + x_1, 3 + x_2, 1 + x_3) \ge d_3(y) = \max(0 + y_1, 1 + y_2, 3 + y_3),$$
$$x \ge (-10, -10, -10), \ y \ge (0, 0, 0).$$

Using the method from Theorem 4.1 we obtain:
$x_j^{(i)}(y) = d_i(y) - c_{ij}$ for all $i \in I$, $j \in J$,

$$\min_{j \in J} f_j(x_j^{(1)}) = \min_{j \in J}(d_1(y) - c_{1j}) = \min(d_1(y) - 4, d_1(y) - 5, d_1(y) - 2)$$
$$= d_1(y) - 5,$$

and similarly we obtain

$$\min_{j \in J} f_j(x_j^{(2)}) = \min_{j \in J}(d_2(y) - c_{2j}) = d_2(y) - 7,$$
$$\min_{j \in J} f_j(x_j^{(3)}) = \min_{j \in J}(d_3(y) - c_{3j}) = d_3(y) - 8.$$

Using the notation from Theorem 4.1, we obtained: $k(1) = 2$, $k(2) = 1$, $k(3) = 1$ so that $P_1 = \{2, 3\}$, $P_2 = \{1\}$, $P_3 = \emptyset$ and therefore the optimal solution $x^*(y)$ has the components

$$x_1^*(y) = \max(d_2(y) - 7, d_3(y) - 8) = \max(0 + y_1, 1 + y_2, 3 + y_3),$$
$$x_2^*(y) = d_1(y) - 5 = \max(-2 + y_1, -1 + y_2, 3 + y_3),$$
$$x_3^*(y) = -10.$$

The optimal value of the objective function is

$$f(x^*(y)) = \max_{j \in J}(x_j^*(y)),$$

so that

$$f(x^*(y)) = \max(\max(0 + y_1, 1 + y_2, 3 + y_3),\ \max(-2 + y_1, -1 + y_2, 3 + y_3), -10)$$
$$= \max(0 + y_1, 1 + y_2, 3 + y_3)\ .$$

The optimal choice of y generating the smallest attainable optimal value (the optimal choice of parameter y in the sense of Remark 4.12) is $y = 0$ with $x^*(0) = (2, 3, -10)$, $f(x^*(0) = \max(2, 3, -10) = 3$.

In the next part of this section we will study the special case of two-sided (max, min)-linear equation system having the form

$$\max_{j \in J}(a_{ij} \wedge x_j) = \max_{j \in J}(b_{ij} \wedge y_j)\ \ i \in I,\ x_j \le \overline{x}_j,\ y_j \le \overline{y}_j\ j \in J, \tag{4.67}$$

where a_{ij}, b_{ij}, $\overline{x}_j$, $\overline{y}_j$ are finite. System (4.67) is a special case of system (4.59)–(4.61) with $r_{ij}(x_j) = x_j$, $q_{ij}(y_j) = y_j$, $i \in I, j \in J$, $I_1 = I_2 = I$, and $\underline{x}_j = \underline{y}_j = -\infty$ for all $j \in J$. Such equation system was investigated in [7]. It follows from the results of this paper that the solution set M of (4.61) is always nonempty and possesses the maximum element $(x^{\max}, y^{\max})$, i.e. element $(x^{\max}, y^{\max}) \in M$ for which the implication $(x, y) \in M \Longrightarrow [x \le x^{\max}$ and $y \le y^{\max}]$ holds. A polynomial algorithm for finding the maximum solution $(x^{\max}, y^{\max})$ of system (4.67) having the complexity $O(mn.(m \wedge n))$ was proposed in [7]. If we include in system (4.67) also finite lower bounds $\underline{x}$, $\underline{y}$, it follows that

$$M \ne \emptyset \iff \underline{x} \le x^{\max} \text{ and } \underline{y} \le y^{\max}.$$

Remark 4.13. Let us note that a system with the same variables on both sides of the equations, i.e. a system

$$\max_{j \in J}(a_{ij} \wedge x_j) = \max_{j \in J}(b_{ij} \wedge x_j),\ \ i \in I$$

can be transformed to the form

$$\max_{j \in J}(a_{ij} \wedge x_j) = \max_{j \in J}(b_{ij} \wedge y_j),\ i \in I,\ x_j = y_j,\ j \in J.$$

By introducing appropriate coefficients this last system will have the form (4.67). Therefore the system with non-separated variables has the same properties as system (4.67), especially the set of its solutions has the maximum element $x^{\max}$, for which it holds $x \leq x^{\max}$ for all solutions x of the equation system.

Taking into account Remark 4.13, we can consider in the sequel systems with variables $x_j,\ j \in J$ on both sides of the equations.

We will solve the following optimization problem:

$$f(x) = \max_{j \in J} f_j(x_j) \longrightarrow \min \tag{4.68}$$

subject to

$$\max_{j \in J}(a_{ij} \wedge x_j) = \max_{j \in J}(b_{ij} \wedge x_j), i \in I,\ \underline{x}_j \leq x_j \leq \overline{x}_j,\ j \in J, \tag{4.69}$$

where $f_j(x_j),\ j \in J$ are continuous convex functions. Let $\hat{x}_j,\ j \in J$ be points at which functions $f_j, j \in J$ attain minimum on interval $[\underline{x}_j, x_j^{\max}]$. Let x^{opt} be the optimal solution of problem (4.68)–(4.69) and let M denote the set of solutions of system (4.69).

Lemma 4.5. *Let $M(\alpha) = \{x \in M \mid f(x) \leq \alpha\}$, $\alpha \in \mathbf{R}$, let functions $f_j,\ j \in J$ be convex. If $M(\alpha) \neq \emptyset$, there exist $\underline{x}(\alpha), \overline{x}(\alpha)$ such that $M(\alpha) = M \cap \{x \in \mathbf{R}^n \mid \underline{x}(\alpha) \leq x \leq \overline{x}(\alpha)\}$. If $\alpha < f(x^{opt})$, then $M(\alpha) = \emptyset$.*

Proof. Let $j \in J$ be arbitrarily chosen. For any $x \in M(\alpha)$ we have $f(x) = \max_{j \in J} f_j(x_j) \leq \alpha$ and therefore it is $f_j(x_j) \leq \alpha$. Since function f_j is convex, there exist $\tilde{x}_j^1(\alpha),\ \tilde{x}_j^2(\alpha) \in \mathbf{R}$ such that $f_j(x_j) \leq \alpha$ if and only if $\tilde{x}_j^1(\alpha) \leq x_j \leq \tilde{x}_j^2(\alpha)$. Using the algorithm of [7] we will find the maximum element of the set $M \cap \{x \in \mathbf{R}^n \mid x_j \leq \tilde{x}_j^2(\alpha) \wedge \overline{x}_j,\ j \in J\}$.

Let $x^{\max}$ denote the maximum element.

If $\max(\underline{x}_j, \tilde{x}_j^1(\alpha)) \leq x_j^{\max},\ j \in J$, we will set for all $j \in J$:

$$\underline{x}_j(\alpha) = \max(\underline{x}_j, \tilde{x}_j^1(\alpha)),\ \overline{x}_j(\alpha) = x_j^{\max}$$

.

Then vectors $\underline{x}(\alpha) = (\underline{x}_1(\alpha), \ldots, \underline{x}_n(\alpha)), \overline{x}(\alpha) = (\overline{x}_1(\alpha), \ldots, \underline{x}_n(\alpha))$ form the lower and upper bounds for the elements of $M(\alpha)$.

If there exists an index $j_0 \in J$ such that $\max(\underline{x}_{j_0}, \tilde{x}_{j_0}^1(\alpha)) > x_{j_0}^{\max}$, then $M(\alpha) = \emptyset$.

The implication $\alpha < f(x^{opt}) \implies M(\alpha) = \emptyset$ is evident. The proof is completed. □

Remark 4.14. Lemma 4.5 can be extended to unimodal continuous functions $f_j,\ j \in J$, i.e. functions, for which there exists a point $x'_j \in [\underline{x}_j(\alpha), x_j^{\max}]$ such that function f_j is strictly decreasing for $x_j \in [\underline{x}_j(\alpha), x'_j]$ and strictly increasing for $x_j \in [x'_j, x_j^{\max}]$.

Lemma 4.5 can be used to construct a search algorithm, which finds an ϵ-optimal solution $x^{opt}(\epsilon)$ of optimization problem (4.68) - (4.69), i.e. an element $x^{opt}(\epsilon) \in M$ such that $|f(x^{opt}(\epsilon)) - f(x^{opt})| < \epsilon$. The algorithm will proceed similarly as the ALGORITHM I for finding τ described in Sect. 4.5 above. We will describe this algorithm in what follows. We will assume that there exists a value $\overline{\alpha}$ such that $M(\overline{\alpha}) \neq \emptyset$ and we have at our disposal a lower bound $\underline{\alpha}$ such that $f(x) \geq \underline{\alpha} \;\; \forall x \in M(\overline{\alpha})$

ALGORITHM II.

$\boxed{1}$ Input $m,\ n,\ I,\ J, a_{ij},\ b_{ij},\ f_j,\ i \in I, j \in J, \epsilon,\ \underline{x}, \overline{x}, \alpha^*$ such that $M(\overline{\alpha}) \neq \emptyset$, $\overline{\alpha}$;

$\boxed{2}$ Set $\alpha^* := \underline{\alpha} + 1/2(\overline{\alpha} - \underline{\alpha})$;

$\boxed{3}$ If $M(\alpha^*) \neq \emptyset$, set $\overline{\alpha} := \alpha^*$ and go to $\boxed{5}$;

$\boxed{4}$ Set $\underline{\alpha} := \alpha^*$ and go to $\boxed{2}$;

$\boxed{5}$ If $|\overline{\alpha} - \underline{\alpha}| < \epsilon$, then any element of $M(\overline{\alpha})$ is an ϵ-optimal solution, STOP;

$\boxed{6}$ Go to $\boxed{2}$

We will illustrate ALGORITHM II by a small numerical example. In this example we use for finding $x^{\max}$ the algorithm proposed in [7].

Example 4.14.

$$f(x) = \max(|x_1 - 6.5|, |x_2 - 8|, |x_3 - 6|, |x_4 - 5|) \longrightarrow \min$$

subject to

$$\begin{aligned}
\max(1 \wedge x_1, 7 \wedge x_2, 15 \wedge x_3, 18 \wedge x_4) &= \max(3 \wedge x_1, 2 \wedge x_2, 5 \wedge x_3, 10 \wedge x_4) \\
\max(0 \wedge x_1, 9 \wedge x_2, 1 \wedge x_3, 0 \wedge x_4) &= \max(4 \wedge x_1, 6 \wedge x_2, 7 \wedge x_3, 3 \wedge x_4) \\
\max(9 \wedge x_1, 6 \wedge x_2, 7 \wedge x_3, 0 \wedge x_4) &= \max(7 \wedge x_1, 3 \wedge x_2, 2 \wedge x_3, 2 \wedge x_4)
\end{aligned}$$

We have therefore in this case $m = 3,\ n = 4,\ \underline{\alpha} = 0$. Let us have further $\epsilon = 0.2$, $\underline{x} = (5, 1, 5, 6)$, $\overline{x} = (7, 7, 10, 9)$ and choose $\overline{\alpha} = 4$. Then $M(\overline{\alpha}) = M(4) \neq \emptyset$ since we have $\overline{x}(\overline{\alpha}) = \overline{x}(4) = x^{\max} = (7, 7, 9, 9)$, $\underline{x}(\overline{\alpha}) = \underline{x}(4) = (5, 4, 5, 6)$, so that $\underline{x}(4) \leq \overline{x}(4) = x^{\max}$, and $f(x^{\max}) = 4$.
Calculations according to ALGORITHM II run as follows:

$\boxed{2}$ $\alpha^* := 2$;

$\boxed{3}$ $M(\alpha^*) \neq \emptyset, \overline{\alpha} := 2$;

$\boxed{5}$ $|\overline{\alpha} - \underline{\alpha}| = 2 > \epsilon$;

$\boxed{2}$ $\alpha^* := 1$;

$\boxed{4}$ $M(\alpha^*) = \emptyset, \underline{\alpha} := 1$
since we have in this case

$$\underline{x}(\alpha^*) = \underline{x}(1) = (5.5, 7, 5, 6) \not\leq x^{\max}(\alpha^*) = x^{\max}(1) = (7, 6, 6, 6);$$

$\boxed{5}$ $|\overline{\alpha} - \underline{\alpha}| = 2 - 1 = 1 > \epsilon$ $\boxed{2}$ $\alpha^* := 3/2$;
$\boxed{3}$ $M(\alpha^*) \neq \emptyset, \overline{\alpha} := 1.25$;
$\boxed{5}$ $|\overline{\alpha} - \underline{\alpha}| = 0.25 > \epsilon$;
$\boxed{2}$ $\alpha^* := 1.125$;
$\boxed{3}$ $M(\alpha^*) = \emptyset, \underline{\alpha} := 1.125$
since $\underline{x}(1.25) = (5.25, 6.75, 5, 6) \not\leq x^{\max}(1.25) = (7, 6.25, 6.25, 6.25)$
$\boxed{5}$ $|\overline{\alpha} - \underline{\alpha}| = 0.125 < \epsilon = 0.2$
the ϵ-optimal solution is any element of $M(\overline{\alpha}) = M(3/2)$, e.g. its maximum element $x^{\max}(3/2) = (7, 6.5, 6.5, 6.5)$ with the value $f(x^{\max}(3/2)) = 3/2$,
STOP

Let us note that we could choose another element of $M(3/2)$ in the last step of the algorithm with possibly a smaller value of the objective function. For instance, we could choose $\tilde{x} = (7, 6.75, 6.75, 6.75) \in M(3/2)$, for which $f(\tilde{x}) = \max(0.5, 1.25, 0.75, 1.25) = 1.25 < f(x^{\max}(3/2)) = 3/2$.

The third special case investigated in this section is the (max, min)-linear inequality system of the form

$$\max_{j \in J}(a_{ij} \wedge x_j) \geq \max_{j \in J}(b_{ij} \wedge x_j), \ \ i \in I, \ \ x_j \leq \overline{x}_j, \ j \in J, \tag{4.70}$$

which is the special case of system (4.57)–(4.58) with $I_1 = I = \{1, 2, \ \dots, \ m\}$, $I_2 = \emptyset$ and with finite upper bounds of the variables $x_j, j \in J$. We will assume that $M^{\geq}$ denotes the set of all solutions of system (4.70). Let us note that set $M^{\geq}$ is always nonempty, because e.g. element $\tilde{x}$ with components

$$\tilde{x}_j = \min_{k \in I} \min_{s \in J}(a_{ks} \wedge b_{ks}), \ j \in J$$

belongs to $M^{\geq}$.

By appropriately chosen m slack variables on the right hand sides, we can transform system (4.70) to an equality system (4.69) with $m + n$ variables and use the method of [7] to find the maximum element of $M^{\geq}$.

In what follows, we will present a direct method for finding the maximum element of $M^{\geq}$. To simplify the notations we will set for any $x \in \mathbf{R}^n$

$$a_i(x) = \max_{j \in J}(a_{ij} \wedge x_j), \ b_i(x) = \max_{j \in J}(b_{ij} \wedge x_j), \ i \in I$$

and introduce the following notations:

$$I^{(1)}(x) = \{i \in I \mid a_i(x) < b_i(x)\},$$

$$\alpha(x) = \min_{i \in I} a_i(x),$$

$$I^{(2)}(x) = \{i \in I^{(1)}(x) \mid a_i(x) = \alpha(x)\},$$

$$J_i(x) = \{j \in J \mid b_{ij} \wedge x_j > \alpha(x)\}, \ i \in I^{(2)}.$$

Note that if $I^{(1)}(x) = \emptyset$, then evidently $x^{\max} = \overline{x}$. If $I^{(1)}(x) \neq \emptyset$, then $\overline{x} \notin M^{\geq}$ and we define $\overline{x}^{(1)} \in \mathbf{R}^n$ as follows:

$$\overline{x}_j^{(1)} = \alpha(\overline{x}),\ j \in \tilde{J},\ \overline{x}_j^{(1)} = \overline{x}_j,\ j \in J \setminus \tilde{J}, \tag{4.71}$$

where

$$\tilde{J} = \bigcup_{i \in I^{(2)}(\overline{x})} J_i(\overline{x}).$$

The following lemmas establish some properties of $\overline{x}_j^{(1)}$.

Lemma 4.6. *If $I^{(1)}(\overline{x}) \neq \emptyset$, then*

$$a_i(\overline{x}^{(1)}) = b_i(\overline{x}^{(1)}),\ i \in I^{(2)}(\overline{x})$$

Proof. Let i be arbitrary index of $I^{(2)}(\overline{x})$ and $j \in J_i$. We have further:

$$\alpha(\overline{x}) = a_i(\overline{x}) = a_i(\overline{x}^{(1)}) = b_{ij} \wedge \overline{x}_j^{(1)},\ j \in J_i.$$

and

$$b_{ij} \wedge \overline{x}_j^{(1)} = b_{ij} \wedge \overline{x}_j,\ j \in J \setminus J_i,$$

so that

$$b_i(\overline{x}^{(1)}) = \alpha(\overline{x})$$

and therefore

$$a_i(\overline{x}^{(1)}) = b_i(\overline{x}^{(1)}) = \alpha(\overline{x}).$$

Since i was an arbitrary index of $I^{(2)}(\overline{x})$, the proof is completed. □

Lemma 4.7. *If $\alpha(\overline{x}) \geq a_i(\overline{x}) \geq b_i(\overline{x})$, then $a_i(\overline{x}^{(1)}) \geq b_i(\overline{x}^{(1)})$.*

Proof. Let i be an arbitrary index such that $\alpha(\overline{x}) \geq a_i(\overline{x}) \geq b_i(\overline{x})$. If $\alpha(\overline{x}) = a_i(\overline{x})$, then it follows from the proof of Lemma 4.6 that

$$a_i(\overline{x}^{(1)}) = a_i(\overline{x}) \geq b_i(\overline{x}) \geq b_i(\overline{x}^{(1)}),$$

where the last inequality follows from (4.71) (it is namely $\overline{x} \geq \overline{x}^{(1)}$).

Let us assume further that $\alpha(\overline{x}) > a_i(\overline{x})$ and let $j_0 \in J$ be any index such that $a_i(\overline{x}) = a_{ij_0} \wedge \overline{x}_{j_0}$. Then $j_0 \notin \tilde{J}$, since if index j_0 belonged to $\tilde{J}$, it would exist an index $i_0 \in I^{(1)}(\overline{x})$ such that $j_0 \in J_{i_0}$ and it would be

$$a_{i_0}(\overline{x}) < \alpha(\overline{x}) = \min_{i \in I^{(1)}(\overline{x})} a_i(\overline{x}),$$

which is a contradiction. Since $j_0 \notin \tilde{J}$, it is according to (4.71) $\overline{x}^{(1)}_{j_0} = \overline{x}_{j_0}$. Since j_0 was an arbitrary "active" index determining the value $a_i(\overline{x})$, it is $a_i(\overline{x}^{(1)}) = a_i(\overline{x})$. Further since, as we have already mentioned above, $\overline{x} \geq \overline{x}^{(1)}$, we obtain:

$$\alpha(\overline{x}) > a_i(\overline{x}^{(1)}) = a_i(\overline{x}) \geq b_i(\overline{x}) \geq b_i(\overline{x}^{(1)}),$$

which completes the proof. □

Remark 4.15. Summarizing Lemma 4.7, we can say that if we replace $\overline{x}$ by $\overline{x}^{(1)}$, all inequalities $a_i(x) \geq b_i(x)$, which were satisfied for $x = \overline{x}$ and for which $\alpha(\overline{x}) \geq a_i(\overline{x})$ remain satisfied also for $x = \overline{x}^{(1)}$.

Lemma 4.8. *Let $x^* \not\leq \overline{x}^{(1)}$. Then $x^* \notin M^{\geq}$.*

Proof. Let $j_0 \in J$ be an arbitrary index, for which the inequality $x^*_{j_0} > \overline{x}^{(1)}_{j_0}$ holds. If $j_0 \notin \tilde{J}$, then it is $x^*_{j_0} > \overline{x}^{(1)}_{j_0} = \overline{x}_{j_0}$ and therefore $x^* \notin M^{\geq}$. Let us assume further that $j_0 \in \tilde{J}$ and $x^* \in M^{\geq}$.

Then there exists an index $i_0 \in I^{(2)}(\overline{x})$ such that $j_0 \in J_{i_0}$ and it is

$$\alpha(\overline{x}) = a_{i_0}(\overline{x}) = a_{i_0}(\overline{x}^{(1)}),$$

where the last equality follows from the proof of Lemma 4.7. It follows further from the definition of $\overline{x}^{(1)}$ that $b_{i_0}(\overline{x}^{(1)}) = \alpha(\overline{x}) = a_{i_0}(\overline{x}^{(1)})$ (see (4.71)). Since $j_0 \in J_{i_0}$, we have

$$b_{i_0}(\overline{x}^{(1)}) = b_{i_0 j_0} \wedge \overline{x}^{(1)}_{j_0} = b_{i_0 j_0} \wedge \alpha(\overline{x}) = \alpha(\overline{x})$$

since $b_{i_0 j_0} > \alpha(\overline{x})$. We have further according to our assumptions $x^*_{j_0} > \overline{x}^{(1)}_{j_0}$ so that

$$b_{i_0}(x^*) \geq b_{i_0 j_0} \wedge x^*_{j_0} > b_{i_0 j_0} \wedge \overline{x}^{(1)}_{j_0} = b_{i_0 j_0} \wedge \alpha(\overline{x}).$$

Since we assumed that $x^* \in M^{\geq}$, it must be $x^* \leq \overline{x}$ and therefore $a_{i_0}(x^*) \leq a_{i_0}(\overline{x}) = \alpha(\overline{x})$. Summarizing the previous inequalities we obtain:

$$a_{i_0}(x^*) \leq a_{i_0}(\overline{x}) = \alpha(\overline{x}) = b_{i_0 j_0} \wedge \overline{x}^{(1)}_{j_0} < b_{i_0 j_0} \wedge x^*_{j_0} \leq b_{i_0}(x^*),$$

so that $a_{i_0}(x^*) < b_{i_0}(x^*)$ and therefore $x^* \notin M^{\geq}$. This contradiction completes the proof. □

It follows from Lemmas 4.6–4.8 that any element of $M^{\geq}$ must be smaller or equal to $\overline{x}^{(1)}$, and it is $\overline{x}^{(1)} \leq \overline{x}$, $\overline{x}^{(1)} \neq \overline{x}$. Therefore $\overline{x}^{(1)}$ can be accepted as a new upper bound for elements of $M^{\geq}$ and if $\overline{x}^{(1)} \notin M^{\geq}$ we can repeat the calculations with this new upper bound. If $\overline{x}^{(1)} \in M^{\geq}$, then $\overline{x}^{(1)}$ is the maximum element of $M^{\geq}$.

In the sequel, we will describe this procedure in an algorithmic form.
ALGORITHM III.

[1] Input $m, n, I, J, a_{ij}, b_{ij}, \overline{x}$.
[2] Find $I^{(1)}(\overline{x})$.
[3] If $I^{(1)}(\overline{x}) = \emptyset$, then $x^{\max} := \overline{x}$, STOP.
[4] Find $\alpha(\overline{x})$, $I^{(2)}(\overline{x})$.
[5] Find J_i for $i \in I^{(2)}(\overline{x})$ and set $\tilde{J} := \bigcup_{i \in I^{(2)}(\overline{x})} J_i$.
[6] $\overline{x}_j := \alpha(\overline{x}),\ j \in \tilde{J}$, go to [2].

Remark 4.16. Let us note that in each iteration at least one inequality, which was unsatisfied in the preceding step becomes satisfied and for any $i \in I$ such that $a_i(\overline{x}) \le \alpha(\overline{x})$ the inequality $a_i(\overline{x}) \ge b_i(\overline{x})$ holds. Therefore we need at most $m \wedge n$ iterations to get $x^{\max}$. In case we have rational or integer coefficients a_{ij}, b_{ij}, the complexity of each iteration is equal to $O(m.n)$ so that we obtain the total complexity equal to $O((m \wedge n)mn)$.

We will illustrate the theoretical results by a small numerical example.

Example 4.15. Let matrices A, B be given as follows:

$$A = \begin{pmatrix} 6, & 4, & 2, & 1 \\ 8, & 5, & 7, & 3 \\ 3, & 0, & 1, & 0 \end{pmatrix}$$

$$B = \begin{pmatrix} 8, & 5, & 6, & 2 \\ 4, & 7, & 7, & 5 \\ 10, & 2, & 10, & 2 \end{pmatrix}$$

We have in this case $m = 3,\ n = 4,\ I = \{1,2,3\},\ J = \{1,2,3,4\}$. Let $\overline{x} = (50, 50, 50, 50)$.

ALGORITHM III. proceeds as follows:

[1] Input $m, n, I, J, A, B, \overline{x}$.

--- Iteration 1

[2] $\overline{x} = (50, 50, 50, 50)$, $I^{(1)}(\overline{x}) := \{1, 3\}$
[3] $I^{(1)}(\overline{x}) \neq \emptyset$.
[4] $\alpha(\overline{x}) := 3,\ I^{(2)}(\overline{x}) = \{3\}$.
[5] $J_3 := \{1, 3\}$, $\tilde{J} := J_3 = \{1, 3\}$.
[6] $\overline{x}_j := 3,\ j \in \tilde{J}$, go to [2].

--- Iteration 2

[2] $\overline{x} = (3, 50, 3, 50)$, $I^{(1)}(\overline{x}) := \{1, 2\}$
[3] $I^{(1)}(\overline{x}) \neq \emptyset$.
[4] $\alpha(\overline{x}) := 4,\ I^{(2)}(\overline{x}) = \{1\}$.
[5] $J_1 := \{2\},\ \tilde{J} := J_1 = \{2\}$.
[6] $\overline{x}_j := 4$ for all $j \in \tilde{J}$, go to [2].

-- Iteration 3

[2] $\overline{x} = (3, 4, 3, 50)$, $I^{(1)}(\overline{x}) := \{2\}$
[3] $I^{(1)}(\overline{x}) \neq \emptyset$.
[4] $\alpha(\overline{x}) := 4,\ I^{(2)}(\overline{x}) = \{2\}$.
[5] $J_2 := \{4\},\ \tilde{J} := J_2 = \{4\}$.
[6] $\overline{x}_j := 4$ for all $j \in \tilde{J}$, go to [2].

-- Iteration 4

[2] $\overline{x} = (3, 4, 3, 4)$, $I^{(1)}(\overline{x}) = \emptyset$
[3] $I^{(1)}(\overline{x}) = \emptyset$, $x^{\max} := \overline{x} = (3, 4, 3, 4)$, STOP.

Remark 4.17. Since we have an effective method for finding the maximum element of set $M^{\geq}$, we can use an iteration algorithm similar to ALGORITHM II. to get an ϵ-optimal solution of the optimization problem

$$f(x) = \max_{j \in J} f_j(x_j) \longrightarrow \min$$

subject to

$$x \in M^{\geq},\ x \geq \underline{x},$$

where $f_j(x),\ j \in J$ are continuous unimodal functions and $\underline{x}$ is a given lower bound of x.

Remark 4.18. Let us note that since each equation can be replaced by two symmetric inequalities, any system of m (max, min)-linear equations can be replaced by an equivalent system of $2m$ (max, min)-linear inequalities. Therefore ALGORITHM II and ALGORITHM III can be used also to solve (max, min)-linear equations and minimization problems solved by ALGORITHM I.

Remark 4.19. Let us note that another special system of (max, min)-linear inequalities with binary entries was solved by a polynomial algorithm in [4].

Remark 4.20. By interchanging the operations max, min we can consider in a similar way (min, max)-linear systems and inequalities of the form

$$\min_{j \in J} \max(a_{ij}, x_j) \ =,\ \leq,\ \geq\ \min_{j \in J} \max(b_{ij}, x_j),\ i \in I.$$

Let us note further that the proposed procedures remain unchanged, if we assume that the constraint coefficients a_{ij}, b_{ij} and components of solution x are elements of any set $\tilde{R}$, $\tilde{R} \subset \mathbf{R}$, which is closed with respect to operations max and min. Besides $\tilde{R} = \mathbf{R}$ as in the theory presented above, we can consider also for instance that $\tilde{R}$ is the set of rational numbers, integer numbers or any finite subset of $\mathbf{R}$.

4.7 Conclusions

The chapter provides a unified approach to studying systems of one- and two-sided (max, min)-separable equation and inequality systems. Further we solve optimization problems, in which the set of feasible solutions is described by such systems and the objective function is a (max)-separable function. While for the one-sided system an explicit formula for solving the optimization problems was presented, for the two-sided systems an effective solution method is presented only for some types of the problems. The unified approach presented in this chapter encompasses so called (max, min)- linear and (max, +)-linear equation and inequality systems as well as optimization problems with a max-separable objective function and (max, min)- linear or (max, +)-linear equation and inequality constraints, which were studied in the literature. Since the max-operation is only a semigroup operation, the transfer of variables from one side of equations or inequalities to the other side is not possible without changing the structure of the relations so that the one-sided systems and two-sided systems have to be studied using different approaches and problems with two-sided equations and/or inequalities could be up to now solved effectively only for some classes of the problems considered in the last part of this chapter.

References

1. Baccelli, F.L., Cohen, G., Olsder, G.J., Quadrat, J.P.: Synchronization and Linearity, An Algebra for Discrete Event Systems. Wiley, Chichester (1992)
2. Butkovic, P.: Max-linear Systems: Theory and Algorithms. Springer, New York (2010)
3. Cao, Z.Q., Kim, K.H., Roush, F.W.: Incline Algebra and Applications. Wiley, New York (1984)
4. Cechlarova, K.: On max-min linear inequalities and coalitional resource games with sharable resources. Lin. Algebra Appl. **433**, 127–135 (2010)
5. Cechlarova, K.: Efficient computation of the greatest eigenvector in fuzzy algebra. Tatra Mt. Math. Publ. **12**, 73–79 (1997)
6. Cuninghame-Green, R.A.: Minimax Algebra, Lecture Notes in Economics and Mathematical Systems, vol. 166. Springer, Berlin (1979)
7. Gavalec, M., Zimmermann, K. : Solving systems of two-sided (max, min)-linear equations. Kybernetika **46**(3), 405–414 (2010)
8. Gondran, M., Minoux, M.: Linear algebra of dioids: A survey of recent results. Ann. Discrete Math. **19**, 147–164 (1984)

9. Kolokoltsov, V.N., Maslov,V.P.: Idempotent Analysis and Its Applications. Kluwer, Dordrecht (1997)
10. Krivulin, N.K.: Methods of Idempotent Algebra in Problems of Modelling and Analysis of Complex Systems. S. Peterburg (2009)
11. Litvinov, G.L., Sergeev, S.N. (eds.): Tropical and Idempotent Mathematics and Applications, Contemporary Mathematics, vol. 616. Providence American Math. Society (2014)
12. Litvinov, G.L., Maslov, V.P., Sergeev, S.N. (eds.): Idempotent and Tropical Mathematics and Problems of Mathematical Physics, vol. I. Independent University Moscow, Moscow (2007)
13. Maslov, V.P., Samborskij, S.N.: Idempotent Analysis, Advances in Soviet Mathematics, vol. 13. AMS, Providence (1992)
14. Plavka, J.: On eigenproblem of circulant matrices in max-algebra. Optimization **50**, 477–483 (5–6)
15. Sanchez, E.: Resolution of eigen fuzzy sets equations. Fuzzy Sets Syst. **1**(1), 69–74 (1978)
16. Sanchez, E.: Inverses of fuzzy relations. Applications to possibility distributions and medical diagnosis, Fuzzy Sets Syst. **2**(1), 75–86 (1979)
17. Vorobjov, N.N.: Extremal algebra of positive matrices, Datenverarbeitung und Kybernetik **3**, 39–71 (1967). (in Russian)
18. Zimmermann, K.: Disjunctive optimization, max-separable problems and extremal algebras. Theor. Comput. Sci. **293**, 45–54 (2003)
19. Zlobec, S.: Characterizing an optimal input in perturbed convex programming. Math. Program. **25**, 109–121 (1983)

Chapter 5
Interval Eigenproblem in Max-Min Algebra

Abstract The eigenvectors of square matrices in max-min algebra correspond to steady states in discrete events system in various application areas, such as design of switching circuits, medical diagnosis, models of organizations and information systems. Imprecise input data lead to considering interval version of the eigenproblem, in which interval eigenvectors of interval matrices in max-min algebra are investigated. Six possible types of an interval eigenvector of an interval matrix are introduced, using various combination of quantifiers in the definition. The previously known characterizations of the interval eigenvectors were restricted to the increasing eigenvectors, see [11]. In this chapter, the results are extended to the non-decreasing eigenvectors, and further to all possible interval eigenvectors of a given max-min matrix. Classification types of general interval eigenvectors are studied and characterization of all possible six types is presented.

5.1 Introduction

The input data in real problems are usually not exact and can be rather characterized by interval values. Considering matrices and vectors with interval coefficients is therefore of great practical importance, see [3–9, 12, 16, 19, 22, 23]. For systems described by interval coefficients the investigation of steady states leads to computing interval eigenvectors. The eigenspace structure of a given interval matrix A in max-min algebra is studied in this chapter.

By *max-min algebra* we understand a triple $(\mathscr{B}, \oplus, \otimes)$, where $\mathscr{B}$ is a linearly ordered set, and $\oplus = \max$, $\otimes = \min$ are binary operations on $\mathscr{B}$. We assume that $\mathscr{B}$ contains the minimal element O and the maximal element I. The notation $\mathscr{B}(m,n)$ $(\mathscr{B}(n))$ denotes the set of all matrices (vectors) of dimension $m \times n$ (dimension n) over $\mathscr{B}$. Operations $\oplus$, $\otimes$ are extended to addition and multiplication of matrices and vectors in the standard way. That is, for matrices $A, B, C \in \mathscr{B}(m,n)$ we write $C = A \oplus B$, if $c_{ij} = a_{ij} \oplus b_{ij}$ for every $i \in M = \{1,2,\ldots,m\}$ and $j \in N = \{1,2,\ldots,n\}$. Similarly, for matrices $A \in \mathscr{B}(m,p), B \in \mathscr{B}(p,n), C \in \mathscr{B}(m,n)$ we write $C = A \otimes B$, if $c_{ij} = \bigoplus_{k=1}^{p} a_{ik} \otimes b_{kj}$ for every $i \in M$ and $j \in N$. It is

The chapter was written by "Martin Gavalec"

M. Gavalec et al., *Decision Making and Optimization*, Lecture Notes in Economics and Mathematical Systems 677, DOI 10.1007/978-3-319-08323-0_5

easy to verify that the usual arithmetic rules, such as commutativity, associativity and distributivity, hold for computation with matrices over $(\mathcal{B}, \oplus, \otimes)$.

The linear order on $\mathcal{B}$ extends componentwise to partial ordering on $\mathcal{B}(m,n)$ and $\mathcal{B}(n)$ and the notation $\vee$ ($\wedge$) is used for the binary join (meet) in these sets. In more detail, for $A, B \in \mathcal{B}(m,n)$ ($x, y \in \mathcal{B}(n)$) we write $A \leq B$ ($x \leq y$), if $a_{ij} \leq b_{ij}$ ($x_j \leq y_j$) for every $i \in M$, $j \in N$. Analogously, we have $(A \vee B)_{ij} = \max(a_{ij}, b_{ij})$ ($(x \vee y)_j = \max(x_j, y_j)$) and $(A \wedge B)_{ij} = \min(a_{ij}, b_{ij})$ ($(x \wedge y)_j = \min(x_j, y_j)$) for every $i \in M$, $j \in N$. Operators $\vee$ ($\wedge$) may also be quantified over a set S, as $\bigvee_{i \in S} x_i$ ($\bigwedge_{i \in S} x_i$) whenever the upper (resp. lower) bound exists. The Boolean notation $\cup$ and $\cap$ respectively will be used for union and intersection of sets.

The *eigenproblem* for a given matrix $A \in \mathcal{B}(n,n)$ in max-min algebra consists of finding a value $\lambda \in \mathcal{B}$ (called the *eigenvalue*) and a vector $x \in \mathcal{B}(n)$ (called the *eigenvector*) such that the equation $A \otimes x = \lambda \otimes x$ holds true. It is well-known that the above problem in max-min algebra can be in principle reduced to solving the equation $A \otimes x = x$. Namely, if $x \in \mathcal{B}(n)$ fulfills the equation $A \otimes x = \lambda \otimes x$, then an easy computation shows that $y = \lambda \otimes x$ fulfills the equation $A \otimes y = y$. The eigenproblem in max-min algebra has been studied by many authors. Interesting results were found in describing the structure of the *eigenspace* (the set of all eigenvectors), and algorithms for computing the greatest eigenvector of a given matrix were suggested, see e.g. [1, 2, 13–15, 17, 18, 20, 21].

The complete structure of the eigenspace has been described in [10], where the eigenspace of a given matrix is presented as a union of intervals of permuted *monotone* (increasing or non-decreasing) *eigenvectors*. Explicit formulas are shown for the lower and upper bounds of the intervals of monotone eigenvectors for any given monotonicity type. By permutation of indices, the formulas are then used for description of the whole eigenspace of a given matrix. The monotonicity approach from [10] has been applied to the *interval eigenproblem* in [11], where a classification consisting of six different types of interval eigenvectors is presented, and detailed characterization of these types is given for increasing interval eigenvectors (which are called 'strictly increasing' in both papers [10, 11]).

The aim of this chapter is to extend the results from [11] to all types of non-decreasing interval eigenvectors (which are simply called 'increasing' in [10] and [11]). Furthermore, permutations of indices are used in the paper for finding characterizations of all six classification types without any restriction. By this, the interval eigenproblem is completely solved.

The content of the chapter is organized as follows. Section 5.1.1 presents applications of max-min eigenvectors and a simple example. In Sect. 5.1.2 the basic results from [10] and the necessary notions are described, which will be used later in Sect. 5.2.1. In Sect. 5.2 we define interval matrices and six basic types of interval eigenvectors, according to [11]. Section 5.2.1 is the main part of the chapter: for a given interval partition D, all six types of D-increasing interval eigenvectors are described. The results are generalized for arbitrary interval eigenvectors in Sect. 5.2.2. Finally, Sect. 5.3 shows all relations between the considered six general types of interval eigenvectors and several examples.

5.1.1 Max-Min Eigenvectors in Applications

Eigenvectors of matrices in max-min algebra are useful in applications such as automata theory, design of switching circuits, logic of binary relations, medical diagnosis, Markov chains, social choice, models of organizations, information systems, political systems and clustering.

As an example of a business application we consider evaluation of projects adopted by a company. Position of a project is characterized by several components, such as its importance for the future of the company, the assigned investments, the intensity of the project, or its impact on market. The level of each component i is described by some value x_i, which is influenced by the levels of all components x_j. The influence is expressed by a constant factor a_{ij}, and the position vector x at time $t+1$ is given by equalities $x_i(t+1) = \max_j \big(\min(a_{ij}, x_j(t))\big) = \bigoplus_j \big(a_{ij} \otimes x_j(t)\big)$ for every i, or shortly $x(t+1) = A \otimes x(t)$, in matrix notation. The steady position of the project is then described by the equation $x(t+1) = x(t)$, i.e. $A \otimes x(t) = x(t)$. In other words, $x(t)$ is then an eigenvector of matrix A, which means that the steady states exactly correspond to eigenvectors. In reality, interval eigenvectors and interval matrices should be considered.

Further examples are related to business applications describing other activities, e.g. planning different forms of advertisement. Similar applications can also be found in social science, biology, medicine, computer nets, and many other areas.

5.1.2 Basics on the Eigenspace Structure of a Max-Min Matrix

Notation $N = \{1, 2, \dots, n\}$ with a given natural number n will be used in Chaps. 5 and 6. For a fixed matrix $A \in \mathscr{B}(n,n)$ we denote analogously as in [10] the set of all *increasing* vectors of dimension n

$$\mathscr{B}^{<}(n) = \{ x \in \mathscr{B}(n) |\, (\forall i, j \in N)[i < j \Rightarrow x_i < x_j] \} ,$$

and the set of all *non-decreasing* vectors

$$\mathscr{B}^{\leq}(n) = \{ x \in \mathscr{B}(n) |\, (\forall i, j \in N)[i < j \Rightarrow x_i \leq x_j] \} .$$

Clearly, every increasing vector is non-decreasing, i.e. $\mathscr{B}^{<}(n) \subset \mathscr{B}^{\leq}(n)$ (the converse inclusion does not hold). Further we denote the eigenspace of a matrix $A \in \mathscr{B}(n,n)$

$$\mathscr{F}(A) = \{ x \in \mathscr{B}(n) |\, A \otimes x = x \} ,$$

and the eigenspaces of all *increasing* eigenvectors (*non-decreasing* eigenvectors)

$$\mathscr{F}^{<}(A) = \mathscr{F}(A) \cap \mathscr{B}^{<}(n) , \quad \mathscr{F}^{\leq}(A) = \mathscr{F}(A) \cap \mathscr{B}^{\leq}(n) .$$

Let P_n denote the set of all permutations on N. It is clear that any vector $x \in \mathcal{B}(n)$ can be permuted to a non-decreasing vector x_φ by some permutation $\varphi \in P_n$. If we denote by $A_{\varphi\varphi}$ the matrix created from a given $n \times n$ max-min matrix A by permuting its rows and columns by φ, then the structure of the eigenspace $\mathcal{F}(A)$ of A can be described by investigating the structure of non-decreasing eigenspaces $\mathcal{F}^{\leq}(A_{\varphi\varphi})$ for all possible permutations φ, in view of the following theorem.

Theorem 5.1 ([10]). *Let $A \in \mathcal{B}(n,n)$, $x \in \mathcal{B}(n)$ and let $\varphi \in P_n$. Then $x \in \mathcal{F}(A)$ if and only if $x_\varphi \in \mathcal{F}(A_{\varphi\varphi})$.*

For the convenience of the reader we present a simple proof of the theorem (see [10]).

Proof. Assume that $x \in \mathcal{F}(A)$, i.e. $A \otimes x = x$. By definition, for every $i \in N$ we have $\bigoplus_{j \in N}(a_{ij} \otimes x_j) = x_i$. As φ is a permutation on N, we have $\bigoplus_{j \in N}(a_{i\varphi(j)} \otimes x_{\varphi(j)}) = x_i$. By the same argument, $\bigoplus_{j \in N}(a_{\varphi(i)\varphi(j)} \otimes x_{\varphi(j)}) = x_{\varphi(i)}$ for every $i \in N$. Hence, $A_{\varphi\varphi} \otimes x_\varphi = x_\varphi$, i.e. $x_\varphi \in \mathcal{F}(A_{\varphi\varphi})$. The converse implication is analogous. □

We shall use notation $\mathcal{B}^{<}_{\varphi}(n) = \{x \in \mathcal{B}(n) |\, x_\varphi \in \mathcal{B}^{<}(n)\}$, $\mathcal{B}^{\leq}_{\varphi}(n) = \{x \in \mathcal{B}(n) |\, x_\varphi \in \mathcal{B}^{\leq}(n)\}$, $\mathcal{F}^{<}_{\varphi}(A) = \{x \in \mathcal{F}(A) |\, x_\varphi \in \mathcal{B}^{<}(n)\}$ and $\mathcal{F}^{\leq}_{\varphi}(A) = \{x \in \mathcal{F}(A) |\, x_\varphi \in \mathcal{B}^{\leq}(n)\}$.

In this notation, the assertions of Theorem 5.1 can be expressed as $\mathcal{F}^{<}_{\varphi}(A) = \{x \in \mathcal{B}(n) |\, x_\varphi \in \mathcal{F}^{<}(A_{\varphi\varphi})\}$ and $\mathcal{F}^{\leq}_{\varphi}(A) = \{x \in \mathcal{B}(n) |\, x_\varphi \in \mathcal{F}^{\leq}(A_{\varphi\varphi})\}$.

For $A \in \mathcal{B}(n,n)$, the structure of the increasing eigenspace $\mathcal{F}^{<}(A)$ (which is a proper subset of $\mathcal{F}^{\leq}(A)$) has been described in [10] as an interval of increasing vectors. The lower and upper bound vectors $m^\star(A)$, $M^\star(A) \in \mathcal{B}(n)$ are defined componentwise for every $i \in N$ by formulas

$$m^\star_i(A) = \max_{j \leq i} \max_{k > j} a_{jk} \ , \qquad M^\star_i(A) = \min_{j \geq i} \max_{k \geq j} a_{jk} \ . \tag{5.1}$$

Theorem 5.2 ([10]). *Let $A \in \mathcal{B}(n,n)$ and let $x \in \mathcal{B}(n)$ be an increasing vector. Then $x \in \mathcal{F}(A)$ if and only if $m^\star(A) \leq x \leq M^\star(A)$. In formal notation,*

$$\mathcal{F}^{<}(A) = [m^\star(A), M^\star(A)] \cap \mathcal{B}^{<}(n) \ .$$

The structure of the whole non-decreasing eigenspace $\mathcal{F}^{\leq}(A)$ is described analogously using *interval partitions* of the index set N. By interval partition we understand a partition of N consisting of integer subintervals of N. The set of all interval partitions on N will be denoted as $\mathcal{D}_n$.

For every interval partition $D \in \mathcal{D}_n$ and for every $i \in N$, we denote by $D[i]$ the integer interval of the partition containing i. In other words, $D[i]$ is the (uniquely determined) integer interval $I \in D$ with $i \in I$. For $i, j \in N$ we write $D[i] < D[j]$ if $i' < j'$ holds for any $i' \in D[i]$, $j' \in D[j]$, and $D[i] \leq D[j]$ if $D[i] < D[j]$ or $D[i] = D[j]$ hold.

For a given partition $D \in \mathscr{D}_n$ and for a vector $x \in \mathscr{B}(n)$ we say that x is D-*increasing* if for any indices $i, j \in N$

$$x_i = x_j \quad \text{if and only if} \quad D[i] = D[j] \ ,$$
$$x_i < x_j \quad \text{if and only if} \quad D[i] < D[j] \ .$$

The set of all D-increasing vectors in $\mathscr{B}(n)$ will be denoted as $\mathscr{B}^{<}(D,n)$). If $A \in \mathscr{B}(n,n)$, then the set of all D-increasing eigenvectors of A is denoted by $\mathscr{F}^{<}(D,A)$.

It is easy to see that for every non-decreasing vector $x \in \mathscr{B}^{\leq}(n)$, there is exactly one interval partition $D \in \mathscr{D}_n$ such that x is D-increasing. In particular, if vector x is increasing on N, then the corresponding partition is equal to the (finest possible) partition consisting of all singletons $D = \{\{i\} | \, i \in N\}$. On the other hand, if x is constant, then the partition D consists of a single class N, i.e. $D = \{N\}$.

Hence, the non-decreasing eigenspace $\mathscr{F}^{\leq}(A)$ can be expressed as the union of D-increasing eigenspaces $\bigcup_{D \in \mathscr{D}_n} \mathscr{F}^{<}(D,A)$. For any fixed interval partition $D \in \mathscr{D}_n$, the bounds given in Theorem 5.2 for increasing eigenvectors, have been generalized in [10] for D-increasing eigenvectors. The bounds $m^{\star}(D,A)$, $M^{\star}(D,A) \in \mathscr{B}(n)$ were defined and the following theorem has been proved. For every $i \in N$

$$m_i^{\star}(D,A) = \max_{j \in N, D[j] \leq D[i]} \ \max_{k \in N, D[k] > D[j]} a_{jk} \ , \tag{5.2}$$

$$M_i^{\star}(D,A) = \min_{j \in N, D[j] \geq D[i]} \ \max_{k \in N, D[k] \geq D[j]} a_{jk} \ . \tag{5.3}$$

Theorem 5.3 ([10]). *Let $A \in \mathscr{B}(n,n)$, $D \in \mathscr{D}_n$ and let $x \in \mathscr{B}(n)$ be a D-increasing vector. Then $x \in \mathscr{F}(A)$ if and only if $m^{\star}(D,A) \leq x \leq M^{\star}(D,A)$. In formal notation,*

$$\mathscr{F}^{<}(D,A) = [m^{\star}(D,A), M^{\star}(D,A)] \cap \mathscr{B}^{<}(D,n) \ .$$

The theorem allows to describe the structure of the non-decreasing eigenspace $\mathscr{F}^{\leq}(A)$ of a given matrix A. In view of Theorem 5.1, the above characterization can be further extended to the whole eigenspace $\mathscr{F}(A)$ using permutations on N. Theorem 5.3 will be used in this paper as an important tool for characterization of interval eigenvectors.

5.2 Interval Eigenvectors Classification

Similarly as in [4, 9, 11, 12], we define interval matrix with bounds $\underline{A}, \overline{A} \in \mathscr{B}(n,n)$ and interval vector with bounds $\underline{x}, \overline{x} \in \mathscr{B}(n)$

$$[\underline{A}, \overline{A}] = \{A \in \mathscr{B}(n,n) | \, \underline{A} \leq A \leq \overline{A}\} \ , \quad [\underline{x}, \overline{x}] = \{x \in \mathscr{B}(n) | \, \underline{x} \leq x \leq \overline{x}\} \ .$$

We assume that an interval matrix $\mathbf{A} = [\underline{A}, \overline{A}]$ and an interval vector $\mathbf{X} = [\underline{x}, \overline{x}]$ are fixed. The interval eigenproblem for $\mathbf{A}$ and $\mathbf{X}$ consists in recognizing whether $A \otimes x = x$ holds true for $A \in \mathbf{A}$, $x \in \mathbf{X}$. In dependence on the applied quantifiers, six types of interval eigenvectors are distinguished (see [11]).

Definition 5.1. For given interval matrix $\mathbf{A}$, the interval vector $\mathbf{X}$ is called

- *strong eigenvector* of $\mathbf{A}$ if $(\forall A \in \mathbf{A})(\forall x \in \mathbf{X})[\, A \otimes x = x\,]$,
- *strongly universal eigenvector* of $\mathbf{A}$ if $(\exists x \in \mathbf{X})(\forall A \in \mathbf{A})\,[A \otimes x = x]$,
- *universal eigenvector* of $\mathbf{A}$ if $(\forall A \in \mathbf{A})(\exists x \in \mathbf{X})[\, A \otimes x = x\,]$,
- *strongly tolerable eigenvector* of $\mathbf{A}$ if $(\exists A \in \mathbf{A})(\forall x \in \mathbf{X})[\, A \otimes x = x\,]$,
- *tolerable eigenvector* of $\mathbf{A}$ if $(\forall x \in \mathbf{X})(\exists A \in \mathbf{A})[\, A \otimes x = x\,]$,
- *weak eigenvector* of $\mathbf{A}$ if $(\exists A \in \mathbf{A})(\exists x \in \mathbf{X})[\, A \otimes x = x\,]$.

The above six types will be referred to as GT1, GT2, ..., GT6 for general interval eigenvectors, and T1, T2, ..., T6 for non-decreasing interval eigenvectors. In the case of increasing eigenvectors $x \in \mathbf{X}$, all six types have been characterized in [11] and relations between them have been described. For D-increasing eigenvectors and for arbitrary eigenvectors $x \in \mathbf{X}$ the characterization of all interval eigenvector types is presented in the following two sections.

5.2.1 Non-decreasing Interval Eigenvectors

We assume that $D \in \mathscr{D}_n$ is a fixed interval partition on N. We define the *D-increasing* interval eigenvectorsof types T1, T2, ..., T6 by restricting the corresponding quantifiers with $x \in \mathbf{X}$ to $x \in \mathbf{X} \cap \mathscr{B}^<(D,n)$. Types T1, T2, T5, T6 are characterized below for arbitrary interval matrix $\mathbf{A}$, types T3, T4 are characterized for a single matrix A (formally: interval matrix $\mathbf{A} = [A, A]$).

Definition 5.2. Let $\mathbf{X} = [\underline{x}, \overline{x}]$ be an interval vector with bounds $\underline{x}$ and $\overline{x}$. The set $\mathbf{X} \cap \mathscr{B}^<(D,n)$ will be denoted as $\mathbf{X}^<(D)$. Further we denote

$$\underline{x}_D = \bigwedge\{x \in \mathbf{X} |\, x \in \mathscr{B}^<(D,n)\} = \bigwedge \mathbf{X}^<(D) \ , \tag{5.4}$$

$$\overline{x}_D = \bigvee\{x \in \mathbf{X} |\, x \in \mathscr{B}^<(D,n)\} = \bigvee \mathbf{X}^<(D) \ . \tag{5.5}$$

Proposition 5.1. *For given* $\mathbf{X} = [\underline{x}, \overline{x}]$, *the following statements hold true*

(i) $\underline{x}_D \in \mathscr{B}^{\leq}(D,n)\,, \quad \overline{x}_D \in \mathscr{B}^{\leq}(D,n)$,
(ii) $\underline{x} \leq \underline{x}_D\,, \quad \overline{x}_D \leq \overline{x}$,
(iii) *if* $x \in \mathscr{B}^<(D,n)$, *then* $x \in \mathbf{X}^<(D)$ *if and only if* $\underline{x}_D \leq x \leq \overline{x}_D$,
(iv) *if* $\mathbf{X}^<(D) \neq \emptyset$, *then* $\underline{x}_D \leq \overline{x}_D$,
(v) *if* $\mathbf{X}^<(D) = \emptyset$, *then* $\underline{x}_D = \{I, I, \ldots, I\}, \overline{x}_D = \{O, O, \ldots, O\}$.

Proof. (i) Let $i, k \in N, D[i] \leq D[k]$. Then $x_i \leq x_k$ for every $x \in \mathscr{B}^{<}(D,n)$. Hence

$$(\underline{x}_D)_i = \bigwedge \{x_i \,|\, x \in \mathbf{X}^{<}(D)\} \leq \bigwedge \{x_k \,|\, x \in \mathbf{X}^{<}(D)\} = (\underline{x}_D)_k \ ,$$

which implies $\underline{x}_D \in \mathscr{B}^{\leq}(D,n)$, and analogously for $\overline{x}_D$.

(ii), (iii) The inequalities follow directly from the definition.

(iv) Let $\mathbf{X}^{<}(D) \neq \emptyset$. Then there is x such that $x \in \mathbf{X}$, $x \in \mathscr{B}^{<}(D,n)$, and assertion (iii) directly implies $\underline{x}_D \leq \overline{x}_D$.

(v) Let $\mathbf{X}^{<}(D) = \emptyset$. Then $\underline{x}_D$ is infimum of the empty set, which is the constant vector with the maximum value I, and dually, $\overline{x}_D$ is supremum of the empty set, which is the constant vector with the minimum value O. □

Theorem 5.4 (T1). *Let interval matrix* $\mathbf{A} = [\underline{A}, \overline{A}]$ *and interval vector* $\mathbf{X} = [\underline{x}, \overline{x}]$ *be given. Then* $\mathbf{X}$ *is strong* D*-increasing eigenvector of* A *if and only if*

$$m^{\star}(D, \overline{A}) \leq \underline{x}_D \ , \quad \overline{x}_D \leq M^{\star}(D, \underline{A}) \ . \tag{5.6}$$

Proof. Let $\mathbf{X}$ be a strong D-increasing eigenvector of $\mathbf{A}$ i.e. $A \otimes x = x$ for every $x \in \mathbf{X}^{<}(D)$, $A \in \mathbf{A}$. In view of Theorem 5.3 we have

$$m^{\star}(D, A) \leq x \leq M^{\star}(D, A) \ . \tag{5.7}$$

Applying (5.7) to particular matrices $\underline{A}$, $\overline{A}$ we get

$$m^{\star}(D, \overline{A}) \leq x \leq M^{\star}(D, \underline{A}) \ . \tag{5.8}$$

In view of Definition 5.2, conditions (5.6) follow directly.

For the converse implication, let us assume, that conditions (5.6) are satisfied. Let $x \in \mathbf{X}^{<}(D)$, $A \in \mathbf{A}$. Then we have

$$m^{\star}(D, A) \leq m^{\star}(D, \overline{A}) \leq \underline{x}_D \leq x \ ,$$

and

$$x \leq \overline{x}_D \leq M^{\star}(D, \underline{A}) \leq M^{\star}(D, A) \ .$$

Therefore, $m^{\star}(D, A) \leq x \leq M^{\star}(D, A)$, which implies $A \otimes x = x$. Hence, $\mathbf{X}$ is a strong D-increasing eigenvector of $\mathbf{A}$. □

Theorem 5.5 (T2). *Let interval matrix* $\mathbf{A} = [\underline{A}, \overline{A}]$ *and interval vector* $\mathbf{X} = [\underline{x}, \overline{x}]$ *be given. Then* $\mathbf{X}$ *is strongly universal* D*-increasing eigenvector of* $\mathbf{A}$ *if and only if*

$$\left[\left(m^{\star}(D, \overline{A}) \vee \underline{x}_D\right), \left(M^{\star}(D, \underline{A}) \wedge \overline{x}_D\right)\right] \cap \mathscr{B}^{<}(D,n) \neq \emptyset \ . \tag{5.9}$$

Proof. Let **X** be strongly universal D-increasing eigenvector of **A** i.e. there is $x \in \mathbf{X}^{<}(D)$ such that $A \otimes x = x$ holds for every $A \in \mathbf{A}$. By Theorem 5.3 we then have

$$m^{\star}(D, A) \leq x \leq M^{\star}(D, A) \ . \tag{5.10}$$

For particular matrices $\underline{A}, \overline{A}$ we get

$$m^{\star}(D, \overline{A}) \leq x \leq M^{\star}(D, \underline{A}) \ . \tag{5.11}$$

By assumption $x \in \mathbf{X}^{<}(D)$ we have

$$\underline{x}_D \leq x \leq \overline{x}_D, \quad x \in \mathscr{B}^{<}(D, n) \ . \tag{5.12}$$

Condition (5.9) then follows from (5.11) and (5.12) by an easy computation.

Conversely, let us assume that condition (5.9) is satisfied, and vector $x \in \mathscr{B}^{<}(D, n)$ belongs to the interval $\left[(m^{\star}(D, \overline{A}) \vee \underline{x}_D), (M^{\star}(D, \underline{A}) \wedge \overline{x}_D)\right]$. Then $\underline{x}_D \leq x \leq \overline{x}_D$, i.e. $x \in \mathbf{X}^{<}(D)$, in view of Proposition 5.1(iii). Moreover we have, for every $A \in \mathbf{A}$,

$$m^{\star}(D, A) \leq m^{\star}(D, \overline{A}) \leq x \leq M^{\star}(D, \underline{A}) \leq M^{\star}(D, A) \ , \tag{5.13}$$

which implies $A \otimes x = x$. Hence, **X** is strongly universal D-increasing eigenvector of **A**. □

Theorem 5.6 (T3). *Let interval matrix* $\mathbf{A} = [\underline{A}, \overline{A}]$ *with* $\underline{A} = \overline{A} = A$ *and interval vector* $\mathbf{X} = [\underline{x}, \overline{x}]$ *be given. Then* **X** *is universal* D*-increasing eigenvector of* **A** *if and only if*

$$[(m^{\star}(D, A) \vee \underline{x}_D), (M^{\star}(D, A) \wedge \overline{x}_D)] \cap \mathscr{B}^{<}(D, n) \neq \emptyset \ . \tag{5.14}$$

Proof. As $\mathbf{A} = [A, A]$ contains a single matrix A, the following assertions are equivalent

(i) **X** is universal D-increasing eigenvector of $[A, A]$,
(ii) **X** is strongly universal D-increasing eigenvector of $[A, A]$.

Assertion of the theorem follows immediately, because formula (5.14) is a particular case of formula (5.9) in Theorem 5.5. □

Theorem 5.7 (T4). *Let interval matrix* $\mathbf{A} = [\underline{A}, \overline{A}]$ *with* $\underline{A} = \overline{A} = A$ *and interval vector* $\mathbf{X} = [\underline{x}, \overline{x}]$ *be given. Then* **X** *is a strongly tolerable* D*-increasing eigenvector of* $[A, A]$ *if and only if*

$$m^{\star}(D, A) \leq \underline{x}_D \ , \quad \overline{x}_D \leq M^{\star}(D, A) \ . \tag{5.15}$$

Proof. Similarly as in the proof above, we use the fact that interval matrix $\mathbf{A} = [A, A]$ contains a single matrix A. Then the following assertions are equivalent

(i) $\mathbf{X}$ is strongly tolerable D-increasing eigenvector of $[A, A]$,
(ii) $\mathbf{X}$ is strong D-increasing eigenvector of $[A, A]$.

Assertion of the theorem now follows, because formula (5.15) is a particular case of formula (5.6) in Theorem 5.4. □

Theorem 5.8 (T5). *Let interval matrix* $\mathbf{A} = [\underline{A}, \overline{A}]$ *and interval vector* $\mathbf{X} = [\underline{x}, \overline{x}]$ *be given. Then* $\mathbf{X}$ *is tolerable* D*-increasing eigenvector of* $\mathbf{A}$ *if and only if*

$$m^\star(D, \underline{A}) \le \underline{x}_D \ , \quad \overline{x}_D \le M^\star(D, \overline{A}) \ . \tag{5.16}$$

Proof. Let $\mathbf{X}$ be tolerable D-increasing eigenvector of $\mathbf{A}$ i.e. for every $x \in \mathbf{X}^<(D)$, there is a matrix $A \in \mathbf{A}$ such that $A \otimes x = x$. Using Theorem 5.3 we get

$$m^\star(D, \underline{A}) \le m^\star(D, A) \le x \le M^\star(D, A) \le M^\star(D, \overline{A}) \ . \tag{5.17}$$

Conditions (5.16) then follow from (5.17), in view of Definition 5.2.

For the converse implication, let us assume that conditions (5.16) are satisfied. Let $x \in \mathbf{X}^<(D)$ be arbitrary, but fixed. Then we have

$$m^\star(D, \underline{A}) \le \underline{x}_D \le x \le \overline{x}_D \le M^\star(D, \overline{A}) \ . \tag{5.18}$$

We define matrix $A^{(x)} \in \mathscr{B}(n, n)$ by putting, for every $j, k \in N$,

$$a_{jk}^{(x)} = \begin{cases} \overline{a}_{jk} \wedge x_j & D[j] < D[k] \ , \\ \overline{a}_{jk} & D[j] \ge D[k] \ . \end{cases} \tag{5.19}$$

For any $j, k \in N$ with $D[j] < D[k]$ we have $\underline{a}_{jk} \le m_j^\star(D, \underline{A}) \le x_j$, in view of (5.2) and (5.17). Further we have $\underline{a}_{jk} \le \overline{a}_{jk}$, which implies $\underline{a}_{jk} \le a_{jk}^{(x)}$. Clearly, $\underline{a}_{jk} \le \overline{a}_{jk} = a_{jk}^{(x)}$ for $j, k \in N$ with $D[j] \ge D[k]$. Thus, $\underline{A} \le A^{(x)}$. The inequality $A^{(x)} \le \overline{A}$ follows directly from definition (5.19), therefore $A^{(x)} \in \mathbf{A}$.
By definition (5.2), we have for every $i \in N$

$$\begin{aligned} m_i^\star\left(D, A^{(x)}\right) &= \max_{j \in N, D[j] \le D[i]} \ \max_{k \in N, D[k] > D[j]} a_{jk}^{(x)} \\ &= \max_{j \in N, D[j] \le D[i]} \ \max_{k \in N, D[k] > D[j]} \left(\overline{a}_{jk} \wedge x_j\right) \\ &\le \max_{j \in N, D[j] \le D[i]} \ \max_{k \in N, D[k] > D[j]} \left(\overline{a}_{jk} \wedge x_i\right) \\ &= \left(\max_{j \in N, D[j] \le D[i]} \ \max_{k \in N, D[k] > D[j]} \overline{a}_{jk}\right) \wedge x_i \\ &= m_i^\star\left(D, \overline{A}\right) \wedge x_i \ . \end{aligned} \tag{5.20}$$

Hence $m^\star\left(D, A^{(x)}\right) \le m^\star(\overline{A}) \wedge x \le x$. Further we have

$$\begin{aligned}
M_i^\star\left(D, A^{(x)}\right) &= \min_{j\in N, D[j]\ge D[i]} \max_{k\in N, D[k]\ge D[j]} a_{jk}^{(x)} \\
&= \min_{j\in N, D[j]\ge D[i]} \left(\max_{k\in N, D[k]= D[j]} a_{jk}^{(x)} \vee \max_{k\in N, D[k]> D[j]} a_{jk}^{(x)} \right) \\
&= \min_{j\in N, D[j]\ge D[i]} \left(\max_{k\in N, D[k]= D[j]} \overline{a}_{jk} \vee \max_{k\in N, D[k]> D[j]} \left(\overline{a}_{jk} \wedge x_j\right) \right) \\
&\ge \min_{j\in N, D[j]\ge D[i]} \left(\max_{k\in N, D[k]= D[j]} \overline{a}_{jk} \vee \max_{k\in N, D[k]> D[j]} \left(\overline{a}_{jk} \wedge x_i\right) \right) \\
&\ge \min_{j\in N, D[j]\ge D[i]} \max_{k\in N, D[k]\ge D[j]} \left(\overline{a}_{jk} \wedge x_i\right) \\
&= \left(\min_{j\in N, D[j]\ge D[i]} \max_{k\in N, D[k]\ge D[j]} \overline{a}_{jk} \right) \wedge x_i \\
&= M_i^\star\left(D, \overline{A}\right) \wedge x_i \ .
\end{aligned} \tag{5.21}$$

By (5.21) we get $M^\star\left(D, A^{(x)}\right) \ge M^\star(D, \overline{A}) \wedge x \ge x$. Therefore, $m^\star\left(D, A^{(x)}\right) \le x \le M^\star\left(D, A^{(x)}\right)$, which implies $A^{(x)} \otimes x = x$. Hence, $\mathbf{X}$ is tolerable D-increasing eigenvector of $\mathbf{A}$. □

Theorem 5.9 (T6). *Let interval matrix* $\mathbf{A} = [\underline{A}, \overline{A}]$ *and interval vector* $\mathbf{X} = [\underline{x}, \overline{x}]$ *be given. Then* $\mathbf{X}$ *is weak* D*-increasing eigenvector of* $\mathbf{A}$ *if and only if*

$$\left[(m^\star(D, \underline{A}) \vee \underline{x}_D) \ , \left(M^\star(D, \overline{A}) \wedge \overline{x}_D\right)\right] \cap \mathscr{B}^<(D, n) \ne \emptyset \ . \tag{5.22}$$

Proof. Let $\mathbf{X}$ be weak D-increasing eigenvector of $\mathbf{A}$, i.e. there are $A \in \mathbf{A}$, $x \in \mathbf{X}^<(D)$ such that $A \otimes x = x$. By Theorem 5.3 we get

$$m^\star(D, \underline{A}) \le m^\star(D, A) \le x \le M^\star(D, A) \le M^\star(D, \overline{A}) \ . \tag{5.23}$$

By assumption $x \in \mathbf{X}^<(D)$ we have

$$\underline{x}_D \le x \le \overline{x}_D \ , \quad x \in \mathscr{B}^<(D, n) \ . \tag{5.24}$$

Condition (5.22) then easily follows from (5.23) and (5.24).

Conversely, let us assume that condition (5.22) is satisfied, and vector $x \in \mathscr{B}^<(D, n)$ belongs to the interval $\left[(m^\star(D, \underline{A}) \vee \underline{x}_D), \left(M^\star(D, \overline{A}) \wedge \overline{x}_D\right)\right]$. Then $\underline{x}_D \le x \le \overline{x}_D$, i.e. $x \in \mathbf{X}^<(D)$, in view of Proposition 5.1(iii). Moreover we have

$$m^\star(D, \underline{A}) \le x \le M^\star(D, \overline{A}) \ . \tag{5.25}$$

Let us consider D-increasing interval vector $\mathbf{X}' = [\underline{x}', \overline{x}']$ with $\underline{x}' = \overline{x}' = x$, which reduces to a single vector. Then inequalities in (5.25) correspond to conditions in (5.16), therefore $\mathbf{X}'$ is tolerable D-increasing vector of $\mathbf{A}$, in view of Theorem 5.8. As a consequence, there exists matrix $A \in \mathbf{A}$, such that $A \otimes x = x$. Hence, $\mathbf{X}$ is weak D-increasing eigenvector of $\mathbf{A}$. □

5.2.2 General Interval Eigenvectors

For given permutation $\varphi \in P_n$ and partition $D \in \mathscr{D}_n$ we denote

$$\mathscr{B}_{\varphi}^{<}(D,n) = \{x \in \mathscr{B}(n) |\, x_{\varphi} \in \mathscr{B}^{<}(D,n)\} \quad .$$

Further we denote, for any interval matrix $\mathbf{A} = [\underline{A}, \overline{A}]$ and interval vector $\mathbf{X} = [\underline{x}, \overline{x}]$

$$\mathbf{A}_{\varphi\varphi} = [\underline{A}_{\varphi\varphi}, \overline{A}_{\varphi\varphi}] \quad , \quad \mathbf{X}_{\varphi} = [\underline{x}_{\varphi}, \overline{x}_{\varphi}] \quad .$$

Proposition 5.2. *Let* $\underline{x}, \overline{x} \in \mathscr{B}(n)$, $\mathbf{X} = [\underline{x}, \overline{x}]$. *Then*

(i) $\mathscr{B}(n) = \bigcup_{\varphi \in P_n} \bigcup_{D \in \mathscr{D}_n} \mathscr{B}_{\varphi}^{<}(D,n)$,
(ii) $\mathbf{X} = \bigcup_{\varphi \in P_n} \bigcup_{D \in \mathscr{D}_n} \mathbf{X} \cap \mathscr{B}_{\varphi}^{<}(D,n)$.

Proof. Assertion (i) is a consequence of Theorem 5.1, and assertion (ii) follows from (i). □

It is shown in the following six theorems that the general structure of the interval eigenvectors of types GT1, GT2, GT5 and GT6 can be described by considering D-increasing interval vectors $\mathbf{X}_{\varphi}$ of interval matrix $\mathbf{A}_{\varphi\varphi} = [\underline{A}_{\varphi\varphi}, \overline{A}_{\varphi\varphi}]$ for $\varphi \in P_n$, $D \in \mathscr{D}_n$. For types GT3 and GT4 the characterization uses D-increasing interval vectors $\mathbf{X}_{\varphi}$ of single matrix $A_{\varphi\varphi}$ (formally: interval matrix $\mathbf{A}_{\varphi\varphi} = [A_{\varphi\varphi}, A_{\varphi\varphi}]$) with $A \in \mathbf{A}$ and $\varphi \in P_n$, $D \in \mathscr{D}_n$.

Theorem 5.10 (GT1). *Let* $\mathbf{A} = [\underline{A}, \overline{A}]$ *and* $\mathbf{X} = [\underline{x}, \overline{x}]$ *be given. Then* $\mathbf{X}$ *is strong eigenvector of* $\mathbf{A}$ *if and only if, for every* $\varphi \in P_n$, $D \in \mathscr{D}_n$, *interval vector* $\mathbf{X}_{\varphi} = [\underline{x}_{\varphi}, \overline{x}_{\varphi}]$ *is strong D-increasing eigenvector of interval matrix* $\mathbf{A}_{\varphi\varphi} = [\underline{A}_{\varphi\varphi}, \overline{A}_{\varphi\varphi}]$.

Proof. It is easy to verify using Proposition 5.2(ii) that the following assertions are equivalent

$\mathbf{X}$ is strong eigenvector of A ,
$(\forall x \in \mathbf{X})(\forall A \in \mathbf{A})\, x \in \mathscr{F}(A)$,
$(\forall \varphi \in P_n)(\forall D \in \mathscr{D}_n)(\forall x \in \mathbf{X} \cap \mathscr{B}_{\varphi}^{<}(D,n))(\forall A \in \mathbf{A})\, x \in \mathscr{F}(A)$,
$(\forall \varphi \in P_n)(\forall D \in \mathscr{D}_n)(\forall x_{\varphi} \in \mathbf{X}_{\varphi} \cap \mathscr{B}^{<}(D,n))(\forall A_{\varphi\varphi} \in \mathbf{A}_{\varphi\varphi})\, x_{\varphi} \in \mathscr{F}(A_{\varphi\varphi})$
for every $\varphi \in P_n$, $D \in \mathscr{D}_n$, $\mathbf{X}_{\varphi}$ is strong D-increasing eigenvector of $\mathbf{A}_{\varphi\varphi}$.

□

Theorem 5.11 (GT2). *Let* $\mathbf{A} = [\underline{A}, \overline{A}]$ *and* $\mathbf{X} = [\underline{x}, \overline{x}]$ *be given. Then* $\mathbf{X}$ *is strongly universal eigenvector of* $\mathbf{A}$ *if and only if there exist* $\varphi \in P_n$, $D \in \mathscr{D}_n$ *such that interval vector* $\mathbf{X}_\varphi = [\underline{x}_\varphi, \overline{x}_\varphi]$ *is strongly universal* D*-increasing eigenvector of interval matrix* $\mathbf{A}_{\varphi\varphi} = [\underline{A}_{\varphi\varphi}, \overline{A}_{\varphi\varphi}]$.

Proof. The following assertions are equivalent

$\mathbf{X}$ is strongly universal eigenvector of A ,
$(\exists x \in \mathbf{X})(\forall A \in \mathbf{A}) \quad x \in \mathscr{F}(A)$,
$(\exists \varphi \in P_n)(\exists D \in \mathscr{D}_n)(\exists x \in \mathbf{X} \cap \mathscr{B}_\varphi^<(D,n))(\forall A \in \mathbf{A}) \quad x \in \mathscr{F}(A)$,
$(\exists \varphi \in P_n)(\exists D \in \mathscr{D}_n)(\exists x_\varphi \in \mathbf{X}_\varphi \cap \mathscr{B}^<(D,n))(\forall A_{\varphi\varphi} \in \mathbf{A}_{\varphi\varphi}) \quad x_\varphi \in \mathscr{F}(A_{\varphi\varphi})$,
there exist $\varphi \in P_n$, $D \in \mathscr{D}_n$ such that $\mathbf{X}_\varphi$ is strongly universal D-increasing eigenvector of $\mathbf{A}_{\varphi\varphi}$.

□

Theorem 5.12 (GT3). *Let* $\mathbf{A} = [\underline{A}, \overline{A}]$ *and* $\mathbf{X} = [\underline{x}, \overline{x}]$ *be given. Then* $\mathbf{X}$ *is universal eigenvector of* $\mathbf{A}$ *if and only if for every* $A \in \mathbf{A}$ *there exist* $\varphi \in P_n$, $D \in \mathscr{D}_n$ *such that interval vector* $\mathbf{X}_\varphi = [\underline{x}_\varphi, \overline{x}_\varphi]$ *is universal* D*-increasing eigenvector of interval matrix* $[A_{\varphi\varphi}, A_{\varphi\varphi}]$.

Proof. The following assertions are equivalent

$\mathbf{X}$ is universal eigenvector of A ,
$(\forall A \in \mathbf{A})(\exists x \in \mathbf{X}) \quad x \in \mathscr{F}(A)$,
$(\forall A \in \mathbf{A})(\exists \varphi \in P_n)(\exists D \in \mathscr{D}_n)(\exists x \in \mathbf{X} \cap \mathscr{B}_\varphi^<(D,n)) \quad x \in \mathscr{F}(A)$,
$(\forall A \in \mathbf{A})(\exists \varphi \in P_n)(\exists D \in \mathscr{D}_n)(\exists x_\varphi \in \mathbf{X}_\varphi \cap \mathscr{B}^<(D,n)) \quad x_\varphi \in \mathscr{F}(A_{\varphi\varphi})$,
for every $A \in \mathbf{A}$ there exist $\varphi \in P_n$ and $D \in \mathscr{D}_n$ such that $\mathbf{X}_\varphi$ is universal D-increasing eigenvector of $[A_{\varphi\varphi}, A_{\varphi\varphi}]$.

□

Theorem 5.13 (GT4). *Let* $\mathbf{A} = [\underline{A}, \overline{A}]$ *and* $\mathbf{X} = [\underline{x}, \overline{x}]$ *be given. Then* $\mathbf{X}$ *is strongly tolerable eigenvector of* $\mathbf{A}$ *if and only if there exists* $A \in \mathbf{A}$ *such that, for every* $\varphi \in P_n$ *and* $D \in \mathscr{D}_n$, *interval vector* $\mathbf{X}_\varphi = [\underline{x}_\varphi, \overline{x}_\varphi]$ *is strongly tolerable* D*-increasing eigenvector of interval matrix* $[A_{\varphi\varphi}, A_{\varphi\varphi}]$.

Proof. The following assertions are equivalent

$\mathbf{X}$ is strongly tolerable eigenvector of A ,
$(\exists A \in \mathbf{A})(\forall x \in \mathbf{X}) \quad x \in \mathscr{F}(A)$,
$(\exists A \in \mathbf{A})(\forall \varphi \in P_n)(\forall D \in \mathscr{D}_n)(\forall x \in \mathbf{X} \cap \mathscr{B}_\varphi^<(D,n)) \quad x \in \mathscr{F}(A)$,
$(\exists A \in \mathbf{A})(\forall \varphi \in P_n)(\forall D \in \mathscr{D}_n)(\forall x_\varphi \in \mathbf{X}_\varphi \cap \mathscr{B}^<(D,n)) \quad x_\varphi \in \mathscr{F}(A_{\varphi\varphi})$,
there exists $A \in \mathbf{A}$ such that, for every $\varphi \in P_n$ and $D \in \mathscr{D}_n$, $\mathbf{X}_\varphi$ is strongly tolerable D-increasing eigenvector of $[A_{\varphi\varphi}, A_{\varphi\varphi}]$.

□

Theorem 5.14 (GT5). *Let* $\mathbf{A} = [\underline{A}, \overline{A}]$ *and* $\mathbf{X} = [\underline{x}, \overline{x}]$ *be given. Then* $\mathbf{X}$ *is tolerable eigenvector of* $\mathbf{A}$ *if and only if, for every* $\varphi \in P_n$ *and* $D \in \mathscr{D}_n$, *interval* *vector* $\mathbf{X}_\varphi = [\underline{x}_\varphi, \overline{x}_\varphi]$ *is tolerable D-increasing eigenvector of interval matrix* $\mathbf{A}_{\varphi\varphi} = [\underline{A}_{\varphi\varphi}, \overline{A}_{\varphi\varphi}]$.

Proof. The following assertions are equivalent

$\mathbf{X}$ is tolerable eigenvector of A ,
$(\forall x \in \mathbf{X})(\exists A \in \mathbf{A}) \quad x \in \mathscr{F}(A)$,
$(\forall \varphi \in P_n)(\forall D \in \mathscr{D}_n)(\forall x \in \mathbf{X} \cap \mathscr{B}^<_\varphi(D,n))(\exists A \in \mathbf{A}) \quad x \in \mathscr{F}(A)$,
$(\forall \varphi \in P_n)(\forall D \in \mathscr{D}_n)(\forall x_\varphi \in \mathbf{X}_\varphi \cap \mathscr{B}^<(D,n))(\exists A_{\varphi\varphi} \in \mathbf{A}_{\varphi\varphi}) \quad x_\varphi \in \mathscr{F}(A_{\varphi\varphi})$,
for every $\varphi \in P_n$ and $D \in \mathscr{D}_n$, $\mathbf{X}_\varphi$ is tolerable D-increasing eigenvector of $\mathbf{A}_{\varphi\varphi}$.

□

Theorem 5.15 (GT6). *Let* $\mathbf{A} = [\underline{A}, \overline{A}]$ *and* $\mathbf{X} = [\underline{x}, \overline{x}]$ *be given. Then* $\mathbf{X}$ *is weak eigenvector of* $\mathbf{A}$ *if and only if there exist* $\varphi \in P_n$ *and* $D \in \mathscr{D}_n$ *such that interval vector* $\mathbf{X}_\varphi = [\underline{x}_\varphi, \overline{x}_\varphi]$ *is weak D-increasing eigenvector of interval matrix* $\mathbf{A}_{\varphi\varphi} = [\underline{A}_{\varphi\varphi}, \overline{A}_{\varphi\varphi}]$.

Proof. The following assertions are equivalent

$\mathbf{X}$ is weak eigenvector of A ,
$(\exists x \in \mathbf{X})(\exists A \in \mathbf{A}) \quad x \in \mathscr{F}(A)$,
$(\exists \varphi \in P_n)(\exists D \in \mathscr{D}_n)(\exists x \in \mathbf{X} \cap \mathscr{B}^<_\varphi(D,n))(\exists A \in \mathbf{A}) \quad x \in \mathscr{F}(A)$,
$(\exists \varphi \in P_n)(\exists D \in \mathscr{D}_n)(\exists x_\varphi \in \mathbf{X}_\varphi \cap \mathscr{B}^<(D,n))(\exists A_{\varphi\varphi} \in \mathbf{A}_{\varphi\varphi}) \quad x_\varphi \in \mathscr{F}(A_{\varphi\varphi})$,
there exist $\varphi \in P_n$ and $D \in \mathscr{D}_n$ such that $\mathbf{X}_\varphi$ is weak D-increasing eigenvector of $\mathbf{A}_{\varphi\varphi}$.

□

5.3 Relations Between Types of Interval Eigenvectors

Relations between the general types GT1, GT2, ..., GT6 are basically the same as the relations between the increasing versions T1, T2, ..., T6, shown in [11]. That is, the implications

$$GT1 \Rightarrow GT2 \Rightarrow GT3 \Rightarrow GT6 \ , \tag{5.26}$$

$$GT1 \Rightarrow GT4 \Rightarrow GT5 \Rightarrow GT6 \tag{5.27}$$

hold, but the converse implications are not valid

$$GT6 \not\Rightarrow GT3 \not\Rightarrow GT2 \not\Rightarrow GT1 \ , \tag{5.28}$$

$$GT6 \not\Rightarrow GT5 \not\Rightarrow GT4 \not\Rightarrow GT1 \ . \tag{5.29}$$

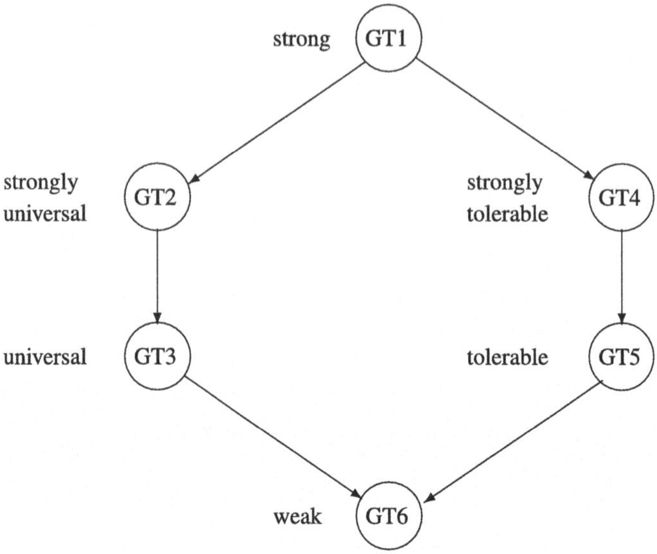

Fig. 5.1 Hasse diagram of the relations between different types of interval eigenvectors

Moreover, neither of the 'cross implications' hold. That is

$$GT2 \not\Rightarrow (GT4 \vee GT5) \ , \tag{5.30}$$

$$GT3 \not\Rightarrow (GT4 \vee GT5) \ , \tag{5.31}$$

$$GT4 \not\Rightarrow (GT2 \vee GT3) \ , \tag{5.32}$$

$$GT5 \not\Rightarrow (GT2 \vee GT3) \ . \tag{5.33}$$

The implications in (5.26) and (5.27) follow directly from definition. On the other hand, the counterexamples in Sect. 5.3.1 below show that the converse implications and the cross implications (5.28)–(5.33) do not hold.

The directed graph of all relations between GT1, GT2, ..., GT6 can be visualized by Hasse diagram in Fig. 5.1 where every arrow indicates an implication, and a non-existing arrow indicates that there is no implication between corresponding types of interval eigenvectors.

5.3.1 Examples and Counterexamples

The first example shows that the converse implication to $GT1 \Rightarrow GT2$ does not hold. That is, if an interval vector **X** is strongly universal eigenvector of an interval matrix **A**, then **X** need not be strong eigenvector of **A**.

Example 5.1 ($GT2 \not\Rightarrow GT1$). Take $\mathbf{A} = [\underline{A}, \overline{A}]$ and $\mathbf{X} = [\underline{x}, \overline{x}]$ as follows

$$\underline{A} = \begin{pmatrix} 1 & 2 \\ 2 & 2 \end{pmatrix}, \quad \overline{A} = \begin{pmatrix} 1 & 3 \\ 3 & 3 \end{pmatrix}, \quad \underline{x} = \begin{pmatrix} 1 \\ 1 \end{pmatrix}, \quad \overline{x} = \begin{pmatrix} 2 \\ 2 \end{pmatrix}.$$

It is easy to see that $A \otimes \underline{x} = \underline{x}$ for every matrix A with $\underline{A} \le A \le \overline{A}$, that is, $\mathbf{X}$ is strongly universal eigenvector of $\mathbf{A}$. On the other hand, $\mathbf{X}$ is not strong eigenvector of $\mathbf{A}$, because $A \otimes x = x$ does not necessarily hold for every $A \in \mathbf{A}$, $x \in \mathbf{X}$. E.g. for vector $x = (1, 1.5)^T$ we have $\underline{A} \otimes x = (1.5, 1.5)^T \neq x$.

The second example shows that ($GT3 \not\Rightarrow GT2$). In other words, if $\mathbf{X}$ is universal eigenvector of $\mathbf{A}$, then $\mathbf{X}$ need not be strongly universal eigenvector of $\mathbf{A}$.

Example 5.2 ($GT3 \not\Rightarrow GT2$). Take $\mathbf{A} = [\underline{A}, \overline{A}]$ and $\mathbf{X} = [\underline{x}, \overline{x}]$ as follows

$$\underline{A} = \begin{pmatrix} 2 & 2 \\ 3 & 4 \end{pmatrix}, \quad \overline{A} = \begin{pmatrix} 2 & 3 \\ 3 & 5 \end{pmatrix}, \quad \underline{x} = \begin{pmatrix} 2 \\ 4 \end{pmatrix}, \quad \overline{x} = \begin{pmatrix} 3 \\ 4 \end{pmatrix}.$$

Then for every $A \in \mathbf{A}$, $x \in \mathbf{X}$ there exist values r, s, k such that

$$A = \begin{pmatrix} 2 & r \\ 3 & s \end{pmatrix}, \quad \begin{matrix} 2 \le r \le 3 \\ 4 \le s \le 5 \end{matrix}, \quad x = \begin{pmatrix} k \\ 4 \end{pmatrix}, \quad 2 \le k \le 3.$$

By easy computation we get $A \otimes x = (r, 4)^T$. That is, x is eigenvector of A if and only if $r = k$. This condition is fulfilled for every $A \in \mathbf{A}$ by some $x \in \mathbf{X}$, which means that $\mathbf{X}$ is universal eigenvector of $\mathbf{A}$. On the other hand, it is obvious that the condition cannot be fulfilled by one single vector for all matrices in $[\underline{A}, \overline{A}]$. Hence, $\mathbf{X}$ is not strongly universal eigenvector for interval matrix $\mathbf{A}$.

The third example shows that ($GT6 \not\Rightarrow GT3$). In other words, if $\mathbf{X}$ is weak eigenvector of $\mathbf{A}$, then $\mathbf{X}$ need not be universal eigenvector of $\mathbf{A}$.

Example 5.3 ($GT6 \not\Rightarrow GT3$). Take $\mathbf{A} = [\underline{A}, \overline{A}]$ and $\mathbf{X} = [\underline{x}, \overline{x}]$ as follows

$$\underline{A} = \begin{pmatrix} 1 & 1 \\ 1 & 1 \end{pmatrix}, \quad \overline{A} = \begin{pmatrix} 1 & 2 \\ 3 & 4 \end{pmatrix}, \quad \underline{x} = \begin{pmatrix} 2 \\ 3 \end{pmatrix}, \quad \overline{x} = \begin{pmatrix} 2 \\ 4 \end{pmatrix}.$$

By easy computation we verify that $\overline{A} \otimes \underline{x} = \underline{x}$. Hence, $\mathbf{X}$ is weak eigenvector of $\mathbf{A}$. On the other hand, for matrix $\underline{A}$ and for any $x \in \mathbf{X}$ we have $\underline{A} \otimes x = (1, 1)^T < \underline{x} \le x$. Therefore, $\mathbf{X}$ is not universal eigenvector of $\mathbf{A}$.

By examples 5.1–5.3 we have proved the non-implications in (5.28). In the next three examples we demonstrate the non-implications in (5.29). First we show that if an interval vector $\mathbf{X}$ is strongly tolerable eigenvector of an interval matrix $\mathbf{A}$, then $\mathbf{X}$ need not be strong eigenvector of $\mathbf{A}$.

Example 5.4 (GT4 $\not\Rightarrow$ GT1). Take $\mathbf{A} = [\underline{A}, \overline{A}]$ and $\mathbf{X} = [\underline{x}, \overline{x}]$ as follows

$$\underline{A} = \begin{pmatrix} 2 & 1 \\ 1 & 3 \end{pmatrix} , \quad \overline{A} = \begin{pmatrix} 2 & 2 \\ 2 & 3 \end{pmatrix} , \quad \underline{x} = \begin{pmatrix} 1 \\ 3 \end{pmatrix} , \quad \overline{x} = \begin{pmatrix} 2 \\ 3 \end{pmatrix} .$$

Every vector $x \in \mathbf{X}$ has the form

$$x = \begin{pmatrix} k \\ 3 \end{pmatrix} , \quad 1 \le k \le 2 .$$

It is easy to see that $\underline{A} \otimes x = x$ for every $x = (k, 3)^T$. That is, $\mathbf{X}$ is strongly tolerable eigenvector of $\mathbf{A}$. On the other hand, $\mathbf{X}$ is not strong eigenvector of $\mathbf{A}$, because $A \otimes x = x$ does not necessarily hold for every $A \in \mathbf{A}$, $x \in \mathbf{X}$. E.g. we have $\overline{A} \otimes \underline{x} = (2, 3)^T \neq \underline{x}$.

The following example demonstrates that if $\mathbf{X}$ is tolerable eigenvector of $\mathbf{A}$, then $\mathbf{X}$ need not be strongly tolerable eigenvector of $\mathbf{A}$.

Example 5.5 (GT5 $\not\Rightarrow$ GT4). Take $\mathbf{A} = [\underline{A}, \overline{A}]$ and $\mathbf{X} = [\underline{x}, \overline{x}]$ as follows

$$\underline{A} = \begin{pmatrix} 2 & 3 \\ 3 & 4 \end{pmatrix} , \quad \overline{A} = \begin{pmatrix} 2 & 4 \\ 3 & 4 \end{pmatrix} , \quad \underline{x} = \begin{pmatrix} 3 \\ 4 \end{pmatrix} , \quad \overline{x} = \begin{pmatrix} 4 \\ 4 \end{pmatrix} ,$$

For every $A \in \mathbf{A}$, $x \in \mathbf{X}$ there exist values r, k such that

$$A = \begin{pmatrix} 2 & r \\ 3 & 4 \end{pmatrix} , \quad 3 \le r \le 4 , \quad x = \begin{pmatrix} k \\ 4 \end{pmatrix} , \quad 3 \le k \le 4 .$$

It is easy to verify that $A \otimes x = (r, 4)^T$. That is, x is eigenvector of A if and only if $r = k$. This condition is fulfilled for every $x \in \mathbf{X}$ by some $A \in \mathbf{A}$, which means that $\mathbf{X}$ is tolerable eigenvector of $\mathbf{A}$. On the other hand, the condition cannot be fulfilled by one single matrix for all vectors in $[\underline{x}, \overline{x}]$. Hence, $\mathbf{X}$ is not strongly tolerable eigenvector of interval matrix $\mathbf{A}$.

The next example shows that a weak interval eigenvector need not be tolerable eigenvector of interval matrix $\mathbf{A}$.

Example 5.6 (GT6 $\not\Rightarrow$ GT5). Take $\mathbf{A} = [\underline{A}, \overline{A}]$ and $\mathbf{X} = [\underline{x}, \overline{x}]$ as follows

$$\underline{A} = \begin{pmatrix} 1 & 1 \\ 2 & 1 \end{pmatrix} , \quad \overline{A} = \begin{pmatrix} 1 & 1 \\ 2 & 2 \end{pmatrix} , \quad \underline{x} = \begin{pmatrix} 1 \\ 1 \end{pmatrix} , \quad \overline{x} = \begin{pmatrix} 2 \\ 1 \end{pmatrix} .$$

By easy computation we verify that $\underline{A} \otimes \underline{x} = \underline{x}$. Hence, $\mathbf{X}$ is weak eigenvector of $\mathbf{A}$. On the other hand, $A \otimes \overline{x} = (1, 2)^T \neq \overline{x}$ for every $A \in \mathbf{A}$. Therefore, $\mathbf{X}$ is not tolerable eigenvector of $\mathbf{A}$.

The last two examples contain the key arguments for the cross non-implications in (5.30)–(5.33). First we show that a strongly universal interval eigenvector need not be tolerable eigenvector of interval matrix **A**.

Example 5.7 ($T2 \not\Rightarrow T5$). Take $\mathbf{A} = [\underline{A}, \overline{A}]$ and $\mathbf{X} = [\underline{x}, \overline{x}]$ as follows

$$\underline{A} = \begin{pmatrix} 2 & 3 \\ 3 & 4 \end{pmatrix}, \quad \overline{A} = \begin{pmatrix} 2 & 3 \\ 3 & 5 \end{pmatrix}, \quad \underline{x} = \begin{pmatrix} 3 \\ 4 \end{pmatrix}, \quad \overline{x} = \begin{pmatrix} 4 \\ 5 \end{pmatrix}.$$

For every $A \in \mathbf{A}$ there exists value r such that

$$A = \begin{pmatrix} 2 & 3 \\ 3 & r \end{pmatrix}, \quad 4 \le r \le 5 .$$

Clearly, $A \otimes \underline{x} = \underline{x}$ for all $A \in \mathbf{A}$, which means that **X** is strongly universal eigenvector of **A**. On the other hand, $A \otimes \overline{x} = (3, r)^T \neq \overline{x}$ for all $A \in \mathbf{A}$. Therefore, **X** is not tolerable eigenvector of interval matrix **A**.

The second key non-implication says that a strongly tolerable interval eigenvector need not be universal eigenvector of interval matrix **A**.

Example 5.8 ($T4 \not\Rightarrow T3$). Take $\mathbf{A} = [\underline{A}, \overline{A}]$ and $\mathbf{X} = [\underline{x}, \overline{x}]$ as follows

$$\underline{A} = \begin{pmatrix} 3 & 2 \\ 3 & 3 \end{pmatrix}, \quad \overline{A} = \begin{pmatrix} 3 & 3 \\ 3 & 5 \end{pmatrix}, \quad \underline{x} = \begin{pmatrix} 3 \\ 4 \end{pmatrix}, \quad \overline{x} = \begin{pmatrix} 3 \\ 5 \end{pmatrix}.$$

Every vector $x \in \mathbf{X}$ has the form

$$x = \begin{pmatrix} 3 \\ k \end{pmatrix}, \quad 4 \le k \le 5 .$$

Then we have $\overline{A} \otimes x = x$ for every $x \in \mathbf{X}$, which means that **X** is strongly tolerable eigenvector of **A**. On the other hand, **X** is not universal eigenvector of **A**, because $\underline{A} \otimes x \neq x$ for every $x \in \mathbf{X}$.

Remark 5.1. Examples 5.7 and 5.8, in combination with implications $GT2 \Rightarrow GT3$ and $GT4 \Rightarrow GT5$ from (5.26) and (5.27), also imply all six remaining non-implications contained in (5.30), (5.31), (5.32) and (5.33).

E.g., assuming $GT2 \Rightarrow GT4$ we get $GT2 \Rightarrow GT5$, because of $GT4 \Rightarrow GT5$, a contradiction with Example 5.7. Hence, $GT2 \not\Rightarrow GT4$. The proofs of the remaining non-implications are analogous.

5.4 Conclusions

The results in this section come from the systematic investigation of the interval version of the eigenproblem in max-min algebra. Interval eigenvectors of interval matrices are motivated by the fact that most of the inputs in real applications are not exact numbers, and the interval approach is one of the possible ways of treating the problems with inexact data. In dependence of the quantifiers used in the definition of an interval eigenvector of an interval matrix, six types of interval eigenvectors can by defined in max-min algebra. The eigenspace structure differs from one type to another, and there are relations between some types. On the other hand, some pairs of eigenvector types are independent. These questions are systematically answered by the presented results.

The interval eigenvectors have been studied in previous paper [11]. The investigation was based on the characterization of increasing eigenvectors of a max-min matrix presented in [10], and the results were restricted to increasing eigenvectors. This section extends the results to non-decreasing eigenvectors, and subsequently to the general case of arbitrary interval eigenvectors. Thus, the basic six types of interval eigenvectors of a given interval matrix in max-min algebra are completely described. Moreover, all relations between interval eigenvectors of various types are shown, and it is demonstrated by examples that no further implications exist. Further research in the area will be oriented to decreasing the computational complexity of the results.

References

1. Cechlárová, K: Eigenvectors in bottleneck algebra. Lin. Algebra Appl. **175**, 63–73 (1992)
2. Cechlárová, K: Efficient computation of the greatest eigenvector in fuzzy algebra. Tatra Mt. Math. Publ. **12**, 73–79 (1997)
3. Cechlárová, K: Solutions of interval linear systems in (max, +)-algebra. In: Proceedings of the 6th International Symposium on Operational Research Preddvor, pp. 321–326, Slovenia (2001)
4. Cechlárová, K., Cuninghame-Green, R.A.: Interval systems of max-separable linear equations. Lin. Algebra Appl. **340**, 215–224 (2002)
5. Cohen, G., Dubois, D., Quadrat, J.P., Viot, M.: A linear-system-theoretic view of discrete event processes and its use for performance evaluation in manufacturing. IEEE Trans. Automat. Contr. **AC-30**, 210–220 (1985)
6. Cuninghame-Green, R.A.: Describing industrial processes with interference and approximating their steady-state behavior. Oper. Res. Quart. **13**, 95–100 (1962)
7. Cuninghame-Green, R.A.: Minimax Algebra, Lecture Notes in Economics and Mathematical Systems, vol. 166. Springer, Berlin (1979)
8. Cuninghame-Green, R.A.: Minimax algebra and application. In: Hawkes, P.W. (ed.), Advances in Imaging and Electron Physics, vol. 90. Academic Press, New York (1995)
9. Fiedler, M., Nedoma, J., Ramík, J., Rohn, J., Zimmermann, K.: Linear Optimization Problems with Inexact Data. Springer, Berlin (2006)
10. Gavalec, M.: Monotone eigenspace structure in max–min algebra. Lin. Algebra Appl. **345**, 149–167 (2002)

11. Gavalec, M., Plavka, J.: Monotone interval eigenproblem in max-min algebra. Kybernetika **46**, 387–396 (2010)
12. Gavalec, M., Zimmermann, K.: Classification of solutions to systems of two-sided equations with interval coefficients. Int. J. Pure Appl. Math. **45**, 533–542 (2008)
13. Gondran, M.: Valeurs propres et vecteurs propres en classification hiérarchique. R. A. I. R. O. Informatique Théorique **10**, 39–46 (1976)
14. Gondran, M., Minoux, M.: Eigenvalues and eigenvectors in semimodules and their interpretation in graph theory. In: Proc. 9th Prog. Symp., pp. 133–148 (1976)
15. Gondran, M., Minoux, M.: Valeurs propres et vecteurs propres en théorie des graphes. Colloques Internationaux, pp. 181–183. CNRS, Paris (1978)
16. Gondran, M., Minoux, M.: Dioïds and semirings: Links to fuzzy sets and other applications. Fuzzy Sets Syst. **158**, 1273–1294 (2007)
17. Olsder, G.: Eigenvalues of dynamic max–min systems. Discrete Events Dynamic Systems, vol. 1, pp. 177–201. Kluwer, Dordrecht (1991)
18. Rashid, I., Gavalec, M., Sergeev, S.: Eigenspace of a three-dimensional max-Łukasiewicz fuzzy matrix. Kybernetika **48**, 309–328 (2012)
19. Rohn, J.: Systems of linear interval equations. Lin. Algebra Appl. **126**, 39–78 (1989)
20. Sanchez, E.: Resolution of eigen fuzzy sets equations. Fuzzy Sets Syst. **1**, 69–74 (1978)
21. Tan, Y.-J.: Eigenvalues and eigenvectors for matrices over distributive lattices. Lin. Algebra Appl. **283**, 257–272 (1998)
22. Tan, Y.-J.: On the powers of matrices over a distributive lattice. Lin. Algebra Appl. **336**, 1–14 (2001)
23. Zimmermann, U.: Linear and Combinatorial Optimization in Ordered Algebraic Structure. In: Ann. Discrete Math., vol. 10. North Holland, Amsterdam (1981)

Chapter 6
Eigenproblem in Max-Drast and Max-Łukasiewicz Algebra

Abstract When the max-min operations on the unit real interval are considered as a particular case of fuzzy logic operations (Gödel operations), then the max-min algebra can be viewed as a specific case of more general fuzzy algebra with operations max and T, where T is a triangular norm (in short: t-norm). Such *max-T algebras* are useful in various applications of the fuzzy set theory. In this chapter we investigate the structure of the eigenspace of a given fuzzy matrix in two specific max-T algebras: the so-called *max-drast algebra*, in which the least t-norm T (often called the drastic norm) is used, and *max-Lukasiewicz algebra* with Łukasiewicz t-norm L. For both of these max-T algebras the necessary and sufficient conditions are presented under which the *monotone eigenspace* (the set of all non-decreasing eigenvectors) of a given matrix is non-empty and, in the positive case, the structure of the monotone eigenspace is described. Using permutations of matrix rows and columns, the results are extended to the whole eigenspace.

6.1 Introduction

Eigenvectors of a fuzzy matrix correspond to steady states of a complex discrete-events system, characterized by the transition matrix and the fuzzy state vectors. In this chapter we investigate the eigenspace structure of a given fuzzy matrix in max-T algebra for two specific triangular norms T: the so-called max-drast algebra, where T is the least t-norm (often called drastic), and max-Łukasiewicz algebra with Łukasiewicz t-norm. The results presented in this chapter have been found within the framework of research aimed at investigating the steady states of systems with a fuzzy transition matrix in max-T algebras with various t-norms T.

Investigation of the eigenspace structure in fuzzy algebras is important for applications. The eigenproblem has been studied by many authors in the case of a max–min algebra (one of the basic fuzzy algebras), see e.g. [4, 5, 15, 18]. Many interesting results were found in describing the structure of the eigenspace. Algorithms have been suggested for computing the maximal eigenvector of a given max–min matrix, see [2, 3, 6, 7]. The problem has also been studied in more general

The chapter was written by "Martin Gavalec"

M. Gavalec et al., *Decision Making and Optimization*, Lecture Notes in Economics and Mathematical Systems 677, DOI 10.1007/978-3-319-08323-0_6

structures such as semi-modules or distributive lattices [11–14,19–21]. The structure of the eigenspace in a max–min algebra, as a union of intervals of monotone eigenvectors, was described in [8]. The approach of [8] can be used for other max-T algebras. It has been applied to the drastic t-norm in [9], where the structure of the space of all increasing eigenvectors of a given max-drast fuzzy matrix was presented.

This chapter starts with investigation of the complete eigenspace structure for matrices in a max-drast algebra without any restriction on the monotonicity. First, necessary and sufficient conditions are presented under which the increasing or non-decreasing eigenspace of a given matrix is non-empty, in other words, that the corresponding system has a stable state. The structure of the corresponding eigenspace is then fully described. Further, using simultaneous row and column permutations of the matrix, a characterization of the general eigenspace structure of a given max-drast fuzzy matrix is presented. In this way all stable states of the corresponding fuzzy system are described, see also [10]. The investigation of stable states in max-drast algebra is useful in the fuzzy systems where the extreme reliability is required, in the sense that any considered sequence of events must not contain two events with the reliability values less than one.

In the rest of the chapter we investigate the eigenspace structure for matrices in max-Łukasiewicz algebra, see [17]. As the eigenspace structure for matrices of higher dimensions is rather complex, we only describe in full the eigenproblem for three-dimensional fuzzy matrices. Similarly as in max-drast case, necessary and sufficient conditions are proved under which the monotone eigenspace of a given matrix is non-empty. Further, the structure of the monotone eigenspace is described and finally, simultaneous row and column permutations of the matrix are used to obtain a complete characterization of the general eigenspace structure.

6.2 Eigenvectors in Max-T Algebra

Let us denote the real unit interval by $\mathscr{I} = \langle 0, 1\rangle$, let T be one of the triangular norms used in fuzzy theory. By a max-T algebra we understand a triple $(\mathscr{I}, \oplus, \otimes_T)$ with $\oplus = \max$ and $\otimes_T = T$, binary operations on $\mathscr{I}$. Further, notation $\mathscr{I}(n,n)$ ($\mathscr{I}(n)$) denotes the set of all square matrices (all vectors) of a given dimension n over $\mathscr{I}$. The operations $\oplus$, $\otimes_T$ are analogously extended to matrices and vectors in a formal way. The subscript T is often omitted, when the operation $\otimes = \otimes_T$ is clear from the context.

The eigenproblem for a given matrix $A \in \mathscr{I}(n,n)$ in a max-T algebra consists in finding an eigenvector $x \in \mathscr{I}(n)$ such that $A \otimes x = x$. The eigenspace of $A \in \mathscr{I}(n,n)$ is denoted by $\mathscr{F}_T(A) = \{x \in \mathscr{I}(n) | \, A \otimes_T x = x\}$. Similarly, as in the case of operation $\otimes_T$, the subscript is often omitted and we write simply $\mathscr{F}(A)$ instead of $\mathscr{F}_T(A)$.

Analogously as in Chap. 5, the eigenspace structure will be investigated by permuting any vector $x \in \mathscr{I}(n)$ to a non-decreasing form. We shall also use further notions and notation form the previous chapter, with the difference that the abstract linearly ordered set $\mathscr{B}$ is replaced here by the unit interval $\mathscr{I}$.

E.g., the *increasing* eigenspace of a matrix $A \in \mathscr{I}(n,n)$ is denoted as

$$\mathscr{F}^{<}(A) = \left\{ x \in \mathscr{I}(n) |\ A \otimes x = x,\ (\forall i, j)[i < j \ \Rightarrow \ x_i < x_j\,] \right\} .$$

and the *non-decreasing* eigenspace as

$$\mathscr{F}^{\leq}(A) = \left\{ x \in \mathscr{I}(n) |\ A \otimes x = x,\ (\forall i, j)[i < j \ \Rightarrow \ x_i \leq x_j\,] \right\} ,$$

Similarly, we use notation $\mathscr{I}^{<}(n)$, $\mathscr{I}^{\leq}(n)$, $\mathscr{I}^{<}_{\varphi}(n) = \left\{ x \in \mathscr{I}(n) |\ x_{\varphi} \in \mathscr{I}^{<}(n) \right\}$, $\mathscr{I}^{\leq}_{\varphi}(n) = \left\{ x \in \mathscr{I}(n) |\ x_{\varphi} \in \mathscr{I}^{\leq}(n) \right\}$, $\mathscr{F}^{<}_{\varphi}(A) = \left\{ x \in \mathscr{F}(A) |\ x_{\varphi} \in \mathscr{I}^{<}(n) \right\}$ and $\mathscr{F}^{\leq}_{\varphi}(A) = \left\{ x \in \mathscr{F}(A) |\ x_{\varphi} \in \mathscr{I}^{\leq}(n) \right\}$. Again, in this notation the assertions of Theorem 5.1 are expressed as $\mathscr{F}^{<}_{\varphi}(A) = \left\{ x \in \mathscr{I}(n) |\ x_{\varphi} \in \mathscr{F}^{<}(A_{\varphi\varphi}) \right\}$ and $\mathscr{F}^{\leq}_{\varphi}(A) = \left\{ x \in \mathscr{I}(n) |\ x_{\varphi} \in \mathscr{F}^{\leq}(A_{\varphi\varphi}) \right\}$.

It has been proved in [8] that if the binary operation $\otimes$ coincides with the minimum operation, then the increasing eigenspace $\mathscr{F}^{<}(A)$ can be described as an interval of increasing eigenvectors, see Theorem 5.2.

Remark 6.1. Theorem 5.1 and Theorem 5.2 are presented in a slightly different formulation, namely for a max–min algebra $(\mathscr{B}, \oplus, \otimes)$ with an arbitrary bounded linearly ordered set $\mathscr{B}$ and with the operations $\oplus = \max$, $\otimes = \min$. Moreover, an analogous description is also given for the non-decreasing eigenspace $\mathscr{F}^{\leq}(A)$, and for constant eigenvectors. That is, in [8] the structure of $\mathscr{F}(A)$ in a max–min algebra $\mathscr{B}$ is completely described for any matrix $A \in \mathscr{B}(n,n)$.

In fuzzy sets theory, various triangular norms are used (see Chap. 1). The most frequent of them are (see also the notation in Chap. 1):

Gödel norm $\quad T_M = G(x,y) = \min(x,y)$,

product norm $\quad T_P = \mathrm{prod}(x,y) = (x \cdot y)$,

drastic norm

$$T_D = \mathrm{drast}(x,y) = \begin{cases} \min(x,y) & \text{if } \max(x,y) = 1 \ , \\ 0 & \text{if } \max(x,y) < 1 \ , \end{cases}$$

Łukasiewicz norm $\quad T_L = L(x,y) = \max(x + y - 1, 0)$.

The max-T fuzzy algebra with the Gödel norm is a special case of the max-min algebra, and by Remark 6.1 the eigenspace for this case is described in [8]. The eigenspace for the max-prod fuzzy algebra with the product norm has been discussed in [16].

6.3 Eigenvectors in Max-Drast Algebra

In this section, the above considerations will be transferred to the case of a max-drast algebra, which is a special case of a max-T fuzzy algebra with the so-called drastic triangular norm. The drastic norm is the basic example of a non-divisible t-norm on any partially ordered set, see [1]. In Sect. 6.2 we work with the max-drast fuzzy algebra $(\mathscr{I}, \oplus, \otimes)$ with binary operations $\oplus = \max$ and $\otimes = \text{drast} = T_D$.

Hence, for $x, y \in \mathscr{I}(n)$ we have

$$(x \oplus y)_i = \max(x_i, y_i) \ ,$$

$$(x \otimes y)_i = \begin{cases} \min(x_i, y_i) & \text{if } \max(x_i, y_i) = 1 \ , \\ 0 & \text{if } \max(x_i, y_i) < 1 \ , \end{cases}$$

for every $i \in N$.

Proposition 6.1. *Let $A \in \mathscr{I}(n,n)$ and $x \in \mathscr{I}(n)$. Then $x \in \mathscr{F}(A)$ if and only if for every $i \in N$ the following hold*

$$\text{drast}(a_{ij}, x_j) \le x_i \quad \textit{for every} \quad j \in N \ , \tag{6.1}$$

$$\text{drast}(a_{ij}, x_j) = x_i \quad \textit{for some} \quad j \in N \ . \tag{6.2}$$

Proof. Conditions (6.1), (6.2) are equivalent to the equality $\bigoplus_{j=1}^{n} (a_{ij} \otimes x_j) = x_i$. Thus, the proposition follows from the definition of the eigenspace $\mathscr{F}(A)$. □

Proposition 6.2. *Let $A \in \mathscr{I}(n,n)$, $x \in \mathscr{I}^<(n)$. Then $x \in \mathscr{F}^<(A)$ if and only if for every $i \in N$ the following hold*

$$\text{drast}(a_{ij}, x_j) \le x_i \quad \textit{for every} \quad j \in N, \quad j \ge i \ , \tag{6.3}$$

$$\text{drast}(a_{ij}, x_j) = x_i \quad \textit{for some} \quad j \in N, \quad j \ge i \ . \tag{6.4}$$

Proof. For $j < i$ we have $x_j < x_i \le 1$, in view of the strict monotonicity of x. Hence $\text{drast}(a_{ij}, x_j) \le x_j < x_i$. The proposition then follows from Proposition 6.1. □

Proposition 6.3. *Let $A \in \mathscr{I}(n,n)$, $x \in \mathscr{I}^<(n)$. Then $x \in \mathscr{F}^<(A)$ if and only if for every $i \in N$ the following hold*

$$a_{ij} < 1 \quad \textit{for every } j \in N, \ j > i \ , \tag{6.5}$$

$$a_{in} \le x_i \quad \textit{if } x_n = 1 \ , \tag{6.6}$$

$$\text{drast}(a_{ij}, x_j) = x_i \quad \textit{for some } j \in \{i, n\} \ . \tag{6.7}$$

Proof. Suppose $x \in \mathscr{F}^<(A)$ and $a_{ij} = 1$ for some $j > i$. Then $\text{drast}(a_{ij}, x_j) = \text{drast}(1, x_j) = x_j > x_i$, which is a contradiction with (6.3). Hence, condition (6.5) is necessarily fulfilled. Suppose $x_n = 1$ and $a_{ij} > x_i$, then we get $\text{drast}(a_{in}, x_n) = \text{drast}(a_{in}, 1) = a_{in} > x_i$, which again is a contradiction with (6.3). Therefore, condition (6.6) must hold. As a consequence of condition (6.5), we have $\text{drast}(a_{ij}, x_j) = 0$ for every $j \in N, i < j < n$. Then condition (6.7) follows from (6.4). On the other hand, it is easy to see that conditions (6.5), (6.6), and (6.7) imply (6.3) and (6.4), hence they are also sufficient for $x \in \mathscr{F}^<(A)$. □

The following theorem describes necessary and sufficient conditions under which a square matrix possesses an increasing eigenvector.

Theorem 6.1. *Let $A \in \mathscr{I}(n,n)$. Then $\mathscr{F}^<(A) \neq \emptyset$ if and only if the following conditions are satisfied*

(i) $a_{ij} < 1$ *for all* $i, j \in N, i < j$,
(ii) $0 < a_{in}$ *for all* $i \in N \setminus \{1\}$ *with* $a_{ii} < 1$,
(iii) $a_{kn} < a_{in}$ *for all* $i, k \in N, k < i$ *with* $a_{ii} < 1$,
(iv) $a_{nn} = 1$.

Proof. First we prove that conditions (i)–(iv) are necessary. If $\mathscr{F}^<(A) \neq \emptyset$, then there exists $x \in \mathscr{F}^<(A)$ such that conditions (6.5) and (6.7) are satisfied for every $i \in N$. Condition (i) is in fact equivalent to condition (6.5).

To prove condition (ii), let us assume that $i \in N \setminus \{1\}$ with $a_{ii} < 1$. Then $\text{drast}(a_{ii}, x_i) < x_i$, and by (6.7) we get $\text{drast}(a_{in}, x_n) = x_i$, which cannot hold, if $a_{in} = 0$. Hence $a_{in} > 0$.

Now, suppose $i, k \in N, k < i$ with $a_{ii} < 1$. Similarly to the above, we get $\text{drast}(a_{in}, x_n) = x_i$, which implies $x_n = 1$, in view of condition (i). Then, by (6.3), we have

$$a_{kn} = \text{drast}(a_{kn}, 1) = \text{drast}(a_{kn}, x_n) \leq x_k < x_i = \text{drast}(a_{in}, x_n) =$$
$$= \text{drast}(a_{in}, 1) = a_{in} ,$$

which proves condition (iii).

Condition (iv) follows from the fact that, by (6.7), the equality $\text{drast}(a_{nn}, x_n) = x_n$ must be satisfied, which is impossible if $a_{nn} < 1$.

For the proof of the converse implication, let us assume that the matrix satisfies conditions (i)–(iv). Then we define an increasing vector $x \in \mathscr{I}(n)$ by recursion using the following rules.

Rule 1. If $i < n$, $a_{ii} < 1$, then put $x_i = a_{in}$.
Rule 2. If $i < n$, $a_{ii} = 1$, then choose $x_i \in \mathscr{I}$ such that

(a) $x_{i-1} < x_i$ (not applied for $i = 1$) ,
(b) $a_{in} \leq x_i < 1$,
(c) $x_i < a_{kn}$ for every $k \in N, k > i$ with $a_{kk} < 1$ (not applied if $a_{kk} = 1$ for all $k > i$) .

Rule 3. For $i = n$, choose $x_n \in \mathscr{I}$ such that

(a) $x_{n-1} < x_n$,
(b) if $a_{11} < 1$, $0 < a_{1n}$, then $x_n = 1$,
(c) if $a_{kk} < 1$ for some $k \in N$, $k > 1$, then $x_n = 1$.

It follows from condition (iii) that the recursion according to the above rules always gives an output $x \in \mathscr{I}^{<}(n)$. Further, it can be easily verified that condition (6.7) holds true, while (6.5) follows directly from condition (i). Hence, we have proved that conditions (i)–(iv) are also sufficient. □

The next theorem characterizes all the eigenvectors of a given matrix. In other words, the theorem completely describes the eigenspace structure.

Theorem 6.2. *Suppose $A \in \mathscr{I}(n,n)$ fulfills the conditions (i)–(iv) from Theorem 6.1, and let $x \in \mathscr{I}^{<}(n)$. Then $x \in \mathscr{F}^{<}(A)$ if and only if the following conditions are satisfied*

(v) $x_i = a_{in}$ *for all* $i \in N$ *with* $a_{ii} < 1$,
(vi) *if* $x_n = 1$, *then* $x_i \geq a_{in}$ *for all* $i \in N \setminus \{n\}$ *with* $a_{ii} = 1$,
(vii) $x_n = 1$ *if* $a_{11} < 1$, $0 < a_{1n}$,
(viii) $x_n = 1$ *if* $a_{ii} < 1$ *for some* $i \in N \setminus \{1\}$.

Proof. The assertion of the theorem follows from Proposition 6.3 and from the arguments in the proof of Theorem 6.1. □

In contrast to the above investigation of increasing eigenvectors, the next two theorems characterize the existence and description of all the constant eigenvectors of a given matrix in a max-drast algebra.

Theorem 6.3. *If $A \in \mathscr{I}(n,n)$, then there is a non-zero constant eigenvector $x \in \mathscr{F}(A)$ if and only if*

(ix) $\max\{ a_{ij} | \, j \in N \} = 1$ *for every* $i \in N$.

Proof. Let $x = (c, c, \ldots, c) \in \mathscr{I}(n)$ with $c > 0$ be a constant eigenvector of A. From Proposition 6.1 we get for every $i \in N$

$$\mathrm{drast}(a_{ij}, c) \leq c \quad \text{for every} \quad j \in N \ , \tag{6.8}$$

$$\mathrm{drast}(a_{ij}, c) = c \quad \text{for some} \quad j \in N \ . \tag{6.9}$$

By the assumption that $c > 0$, the condition (6.9) holds for fixed $i \in N$ if and only if $a_{ij} = 1$ for some $j \in N$, in other words, if $\max\{ a_{ij} | \, j \in N \} = 1$. Hence, condition (ix) is necessary. On the other hand, it is easy to see that (ix) is also sufficient for the existence of a non-zero constant eigenvector. An even stronger formulation is contained in theorem below. □

Theorem 6.4. *Suppose that $A \in \mathscr{I}(n,n)$ fulfills condition (ix) from Theorem 6.3. Then the constant vector $x = (c, c, \ldots, c) \in \mathscr{I}(n)$ belongs to $\mathscr{F}(A)$ for every $c \in \mathscr{I}$*

Proof. For $c = 0$, the constant vector fulfills conditions (6.8) and (6.9) for an arbitrary matrix. For $c > 0$, the assertion of the theorem follows from Theorem 6.3. □

Example 6.1. The utility of the theorems of this section will be demonstrated by computing the increasing eigenspace of

$$A = \begin{pmatrix} 0.7 & 0.3 \\ 0.2 & 1 \end{pmatrix} ,$$

and its permuted counterpart,

$$A_{\varphi\varphi} = \begin{pmatrix} 1 & 0.2 \\ 0.3 & 0.7 \end{pmatrix} .$$

The permuted matrix is used to compute the decreasing eigenvectors of A. As a result of the computation, we get that the eigenspace $\mathcal{F}(A)$ contains

- exactly one increasing eigenvector $x = (0.3, 1)$,
- no decreasing eigenvectors ,
- the constant vector $x = (0, 0)$.

Example 6.2. In this example, we compute the increasing eigenspace of

$$B = \begin{pmatrix} 1 & 0.3 \\ 0.2 & 1 \end{pmatrix} ,$$

and its permuted counterpart

$$B_{\varphi\varphi} = \begin{pmatrix} 1 & 0.2 \\ 0.3 & 1 \end{pmatrix} .$$

The eigenspace $\mathcal{F}(B)$ consists of

- increasing vectors with $0 \le x_1 < x_2 < 1$,
- increasing vectors with $0.3 \le x_1 < x_2 = 1$,
- decreasing vectors with $0 \le x_2 < x_1 < 1$,
- decreasing vectors with $0.2 \le x_2 < x_1 = 1$,
- constant vectors $x = (c, c)$ with $0 \le c \le 1$.

The eigenspaces of A and B are shown in Fig. 6.1.

Remark 6.2. The description of the non-decreasing eigenvectors is necessary for computing the eigenspace when the dimension is $n > 2$. This extension of the presented theory is done in the next subsection by an analogous method to that for max–min matrices in [8].

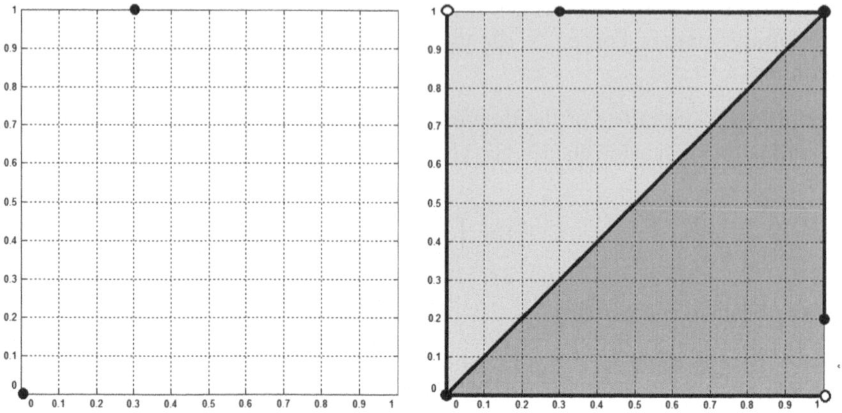

Fig. 6.1 Eigenspaces of A and B from Examples 6.1 and 6.2

6.3.1 Non-decreasing Eigenvectors

In this subsection, the max-drast eigenproblem is investigated for non-decreasing eigenvectors. We use the notation introduced in [8]. The structure of the non-decreasing eigenspace $\mathcal{F}^{\leq}(A)$ is described by considering every non-decreasing eigenvector as increasing, with respect to some *interval partition*, D, of the index set N., i.e., a set of disjoint integer intervals in N, whose union is $\bigcup D = N$. The set of all interval partitions of N will be denoted by $\mathcal{D}_n$. For every interval partition $D \in \mathcal{D}_n$ and for every $i \in N$, we denote by $D[i]$ the partition class containing i, in other words, $D[i]$ is an integer interval $I \in D$ with $i \in I$. For $i, j \in N$ we write $D[i] < D[j]$ if $i' < j'$ for any $i' \in D[i]$ and $j' \in D[j]$. If $D[i] < D[j]$ or $D[i] = D[j]$, then we write $D[i] \leq D[j]$.

For a given partition $D \in \mathcal{D}_n$, we say that vector $x \in \mathcal{I}(n)$ is D-increasing, D-increasing vector if for any indices $i, j \in N$

$$x_i = x_j \quad \text{if and only if} \quad D[i] = D[j] \ ,$$

$$x_i < x_j \quad \text{if and only if} \quad D[i] < D[j] \ .$$

The set of all D-increasing vectors in $\mathcal{I}(n)$ will be denoted by $\mathcal{I}^{<}(D, n)$. If $A \in \mathcal{I}(n, n)$, then the set of all D-increasing eigenvectors of A is denoted by $\mathcal{F}^{<}(D, A)$.

Remark 6.3. It is easy to see that for every non-decreasing vector $x \in \mathcal{I}^{\leq}(n)$ there is exactly one interval partition $D \in \mathcal{D}_n$ such that x is D-increasing. If x is increasing on N, then the corresponding partition is equal to the finest partition consisting of all singletons $D = \{\{i\} | i \in N\}$. On the other hand, if x is constant, then $D = N$.

Proposition 6.4. *Let $A \in \mathscr{I}(n,n)$ and $x \in \mathscr{I}^<(D,n)$. Then $x \in \mathscr{F}^<(D,A)$ if and only if for every $i \in N$ the following hold*

$$\mathrm{drast}(a_{ij}, x_j) \le x_i \quad \textit{for every} \quad j \in N, \quad D[j] \ge D[i] \ , \tag{6.10}$$

$$\mathrm{drast}(a_{ij}, x_j) = x_i \quad \textit{for some} \quad j \in N, \quad D[j] \ge D[i] \ . \tag{6.11}$$

Proof. For $D[j] < D[i]$ we have $x_j < x_i \le 1$, in view of the D-monotonicity of x. Hence $\mathrm{drast}(a_{ij}, x_j) \le x_j < x_i$. The assertion of the proposition then follows from Proposition 6.1. □

Proposition 6.5. *Let $A \in \mathscr{I}(n,n)$ and $x \in \mathscr{I}^<(D,n)$. Then $x \in \mathscr{F}^<(D,A)$ if and only if for every $i \in N$ the following hold*

$$a_{ij} < 1 \quad \textit{for every } j \in N, \ D[j] > D[i] \ , \tag{6.12}$$

$$a_{ij} \le x_i \quad \textit{for every } j \in N, \ x_j = 1 \ , \tag{6.13}$$

$$\mathrm{drast}(a_{ij}, x_j) = x_i \quad \textit{for some } j \in D[i] \cup D[n] \ . \tag{6.14}$$

Proof. If $x \in \mathscr{F}^<(D,A)$, $a_{ij} = 1$, and $D[j] > D[i]$ for some $i, j \in N$, then $\mathrm{drast}(a_{ij}, x_j) = \mathrm{drast}(1, x_j) = x_j > x_i$, which is a contradiction with (6.10). Hence condition (6.12) is necessarily fulfilled. If $x_j = 1$ and $a_{ij} > x_i$, then we get $\mathrm{drast}(a_{ij}, x_j) = \mathrm{drast}(a_{ij}, 1) = a_{ij} > x_i$, which is in contradiction with (6.1). Thus, condition (6.13) must hold true. As a consequence of condition (6.12), we have $\mathrm{drast}(a_{ij}, x_j) = 0$ for every $j \in N$ with $D[i] < D[j] < D[n]$. Therefore, condition (6.14) follows from (6.11).

For the proof of the converse implication, assume first that $i, j \in N$, $D[j] \ge D[i]$. If $D[j] = D[i]$, then $x_j = x_i$ and $\mathrm{drast}(a_{ij}, x_j) = \mathrm{drast}(a_{ij}, x_i) \le x_i$. If $D[j] > D[i]$ and $x_j < 1$, then $\mathrm{drast}(a_{ij}, x_j) = 0 \le x_i$ in view of (6.12). Finally, if $D[j] > D[i]$ and $x_j = 1$, then $\mathrm{drast}(a_{ij}, x_j) \le x_i$ in view of (6.13). Hence the condition (6.10) holds true. Second, the condition (6.11) follows directly from (6.14). Now $x \in \mathscr{F}^<(D,A)$ by Proposition 6.4. □

Proposition 6.6. *Suppose $A \in \mathscr{I}(n,n)$ and $x \in \mathscr{F}^<(D,A)$. Then*

(i) *if $x_1 = 0$, $x_n = 1$, then $\max\limits_{k \in D[n]} a_{ik} = 0$ for every $i \in D[1]$,*
(ii) *if $\max\limits_{j \in D[i]} a_{ij} < 1$ for some $i \in N$, $x_i > 0$, $D[i] < D[n]$, then $\max\limits_{k \in D[n]} a_{ik} = x_i$ and $x_n = 1$,*
(iii) *if $\max\limits_{j \in D[i]} a_{ij} < 1$ and $\max\limits_{k \in D[n]} a_{ik} = 0$ for some $i \in D[1] < D[n]$, then $x_1 = 0$.*

Proof. Let $x \in \mathscr{F}^<(D,A)$.

(i) If $x_1 = 0$ and $x_n = 1$, then for every $i \in D[1]$, $k \in D[n]$ we have $x_i = 0$, $x_k = 1$. In view of (6.10) we then have $a_{ik} \le x_i = 0$, i.e. $\max_{k \in D[n]} a_{ik} = 0$.
(ii) Assume that $\max_{j \in D[i]} a_{ij} < 1$, $i \in N$, $x_i > 0$, $D[i] < D[n]$. Then $a_{ij} < 1$, $x_j = x_i < 1$, and $\mathrm{drast}(a_{ij}, x_j) = 0 < x_i$ for every $j \in D[i]$. By (6.9), there

exists $k \in D[n]$ such that $\mathrm{drast}(a_{ik}, x_k) = x_i > 0$, which is only possible if $x_k = x_n = 1$ and $a_{ik} = x_i$, i.e. $\max_{k \in D[n]} a_{ik} = x_i$, in view of (6.10).

(iii) Suppose $\max_{j \in D[i]} a_{ij} < 1$ and $\max_{k \in D[n]} a_{ik} = 0$ for some $i \in D[1] < D[n]$. Assume that $x_1 = x_i > 0$. By (ii), we then get $0 = \max_{k \in D[n]} a_{ik} = x_i$, which is a contradiction.

□

The following theorem describes necessary and sufficient conditions under which a square matrix possesses a D-increasing eigenvector.

Theorem 6.5. *Let $A \in \mathscr{I}(n,n)$. Then $\mathscr{F}^{<}(D, A) \neq \emptyset$ if and only if the following conditions are satisfied*

(i) $a_{ij} < 1$ *for all* $i, j \in N$, $D[i] < D[j]$,

(ii) $0 < \max_{k \in D[n]} a_{ik}$ *for all* $i \in N \setminus D[1]$ *with* $\max_{j \in D[i]} a_{ij} < 1$,

(iii) $\max_{k \in D[n]} a_{hk} < \max_{k \in D[n]} a_{ik}$ *for all* $i, h \in N$, $D[h] < D[i]$ *with* $\max_{j \in D[i]} a_{ij} < 1$,

(iv) $\max_{k \in D[n]} a_{hk} = \max_{k \in D[n]} a_{ik}$ *for all* $i, h \in N$, $D[h] = D[i]$
with $\max_{j \in D[h]} a_{hj} < 1$, $\max_{j \in D[i]} a_{ij} < 1$,

(v) $\max_{k \in D[n]} a_{hk} \leq \max_{k \in D[n]} a_{ik}$ *for all* $i, h \in N$, $D[h] = D[i]$
with $\max_{j \in D[h]} a_{hj} = 1$, $\max_{j \in D[i]} a_{ij} < 1$,

(vi) $\max_{k \in D[n]} a_{ik} = 1$ *for all* $i \in N$, $D[i] = D[n]$.

Proof. First we prove that conditions (i)–(vi) are necessary. If $\mathscr{F}^{<}(A) \neq \emptyset$, then there exists $x \in \mathscr{F}^{<}(A)$ such that conditions (6.8)–(6.9) are satisfied for every $i \in N$. Condition (i) is in fact equivalent to condition (6.8). For the proof of condition (ii), fix an index $i \in N \setminus D[1]$ such that $\max_{j \in D[i]} a_{ij} < 1$. By this assumption, we get $\mathrm{drast}(a_{ij}, x_j) < x_j = x_i$ for every $j \in D[i]$. Therefore, in view of (6.9), there exists $k \in D[n]$ such that $\mathrm{drast}(a_{ik}, x_k) = x_i \geq x_1 > 0$, which is only possible if $a_{ik} > 0$. Hence $0 < \max_{k \in D[n]} a_{ik}$.

Condition (vi) follows from the fact that for every $i \in N$, $D[i] = D[n]$ we have $x_i = x_n$ and so, in view of (6.9), there is a $k \in D[n]$ such that $\mathrm{drast}(a_{ik}, x_k) = \mathrm{drast}(a_{ik}, x_n) = x_i = x_n$. This is only possible when $a_{ik} = 1$. Therefore, $\max_{k \in D[n]} a_{ik} = 1$.

To prove condition (iii), we consider fixed indices $i, h \in N$ such that $D[h] < D[i]$ and $\max_{j \in D[i]} a_{ij} < 1$. Then we have, in view of condition (vi), $D[i] < D[n]$, i.e. $x_h < x_i < x_n$. It follows from (6.9) and from the assumption $\max_{j \in D[i]} a_{ij} < 1$ that there is an $l \in D[n]$ such that $\mathrm{drast}(a_{il}, x_l) = \mathrm{drast}(a_{il}, x_n) = x_i < x_n$. This is only possible if $a_{il} = x_i$ and $x_l = x_n = 1$. Therefore, $\max_{k \in D[n]} a_{ik} = x_i$. On the other hand, $\mathrm{drast}(a_{hk}, x_k) = \mathrm{drast}(a_{hk}, x_n) = \mathrm{drast}(a_{hk}, 1) = a_{hk} \leq x_h < x_i$ for every $k \in D[n]$, which implies that $\max_{k \in D[n]} a_{hk} < x_i$.

For the proof of conditions (iv) and (v), fix indices $i, h \in N$ such that $D[1] < D[h] = D[i]$ and $\max_{j \in D[i]} a_{ij} < 1$. By this assumption and by (vi) and (6.9) we get, similarly to the above, $\max_{k \in D[n]} a_{ik} = x_i$. If, moreover, $\max_{j \in D[h]} a_{hj} < 1$, then also $\max_{k \in D[n]} a_{ik} = x_h = x_i$, which proves condition (iv). Finally, if $\max_{j \in D[h]} a_{hj} = 1$, then we only have the inequality $\max_{k \in D[n]} a_{ik} \leq x_h = x_i$, implying condition (v).

For the proof of the converse implication, we assume that the matrix satisfies conditions (i)–(vi) and write $S = \{i \in N \mid \max_{j \in D[i]} a_{ij} < 1\}$. We shall consider two disjoint cases.

Case 1. Suppose $S = \emptyset$, i.e. $\max_{j \in D[i]} a_{ij} = 1$ for every $i \in N$. We choose a vector $x \in \mathcal{I}^{<}(D, n)$ with $x_n < 1$ and show that $x \in \mathcal{F}^{<}(D, A)$. It is sufficient to verify that conditions (6.12)–(6.14) are satisfied for every $i \in N$. Condition (6.12) is equivalent to (i), condition (6.13) is trivially fulfilled because $x_j \leq x_n < 1$ for every $j \in N$. Finally, by assumption $\max_{j \in D[i]} a_{ij} = 1$, there is a $j \in D[i]$ such that $a_{ij} = 1$, i.e. $\text{drast}(a_{ij}, x_j) = \text{drast}(1, x_j) = \text{drast}(1, x_i) = x_i$ for every $i \in N$. Hence, condition (6.14) holds true as well.

Case 2. Let $S \neq \emptyset$, i.e. $\max_{j \in D[i]} a_{ij} < 1$ for some $i \in N$. In this case we choose an $x \in \mathcal{I}^{<}(D, n)$ such that $x_n = 1$. Moreover, we assume that $\max_{k \in D[n]} a_{hk} \leq x_h$ for every $h \in N$ and $\max_{k \in D[n]} a_{ik} = x_i$ for every $i \in S$. Such a vector x always exists in view of conditions (ii)–(v). Namely, x can be defined by putting $x_n = 1$ (this value follows from statement (ii) in Proposition 6.5), and then by decreasing recursion on $h \in N$ where we put

$$x_h = x_i = \max_{k \in D[n]} a_{ik} \ ,$$

if there is an index $i \in D[h] \cap S$, and we choose any value $x_h \in \mathcal{I}$ such that

$$\max_{i \in S, D[i] \leq D[h]} \max_{k \in D[n]} a_{ik} \leq x_h \leq \min_{i \in N, D[i] > D[h]} x_i \ ,$$

if $D[h] \cap S = \emptyset$. □

Remark 6.4. In fact, all D-increasing eigenvectors with one exception are described in the second part of the above proof. The exception is when $\max_{j \in D[i]} a_{ij} = 1$ for every $i \in N$ and $x_n = 1$.

The next theorem characterizes all eigenvectors of a given matrix. In other words, the theorem completely describes the eigenspace structure.

Theorem 6.6. *Suppose $A \in \mathcal{I}(n, n)$ fulfills conditions (i)–(vi) from Theorem 6.5. Then $\mathcal{F}^{<}(D, A)$ consists exactly of all vectors $x \in \mathcal{I}^{<}(D, n)$ fulfilling the conditions*

(i) *if $\max_{j \in D[i]} a_{ij} = 1$ for every $i \in N$, then either $x_n < 1$, or*

$$x_n = 1 \quad \text{and} \quad \max_{h \in S, D[h] \leq D[i]} \max_{k \in D[n]} a_{hk} \leq x_i \quad \text{for every } i \in N \ ,$$

(ii) *if $\max_{j \in D[i]} a_{ij} < 1$ for some $i \in N$, then for every $h \in N$*

$$x_h = x_i = \max_{k\in D[n]} a_{ik}\,, \quad \textit{if } \max_{j\in D[i]} a_{ij} < 1 \textit{ for some } i \in D[h]\;,$$

$$\max_{i\in S, D[i]\le D[h]} \max_{k\in D[n]} a_{ik} \le x_h\,, \quad \textit{if } \max_{j\in D[i]} a_{ij} = 1 \textit{ for every } i \in D[h]\;.$$

Proof. The assertion of the theorem follows from Proposition 6.5 and from arguments in the proof of Theorem 6.5. □

6.3.2 General Eigenvectors in Max-Drast Algebra

The results of our investigation made in Sect. 6.3.1 for non-decreasing eigenvectors can be easily extended for the general eigenproblem in max-drast fuzzy algebra. As we mentioned in Sect. 6.2, every vector $x \in \mathscr{I}(n)$ can be permuted to a non-decreasing vector x_φ by some permutation $\varphi \in P_n$. That is, $x_\varphi \in \mathscr{I}^{\le}(n)$ and $x \in \mathscr{I}^{\le}_{\varphi}(n)$ in the notation introduced in Sect. 6.2.

By Remark 6.3.1, for every non-decreasing vector $x_\varphi \in \mathscr{I} \le$ there is an interval partition $D \in \mathscr{D}_n$ such that x_φ is D-increasing. As a consequence, for every vector $x \in \mathscr{I}(n)$ there is a permutation $\varphi \in P_n$ and an interval partition $D \in \mathscr{D}_n$ such that $x \in \mathscr{I}^{<}_{\varphi}(D,n)$. Hence, in the notation introduced in Sect. 6.3.1,

$$\mathscr{I}(n) = \bigcup_{\varphi\in P_n} \mathscr{I}^{\le}_{\varphi}(n) = \bigcup_{\varphi\in P_n} \bigcup_{D\in\mathscr{D}_n} \mathscr{I}^{<}_{\varphi}(D,n)\;, \tag{6.15}$$

and

$$\mathscr{F}_{\mathrm{drast}}(A) = \mathscr{F}(A) = \bigcup_{\varphi\in P_n} \mathscr{F}^{\le}_{\varphi}(A) = \bigcup_{\varphi\in P_n} \bigcup_{D\in\mathscr{D}_n} \mathscr{F}^{<}_{\varphi}(D,A)\;. \tag{6.16}$$

We recall that $x \in \mathscr{F}^{\le}_{\varphi}(A)$ is by definition equivalent to $x_\varphi \in \mathscr{F}^{\le}(A_{\varphi\varphi})$. Therefore, formula (6.16) together with Theorem 6.5 and Theorem 6.6 describe the whole eigenspace of A in max-drast algebra. The computation of $\mathscr{F}(A)$ is shown below in Example 6.3 and Example 6.4 and illustrated by pictures.

6.3.3 Examples

The examples presented in this subsection have dimension $n = 3$. In this case, the above general results can be expressed in simpler way. Theorems 6.1 and 6.5 for increasing and for non-decreasing eigenvectors have in the particular cases with interval partitions $D_{12} = \{\{1,2\},\{3\}\}$ and $D_{23} = \{\{1\},\{2,3\}\}$ the following form.

Theorem 6.7. *Let $A \in \mathscr{I}(3,3)$. Then $\mathscr{F}^{<}(A) \neq \emptyset$ if and only if the following conditions are satisfied*

(i) $a_{12} < 1, a_{13} < 1, a_{23} < 1$,
(ii) $a_{22} = 1,$ *or* $a_{13} < a_{23}$,
(iii) $a_{33} = 1$.

Theorem 6.8. *Let $A \in \mathscr{I}(3,3)$. Then $\mathscr{F}^{<}(D_{12}, A) \neq \emptyset$ if and only if the following conditions are satisfied*

(i) $a_{13} < 1, a_{23} < 1$,
(ii) $\max(a_{11}, a_{12}) = 1,$ *or* $a_{13} \geq a_{23}$,
(iii) $\max(a_{21}, a_{22}) = 1,$ *or* $a_{13} \leq a_{23}$,
(iv) $a_{33} = 1$.

Theorem 6.9. *Let $A \in \mathscr{I}(3,3)$. Then $\mathscr{F}^{<}(D_{23}, A) \neq \emptyset$ if and only if the following conditions are satisfied*

(i) $a_{12} < 1, a_{13} < 1$,
(ii) $\max(a_{22}, a_{23}) = 1$,
(iii) $\max(a_{32}, a_{33}) = 1$.

The forms taken by Theorems 6.2 and 6.6 for dimension three are presented in the following three theorems.

Theorem 6.10. *Suppose $A \in \mathscr{I}(3,3)$ and (i)–(iii) of Theorem 6.7 are satisfied. Then $\mathscr{F}^{<}(A)$ consists exactly of all vectors $x \in \mathscr{I}^{<}(3)$ satisfying the conditions*

$$\text{if } a_{11} = 1,\ a_{22} = 1,\ \text{then } x_3 < 1,\ \text{or } x_3 = 1,\ x_1 \geq a_{13},\ x_2 \geq a_{23}\ , \tag{6.17}$$

$$\text{if } a_{11} < 1,\ a_{22} = 1,\ \text{then } x_3 = 1,\ x_1 = a_{13},\ x_2 \geq a_{23}\ , \tag{6.18}$$

$$\text{if } a_{11} = 1,\ a_{22} < 1,\ \text{then } x_3 = 1,\ x_1 \geq a_{13},\ x_2 = a_{23}\ , \tag{6.19}$$

$$\text{if } a_{11} < 1,\ a_{22} < 1,\ \text{then } x_3 = 1,\ x_1 = a_{13},\ x_2 = a_{23}\ . \tag{6.20}$$

Theorem 6.11. *Suppose $A \in \mathscr{I}(3,3)$ and that (i)–(iv) of Theorem 6.8 are satisfied. Then $\mathscr{F}^{<}(D_{12}, A)$ consists exactly of all vectors $x \in \mathscr{I}^{<}(D_{12}, 3)$ satisfying the conditions*

$$\text{if } \max(a_{11}, a_{12}) = 1,\ \max(a_{21}, a_{22}) = 1,\ \text{then}$$

$$x_3 < 1,\ \text{or } x_3 = 1,\ x_1 = x_2 \geq \max(a_{13}, a_{23})\ , \tag{6.21}$$

$$\text{if } \max(a_{11}, a_{12}) < 1,\ \text{or } \max(a_{21}, a_{22}) < 1,\ \text{then}$$

$$x_3 = 1,\ x_1 = x_2 = \max(a_{13}, a_{23})\ . \tag{6.22}$$

Theorem 6.12. *If $A \in \mathscr{I}(3,3)$ and (i)–(iii) of Theorem 6.9 hold, then $\mathscr{F}^{<}(D_{23}, A)$ consists exactly of all vectors $x \in \mathscr{I}^{<}(D_{23}, 3)$ satisfying the conditions*

$$\textit{if } a_{11} = 1, \ \textit{then } x_2 = x_3 < 1, \ \textit{or } x_2 = x_3 = 1, \ x_1 \geq \max(a_{12}, a_{13}) \ , \tag{6.23}$$

$$\textit{if } a_{11} < 1, \ \textit{then } x_2 = x_3 = 1, \ x_1 = \max(a_{12}, a_{13}) \ . \tag{6.24}$$

The computation of the complete eigenspace in a max-drast fuzzy algebra will now be illustrated for two three-dimensional matrices.

Example 6.3. Consider the matrix

$$A = \begin{pmatrix} 0.6 & 0.8 & 0.3 \\ 0.5 & 0.9 & 0.4 \\ 0.3 & 0.7 & 1 \end{pmatrix} .$$

It satisfies conditions (i), (ii) and (iii) of Theorem 6.7, hence $\mathscr{F}^{<}(A) \neq \emptyset$. Condition (6.20) in Theorem 6.10 implies that there is exactly one increasing eigenvector of A, the vector $(0.3, 0.4, 1)$.

Condition (ii) in Theorem 6.8 and condition (ii) in Theorem 6.9 are not satisfied, which implies that $\mathscr{F}^{<}(D_{12}, A) = \emptyset$ and $\mathscr{F}^{<}(D_{23}, A) = \emptyset$, i.e. A has no non-decreasing eigenvectors. By Theorem 6.3, we have that A has only one constant eigenvector $(0, 0, 0)$. By analogous considerations of all matrices $A_{\varphi\varphi}$ for permutations $\varphi \in P_3$ we find, in view of Theorem 5.1, that A has no further eigenvectors. Summarizing these results, we have

$$\mathscr{F}(A) = \mathscr{F}^{\leq}(A) = \{(0,0,0), (0.3, 0.4, 1)\} .$$

The eigenspace $\mathscr{F}(A)$ is shown in Fig. 6.2.

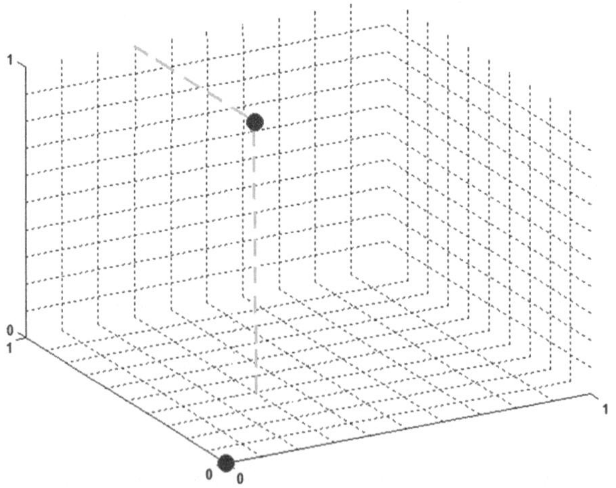

Fig. 6.2 Eigenspace of A from Example 6.3

Example 6.4. In this example we change the entry $a_{22} = 0.9$ to $b_{22} = 1$, and leave the other entries unchanged. First we compute the increasing eigenspace $\mathscr{F}^{\leq}(B) = \mathscr{F}^{<}(B) \cup \mathscr{F}^{<}(D_{12}, B) \cup \mathscr{F}^{<}(D_{23}, B) \cup \mathscr{F}^{=}(B)$ of

$$B = \begin{pmatrix} 0.6 & 0.8 & 0.3 \\ 0.5 & 1 & 0.4 \\ 0.3 & 0.7 & 1 \end{pmatrix} .$$

Similarly to the above, conditions (i), (ii) and (iii) of Theorem 6.7 are satisfied by B, hence $\mathscr{F}^{<}(B) \neq \emptyset$. Condition (6.18) in Theorem 6.10 implies that the increasing eigenvectors of B are exactly the increasing vectors with $x_3 = 1$, $x_1 = a_{13}$, $x_2 \geq a_{23}$. Thus,

$$\mathscr{F}^{<}(B) = \{(0.3, x_2, 1) |\, 0.4 \leq x_2 < 1\} .$$

Condition (ii) in Theorem 6.8 is not satisfied, hence $\mathscr{F}^{<}(D_{12}, B) = \emptyset$. On the other hand, all three conditions (i), (ii), (iii) in Theorem 6.9 hold true, i.e. $\mathscr{F}^{<}(D_{23}, B) \neq \emptyset$ and by condition (6.49) in Theorem 6.12, we get

$$\mathscr{F}^{<}(D_{23}, B) = \{(0.8, 1, 1)\} .$$

Finally, Theorem 6.3 implies that $\mathscr{F}^{=}(B) = \{(0, 0, 0)\}$. Summarizing, we have

$$\mathscr{F}^{\leq}(B) = \{(0.3, x_2, 1) |\, 0.4 \leq x_2 < 1\} \cup \{(0, 0, 0), (0.8, 1, 1)\} .$$

Further eigenvectors of matrix B can be found using Theorem 5.1. Applying the permutation $\varphi = \begin{pmatrix} 1 & 2 & 3 \\ 1 & 3 & 2 \end{pmatrix}$ to the rows and columns of B, we get the matrix $B_{\varphi\varphi}$, which satisfies conditions (i), (ii) and (iii) of Theorem 6.7:

$$B_{\varphi\varphi} = \begin{pmatrix} 0.6 & 0.3 & 0.8 \\ 0.3 & 1 & 0.7 \\ 0.5 & 0.4 & 1 \end{pmatrix} .$$

For simplicity, we use the notation in which the permuted vector x_φ is denoted by $y = (y_1, y_2, y_3) = (x_{\varphi(1)}, x_{\varphi(2)}, x_{\varphi(3)}) = (x_1, x_3, x_2)$.

Using Theorems 6.10–6.12 and Theorem 6.3, we find, analogously as with B, that

$$\mathscr{F}^{<}(B_{\varphi\varphi}) = \{(y_1, y_2, y_3) \in \mathscr{I}^{<}(3) |\, y_1 = 0.8 < y_2 < 1 = y_3\} ,$$

$$\mathscr{F}^{<}(D_{12}, B_{\varphi\varphi}) = \{(y_1, y_2, y_3) \in \mathscr{I}^{<}(D_{12}, 3) |\, y_1 = y_2 = 0.8, y_3 = 1\} ,$$

$$\mathscr{F}^{<}(D_{23}, B_{\varphi\varphi}) = \{(y_1, y_2, y_3) \in \mathscr{I}^{<}(D_{23}, 3) |\, y_1 = 0.8, y_2 = y_3 = 1\} ,$$

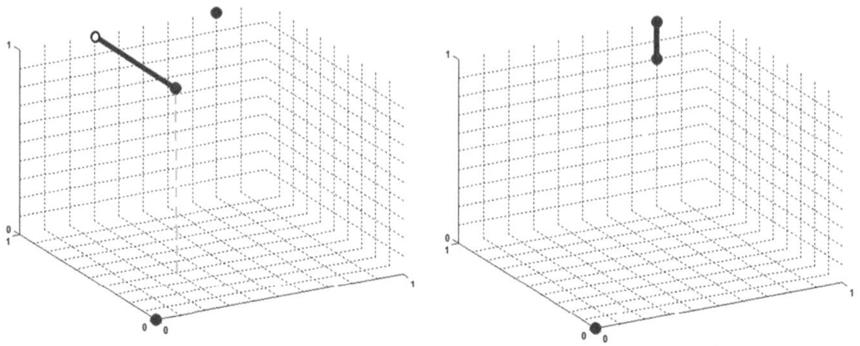

Fig. 6.3 Partial eigenspaces $\mathcal{F}^{\leq}(B)$ and $\mathcal{F}_{\varphi}^{\leq}(B)$ from Example 6.4

$$\mathcal{F}^{=}(B_{\varphi\varphi}) = \{(0,0,0)\} ,$$

$$\mathcal{F}^{\leq}(B_{\varphi\varphi}) = \{(y_1, y_2, y_3) \in \mathcal{I}^{\leq}(3) | y_1 = 0.8 \leq y_2 \leq 1 = y_3\} \cup \{(0,0,0)\} .$$

In the original notation, $x_\varphi = (x_1, x_3, x_2) = (y_1, y_2, y_3)$, we have

$$\mathcal{F}_{\varphi}^{\leq}(B) = \{(0.8, 1, x_3) \in \mathcal{I}_{\varphi}^{\leq}(3) | 0.8 \leq x_3 \leq 1\} \cup \{(0,0,0)\} .$$

The eigenspaces $\mathcal{F}^{\leq}(B)$ and $\mathcal{F}_{\varphi}^{\leq}(B)$ are shown in Fig. 6.3.

If the permutation $\psi = \begin{pmatrix} 1 & 2 & 3 \\ 3 & 1 & 2 \end{pmatrix}$ is applied to the rows and columns of B, then conditions (i), (ii) and (iii) of Theorem 6.7 are satisfied by the resulting matrix

$$B_{\psi\psi} = \begin{pmatrix} 1 & 0.3 & 0.7 \\ 0.3 & 0.6 & 0.8 \\ 0.4 & 0.5 & 1 \end{pmatrix} .$$

Analogously to the above, we use the notation in which the permuted vector x_ψ is denoted by $z = (z_1, z_2, z_3) = (x_{\psi(1)}, x_{\psi(2)}, x_{\psi(3)}) = (x_3, x_1, x_2)$. Here, $B_{\psi\psi}$ does not satisfy condition (ii) in Theorem 6.9, therefore $\mathcal{F}^{<}(D_{23}, B_{\psi\psi}) = \emptyset$. Furthermore, we obtain, using Theorems 6.10–6.11 and Theorem 6.3, that

$$\mathcal{F}^{<}(B_{\psi\psi}) = \{(z_1, z_2, z_3) \in \mathcal{I}^{<}(3) | 0.7 \leq z_1 < z_2 = 0.8, z_3 = 1\} ,$$

$$\mathcal{F}^{<}(D_{12}, B_{\psi\psi}) = \{(z_1, z_2, z_3) \in \mathcal{I}^{<}(D_{12}, 3) | z_1 = z_2 = 0.8, z_3 = 1\} ,$$

$$\mathcal{F}^{=}(B_{\psi\psi}) = \{(0,0,0)\} ,$$

which implies

$$\mathscr{F}^{\leq}(B_{\psi\psi}) = \{(z_1, z_2, z_3) \in \mathscr{I}^{\leq}(3) | 0.7 \leq z_1 \leq 0.8 = z_2, z_3 = 1\} \cup \{(0,0,0)\} \ .$$

In the original notation, $x_\psi = (x_3, x_1, x_2) = (z_1, z_2, z_3)$, we get

$$\mathscr{F}^{\leq}_{\psi}(B) = \{(0.8, 1, x_3) \in \mathscr{I}^{\leq}_{\psi}(3) | 0.7 \leq x_3 \leq 0.8\} \cup \{(0,0,0)\} \ .$$

It is easy to verify that none of the remaining permutations

$$\begin{pmatrix} 1 & 2 & 3 \\ 2 & 3 & 1 \end{pmatrix}, \begin{pmatrix} 1 & 2 & 3 \\ 3 & 2 & 1 \end{pmatrix}, \begin{pmatrix} 1 & 2 & 3 \\ 2 & 1 & 3 \end{pmatrix}$$

fulfills the conditions in Theorems 6.7–6.9. Therefore no further non-zero eigenvectors can be found using these permutations. As a consequence, the eigenspace of B in the max-drast fuzzy algebra is equal to

$$\mathscr{F}(B) = \mathscr{F}^{\leq}(B) \cup \mathscr{F}^{\leq}_{\varphi}(B) \cup \mathscr{F}^{\leq}_{\psi}(B) \ . \tag{6.25}$$

The eigenspace $\mathscr{F}^{\leq}_{\psi}(B)$ and the complete eigenspace $\mathscr{F}(B)$ are shown in Fig. 6.4. In addition, several particular max-drast fuzzy eigenvectors of matrix B are listed below as an illustration of the final formula (6.25)

$\mathscr{F}^{\leq}(B)$ $x_1 \leq x_2 \leq x_3$	$\mathscr{F}^{\leq}_{\varphi}(B)$ $x_1 \leq x_3 \leq x_2$	$\mathscr{F}^{\leq}_{\psi}(B)$ $x_3 \leq x_1 \leq x_2$
$(0, 0, 0)$	$(0, 0, 0)$	$(0, 0, 0)$
$(0.3, 0.4, 1)$	$(0.8, 1, 0.8)$	$(0.8, 1, 0.7)$
$(0.3, 0.99, 1)$	$(0.8, 1, 0.99)$	$(0.8, 1, 0.79)$
$(0.8, 1, 1)$	$(0.8, 1, 1)$	$(0.8, 1, 0.8)$

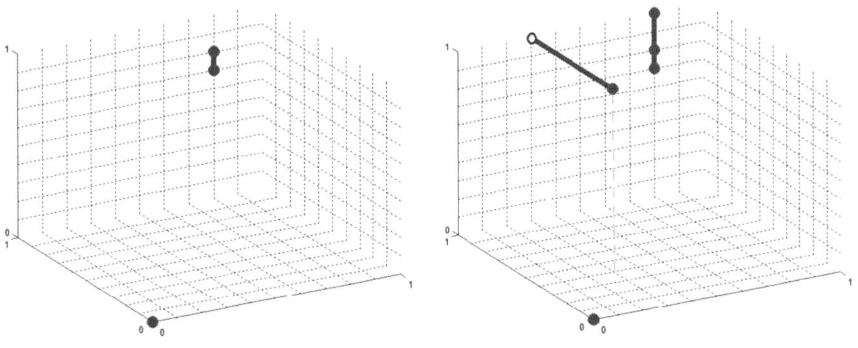

Fig. 6.4 Partial eigenspace $\mathscr{F}^{\leq}_{\psi}(B)$ and complete eigenspace $\mathscr{F}(B)$ from Example 6.4

6.4 Eigenvectors in Max-Łukasiewicz Algebra

In this section, the above considerations will be transferred to the case of max-Łukasiewicz algebra, which is a special case of max-T fuzzy algebra with Łukasiewicz triangular norm. Thus, in Sect. 6.4 we work with the max-L fuzzy algebra $(\mathscr{I}, \oplus, \otimes_l)$ with binary operation $\oplus = \max$ and $\otimes_l = L = T_L$. That is, for every $x, y \in \mathscr{I}(n)$ and for every $i \in N$ we have

$$(x \oplus y)_i = \max(x_i, y_i) \ ,$$

$$(x \otimes_l y)_i = L(x_i, y_i) = \begin{cases} x_i + y_i - 1 & \text{if } \min(x_i + y_i - 1, 0) = 0 \ , \\ 0 & \text{otherwise} \ . \end{cases}$$

The following proposition contains several logical consequences of the definition of Łukasiewicz triangular norm which will be used in development of the further theory.

Proposition 6.7. *Let $a, b, c \in [0, 1]$. Then*

(i) $a \otimes_l b = b$ *if and only if* $a = 1$ *or* $b = 0$,
(ii) $a \otimes_l c = b$ *if and only if* $a = 1 + b - c$ *or* ($a \le 1 - c$ *and* $b = 0$) ,
(iii) $a \otimes_l c \le b$ *if and only if* $a \le 1 + b - c$,
(iv) $a \otimes_l c > b$ *if and only if* $a > 1 + b - c$,
(v) *If* $c < b$ *then* $a \otimes_l c < b$.

Proof. (i) Let $a \otimes_l b = b$. We have either $0 \le a + b - 1$ or $a + b - 1 \le 0$. That is, by definition of Łukasiewicz triangular norm, either $a = 1$ or $b = 0$. For the converse implication, first consider that $a = 1$. Then $a + b - 1 = 1 + b - 1 = b \ge 0$ implies that $a \otimes_l b = b$. If $b = 0$, then $a + b - 1 = a - 1 \le 0$. Hence $a \otimes_l b = b$ follows again from the definition.

(ii) Let $a \otimes_l c = b$. By definition, either $0 \le a + c - 1 = b$ or $a + c - 1 \le 0 (= b)$. That is, either $a = 1 + b - c$ or ($a \le 1 - c$ and $b = 0$). Conversely, let $a = 1 + b - c$ or ($a \le 1 - c$ and $b = 0$). Then either $a + c - 1 = b \ge 0$ or ($a + c - 1 \le 0$ and $b = 0$). It follows directly from the definition that in either case $a \otimes_l c = b$.

(iii) By definition $a \otimes_l c \le b$ implies that, either $a + c - 1 \le 0 (\le b)$ or $0 \le a + c - 1 \le b$. It follows directly that in either case $a \le 1 + b - c$. Conversely, $a \le 1 + b - c$ implies $a + b - 1 \le b$. Also, $0 \le b$. Then by definition $a \otimes_l c \le b$.

(iv) The proof is analogous to (iii).

(v) Let $0 \le c < b$. That is $b > 0$. As $a \le 1$ then $a + c < 1 + b$ implies that $a + c - 1 < b$. Hence, $a \otimes_l c < b$ follows from the definition.

□

Proposition 6.8. *Let $A \in \mathscr{I}(n, n)$, $x \in \mathscr{I}(n)$. Then $x \in \mathscr{F}(A)$ if and only if for every $i \in N$ the following hold*

$$a_{ij} \otimes_l x_j \leq x_i \quad \textit{for every} \quad j \in N \ , \tag{6.26}$$

$$a_{ij} \otimes_l x_j = x_i \quad \textit{for some} \quad j \in N \ . \tag{6.27}$$

Proof. By definition, $x \in \mathscr{F}(A)$ is equivalent with the condition $\max_{j \in N} a_{ij} \otimes_l x_j \leq x_i$ for every $i \in N$, which is equivalent to (1) and (2). □

Proposition 6.9. *Let $A \in \mathscr{I}(n,n)$, $x \in \mathscr{I}^<(n)$. Then $x \in \mathscr{F}^<(A)$ if and only if for every $i \in N$ the following hold*

$$a_{ij} \otimes_l x_j \leq x_i \quad \textit{for every} \quad j \in N, \quad j \geq i \ , \tag{6.28}$$

$$a_{ij} \otimes_l x_j = x_i \quad \textit{for some} \quad j \in N, \quad j \geq i \ . \tag{6.29}$$

Proof. By Proposition 6.7(v), $a_{ij} \otimes_l x_j < x_i$ for every $j < i$, $x_j < x_i$. Hence the terms with $j < i$ in (1) and (2) of Proposition 6.8 can be left out. □

Theorem 6.13. *Let $A \in \mathscr{I}(n,n)$ and $x \in \mathscr{I}^<(n)$. Then $x \in \mathscr{F}^<(A)$ if and only if for every $i \in N$ the following hold*

(i) $a_{ij} \leq 1 + x_i - x_j$ *for every* $j \in N$, $j \geq i$,
(ii) *if* $i = 1$, *then* $x_1 = 0$ *and* $a_{1j} \leq 1 - x_j$, *or there exists* $j \in N$ *such that* $a_{1j} = 1 + x_1 - x_j$,
(iii) *if* $i > 1$, *then* $a_{ij} = 1 + x_i - x_j$ *for some* $j \in N$, $j \geq i$.

Proof. Suppose that $x \in \mathscr{F}^<(A)$, that is $A \otimes_l x = x$. Then $a_{ij} \otimes_l x_j \leq x_i$, for every $j \in N, j \geq i$ by Proposition 6.7(iii) gives $a_{ij} \leq 1 + x_i - x_j$. If $i = 1$ then $a_{1j} \otimes_l x_j = x_1$, for some $j \in N$. Then by Proposition 6.7(ii), we have $x_1 = 0$ and $a_{1j} \leq 1 - x_j$, or $a_{1j} = 1 + x_1 - x_j$, for some $j \in N$. To prove (iii), consider $a_{ij} \otimes_l x_j = x_i$ for some $j \in N$, $j \geq i > 1$. By definition we have, either $a_{ij} + x_j - 1 \leq 0$ or $0 \leq a_{ij} + x_j - 1$. The case $a_{ij} \otimes_l x_j = 0 = x_i$ is not possible, as for $i > 1$, $x_i > 0$. Then we must have $0 \leq a_{ij} + x_j - 1 = x_i$, that is $a_{ij} = 1 + x_i - x_j$.

Conversely, suppose that conditions (i), (ii), (iii) hold true. We show that $x \in \mathscr{F}^<(A)$, that is $A \otimes_l x = x$. In other words, $\max_{j \in N} a_{ij} \otimes_l x_j = x_i$ for every $i \in N$. Let $i \in N$ be fixed. By (i) and Proposition 6.7(iii), $a_{ij} \otimes_l x_j \leq x_i$, for every $j \in N, j \geq i$. If $i = 1$ then by (ii) and Proposition 6.7(ii), $a_{1j} \otimes_l x_j = x_1$ for some $j \in N$. If $i > 1$ then by (iii) $a_{ij} = 1 + x_i - x_j$, for some $j \in N$, $j \geq i > 1$, we have $a_{ij} + x_j - 1 = x_i > 0$, because $i > 1$. This implies $\max(0, a_{ij} + x_j - 1) = x_i$, that is $a_{ij} \otimes_l x_j = x_i$. Hence $\max_{j \in N} a_{ij} \otimes_l x_j = x_i$ for every $i \in N$, that is $A \otimes_l x = x$ or $x \in \mathscr{F}^<(A)$. □

The following theorem describes necessary conditions under which a given square matrix can have a increasing eigenvector.

Theorem 6.14. *Let $A \in \mathscr{I}(n,n)$. If $\mathscr{F}^<(A) \neq \emptyset$, then the following conditions are satisfied*

(i) $a_{ij} < 1$ *for all* $i, j \in N, i < j$,
(ii) $a_{nn} = 1$.

Proof. Let $\mathscr{F}^{<}(A) \neq \emptyset$. That is, there exists $x \in \mathscr{F}^{<}(A)$ such that conditions of Theorem 6.13 hold true. Condition $a_{ij} < 1$ follows directly from (i) and $a_{nn} = 1$ from (iii) of Theorem 6.13. □

Remark 6.5. Generally speaking, the conditions in Theorem 6.14 are only necessary. It can be easily seen that in the case $n = 2$ the necessary conditions in Theorem 6.14 are also sufficient. Namely, if $n = 2$, then we have two conditions: $a_{12} < 1$ and $a_{22} = 1$. Then, arbitrary vector with $x_1 = 0$ and $0 < x_2 \leq 1 - a_{12}$ fulfills the conditions of Theorem 6.13. Further increasing eigenvectors in the case $a_{11} < 1$ are of the form (x_1, x_2) with $0 < x_1 \leq a_{12}$ and $x_2 = x_1 + 1 - a_{12}$. In the case when $a_{11} = 1$, variable x_1 can take arbitrary values from the interval $(0, 1)$ and $x_1 < x_2 \leq \min(1, 1 + x_1 - a_{12})$. The result is completely formulated in the following theorem.

Theorem 6.15. *Let* $A \in \mathscr{I}(2,2)$. *Then* $\mathscr{F}^{<}(A) \neq \emptyset$ *if and only if the* $a_{12} < 1$ *and* $a_{22} = 1$. *If this is the case, then either*

(i) $a_{11} < 1$ *and* $\mathscr{F}^{<}(A) = \{(x_1, x_2) \in \mathscr{I}(2,2) | \, x_1 = 0, \; x_2 \in (0, 1 - a_{12}]\}$
$\cup \{(x_1, x_2) \in \mathscr{I}(2,2) | \, x_1 \in (0, a_{12}], \; x_2 = 1 + x_1 - a_{12}\}$,

or

(ii) $a_{11} = 1$ *and* $\mathscr{F}^{<}(A) = \{(x_1, x_2) \in \mathscr{I}(2,2) | \, x_1 = 0, \; x_2 \in (0, 1 - a_{12}]\}$
$\cup \{(x_1, x_2) \in \mathscr{I}(2,2) | \, x_1 \in (0, 1], \; x_1 < x_2 \leq \min(1, 1 + x_1 - a_{12})\}$.

Proof. The statement follows from the arguments in Remark 6.5. □

In the next theorem, a necessary and sufficient condition for the existence of a non-zero constant eigenvector is presented. The set of all constant eigenvectors of a matrix A is denoted by $\mathscr{F}^{=}(A)$.

Theorem 6.16. *Let* $A \in \mathscr{I}(n,n)$, *then there is a non-zero constant eigenvector* $x \in \mathscr{F}^{=}(A)$ *if and only if*

(i) $\max\{a_{ij} | \, j \in N\} = 1$ *for every* $i \in N$.

Proof. Let $x = (c, c, \ldots, c) \in \mathscr{I}(n)$ with $c > 0$ be a constant eigenvector of A. Then (i) follows from the conditions (6.26), (6.27) in Proposition 6.8. Conversely, if (i) is satisfied, then clearly the conditions (6.26), (6.27) hold true for the unit constant vector $u = (1, 1, \ldots, 1) \in \mathscr{I}(n)$. Hence $u \in \mathscr{F}^{=}(a)$. □

Theorem 6.17. *Let* $A \in \mathscr{I}(n,n)$. *If the condition (i) of Theorem 6.16 is satisfied, then* $\mathscr{F}^{=}(A) = \{(c, c, \ldots, c) | \, c \in \mathscr{I}\}$. *If* A *does not satisfy the condition (i), then* $\mathscr{F}^{=}(A) = \{(0, 0, \ldots, 0)\}$.

Proof. It is easy to verify that if A satisfies the condition (i), then the conditions (6.26), (6.27) hold true for an arbitrary constant vector $(c, c, \ldots, c)$ with $c \in \mathscr{I}$, hence $\mathscr{F}^{=}(A) = \{(c, c, \ldots, c) | \, c \in \mathscr{I}\}$.

On the other hand, let us assume that the condition (i) is not satisfied. Clearly, the zero constant vector $(0, 0, , \ldots, 0)$ fulfills the conditions (6.26), (6.27). Hence we have $\{(0, 0, , \ldots, 0)\} \subseteq \mathscr{F}^{=}(A)$. The equality $\mathscr{F}^{=}(A) = \{(0, 0, , \ldots, 0)\}$ follows from Theorem 6.16. □

Example 6.5. An application of Theorems 6.15 and Theorem 6.17 is presented by computing the increasing eigenspace of

$$A = \begin{pmatrix} 0.7 & 0.3 \\ 0.2 & 1 \end{pmatrix} ,$$

and its permuted counterpart,

$$A_{\varphi\varphi} = \begin{pmatrix} 1 & 0.2 \\ 0.3 & 0.7 \end{pmatrix} .$$

The permuted matrix $A_{\varphi\varphi}$ is used to compute the decreasing eigenvectors of A. As a result of the computation, we get that the eigenspace $\mathscr{F}(A)$ contains

- increasing vectors with $0 = x_1 < x_2 \leq 0.7$,
- increasing vectors with $0 < x_1 \leq 0.3, x_2 = 0.7 + x_1$,
- no decreasing vectors ,
- the constant vector $x = (0,\ 0)$.

Example 6.6. In this example, we compute the increasing eigenspace of

$$B = \begin{pmatrix} 1 & 0.3 \\ 0.2 & 1 \end{pmatrix} ,$$

and its permuted counterpart

$$B_{\varphi\varphi} = \begin{pmatrix} 1 & 0.2 \\ 0.3 & 1 \end{pmatrix} .$$

The eigenspace $\mathscr{F}(B)$ consists of

- increasing vectors with $0 = x_1 < x_2 \leq 0.7$,
- increasing vectors with $0 < x_1 \leq 1, x_1 < x_2 \leq \min(1, 0.7 + x_1)$,
- decreasing vectors with $0 = x_2 < x_1 \leq 0.8$,
- decreasing vectors with $0 < x_2 \leq 1, x_2 < x_1 \leq \min(1, 0.8 + x_2)$,
- constant vectors $x = (c,\ c)$ with $0 \leq c \leq 1$.

The eigenspaces of A and B are shown in Fig. 6.5.

Remark 6.6. Description of non-decreasing eigenvectors is necessary for computing of the general eigenspace of dimension $n = 3$ and higher. This extension of the presented theory is given in the next two sections for three-dimensional matrices.

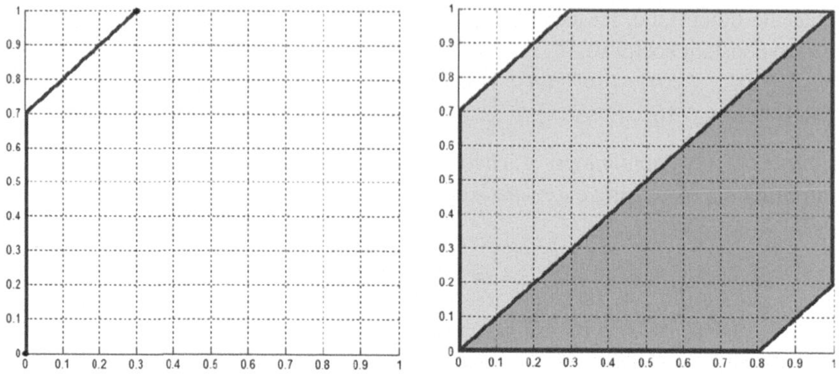

Fig. 6.5 Eigenspaces of A and B from Example 6.5 and 6.6

Moreover, the necessary conditions from Theorem 6.14 are extended to necessary and sufficient ones.

Remark 6.7. The max-L fuzzy algebra is closely related to the max-plus algebra (also called: tropical algebra) with operations $\oplus = \max$ and $\otimes = +$ defined on the set of all real numbers. If we denote by E the $n \times n$ matrix with all inputs equal to 1, and symbols $\mathscr{F}_+$ ($\mathscr{F}_l$) denote the eigenspace of a given matrix in max-plus (max-L) algebra, then the following theorem holds true.

Theorem 6.18. *Let $A \in \mathscr{I}(n,n)$. If a vector $x \in \mathscr{I}(n)$ is max-plus eigenvector of the matrix $A - E$, then x is max-L eigenvector of A. In formal notation,*

$$\mathscr{F}_+(A - E) \cap \mathscr{I}(n) \subseteq \mathscr{F}_L(A) \ .$$

Proof. Assume that $x \in \mathscr{I}(n)$ satisfies the equation $(A - E) \otimes x = x$, i.e. the max-plus equality

$$\bigoplus_{j \in N} (a_{ij} - 1) \otimes x_j = x_i$$

holds for every $i \in N$. Adding the expression $\oplus\ 0$ to both sides, and expressing the operation $\otimes$ explicitly as addition, we get by easy computation

$$\bigoplus_{j \in N} (a_{ij} - 1 + x_j) \oplus 0 = \bigoplus_{j \in N} \big((a_{ij} - 1 + x_j) \oplus 0\big) = \bigoplus_{j \in N} a_{ij} \otimes_l x_j = x_i \oplus 0 = x_i \ .$$

Hence, $A \otimes_L x = x$. □

Now we consider increasing eigenvectors in the three-dimensional eigenproblem in the max-Łukasiewicz fuzzy algebra. In other words, we assume $n = 3$ and we work with matrices in $\mathscr{I}(3,3)$ and vectors in $\mathscr{I}(3)$. The previous results are extended and a complete description of the increasing eigenspace is given.

The following theorem describes the necessary and sufficient conditions under which a three-dimensional fuzzy matrix has an increasing eigenvector.

Theorem 6.19. *Let $A \in \mathscr{I}(3,3)$. Then $\mathscr{F}^<(A) \neq \emptyset$ if and only if the following conditions are satisfied*

(i) $a_{12} < 1$, $a_{13} < 1$, $a_{23} < 1$,
(ii) $a_{22} = 1$, *or* $a_{13} < a_{23}$,
(iii) $a_{33} = 1$.

Proof. Let $\mathscr{F}^<(A) \neq \emptyset$, i.e. there exists $x \in \mathscr{F}^<(A)$. The conditions (i) and (iii) follow directly from Theorem 6.13. To prove the condition (ii), let us assume that $a_{22} < 1$. Then by (iii) of Theorem 6.13 we get $a_{22} = 1+x_2-x_2$ or $a_{23} = 1+x_2-x_3$. The first equation implies $a_{22} = 1$, which is a contradiction. Therefore we must have $a_{23} = 1 + x_2 - x_3 < 1$. By (i) of Theorem 6.13, we have $a_{13} \leq 1 + x_1 - x_3 = 1 + (x_1 - x_2) + (x_2 - x_3) = (1 + x_2 - x_3) + (x_1 - x_2) < 1 + x_2 - x_3 = a_{23}$. This implies $a_{13} < a_{23}$.

Conversely, suppose that conditions (i), (ii) and (iii) are satisfied. To show that $\mathscr{F}^<(A) \neq \emptyset$, we consider two cases.

Case 1. If $a_{22} < 1$, then put $x_1 = 0$ and choose $x_2 \leq \min(1 - a_{12}, a_{23} - a_{13})$ and $0 < x_2 < 1 - a_{13}$, and put $x_3 = x_2 + (1 - a_{23})$. By our assumption, $1-a_{12} > 0$, $a_{23}-a_{13} > 0$ and $1-a_{13} > 0$. Therefore the choice of x_2 fulfilling the conditions $x_2 \leq \min(1-a_{12}, a_{23}-a_{13})$ and $0 < x_2 < 1-a_{13}$ is always possible. Also by assumption, $a_{23} < 1$ implies that $1-a_{23} > 0$. Therefore $x_3 = x_2+(1-a_{23}) > x_2$. Moreover, $x_2 \leq \min(1 - a_{12}, a_{23} - a_{13}) \leq a_{23} - a_{13} \leq a_{23}$, i.e. $x_2 \leq a_{23}$. From this we have $x_3 = x_2 + (1 - a_{23}) \leq 1$, i.e. $x_3 \leq 1$. This shows that $x \in \mathscr{I}^<(3)$. To show that $x \in \mathscr{F}^<(A)$, consider $x_2 \leq \min(1 - a_{12}, a_{23} - a_{13})$ and $0 < x_2 < 1 - a_{13}$. This implies that $x_2 \leq 1 - a_{12}$ and $0 < x_2 < 1 - a_{13}$, or equivalently, $a_{12} \leq 1 - x_2$ and $a_{13} < 1 - x_2$, which implies condition (i) of Theorem 6.13. Choice of $x_1 = 0$ satisfies the condition (ii) of Theorem 6.13. Also, $x_3 = x_2 + (1 - a_{23})$, $a_{23} = 1 + x_2 - x_3$ imply condition (iii) of Theorem 6.13. Hence $\mathscr{F}^<(A) \neq \emptyset$.

Case 2. If $a_{22} = 1$, then put $x_1 = 0$, choose $0 < x_2 < \min(1 - a_{12}, 1 - a_{13})$ and choose x_3 such that $x_2 < x_3 \leq \min(1 - a_{13}, x_2 + (1 - a_{23}))$. The choice $0 < x_2 < \min(1 - a_{12}, 1 - a_{13})$ is always possible because by our assumption $1 - a_{12} > 0$ and $1 - a_{13} > 0$. Also $1 - a_{23} > 0$ by the same argument and therefore $x \in \mathscr{I}^<(3)$. Consider $x_2 < \min(1-a_{12}, 1-a_{13}) \leq 1-a_{12}$. Then $x_2 < 1-a_{12}$ implies that $a_{12} < 1 - x_2$. Similarly $a_{13} < 1 - x_2$ and by $x_3 \leq \min(1 - a_{13}, x_2 + (1 - a_{23}))$ we have $a_{23} \leq 1 + x_2 - x_3$, showing that condition (i) of Theorem 6.13 is satisfied. Conditions (ii) and (iii) of Theorem 6.13 are satisfied by the choice of $x_1 = 0$ and assumption $a_{22} = 1$, respectively. Hence $\mathscr{F}^<(A) \neq \emptyset$. □

Theorem 6.20. *Let $A \in \mathscr{I}(3,3)$ satisfy conditions (i), (ii) and (iii) of Theorem 6.19. Then $x \in \mathscr{F}^<(A)$ if and only if $x \in \mathscr{I}^<(3)$ and either $x_1 = 0$ and conditions* *if* $a_{22} < 1$, *then*

$$0 < x_2 \le \min(1 - a_{12}, a_{23} - a_{13}),\ x_2 < 1 - a_{13},\ x_3 = x_2 + (1 - a_{23})\ , \tag{6.30}$$

if $a_{22} = 1$, *then*

$$0 < x_2 < \min(1 - a_{12}, 1 - a_{13}),\ x_2 < x_3 \le \min\bigl(1 - a_{13}, x_2 + (1 - a_{23})\bigr)\ , \tag{6.31}$$

are satisfied, or $x_1 > 0$ *and conditions*

if $a_{11} = 1, a_{22} = 1$, *then* $x_1 < x_2 < 1$,

$$x_2 \le x_1 + (1 - a_{12}),\ x_3 \le \min\bigl(x_1 + (1 - a_{13}), x_2 + (1 - a_{23}), 1\bigr)\ , \tag{6.32}$$

if $a_{11} = 1, a_{22} < 1$, *then* $x_1 < a_{23}$,

$$x_2 \le \min\bigl(a_{23}, x_1 + (1 - a_{12}), x_1 + (a_{23} - a_{13})\bigr),\ x_3 = x_2 + (1 - a_{23})\ , \tag{6.33}$$

if $a_{11} < 1, a_{22} = 1$, *then*

$$0 < x_1 < a_{12},\ x_2 = x_1 + (1 - a_{12})\ ,$$
$$x_3 \le \min\bigl(x_1 + 2 - (a_{12} + a_{23}), x_1 + (1 - a_{13}), 1\bigr),\ \textit{or}$$
$$0 < x_1 \le a_{13},\ x_1 + (a_{23} - a_{13}) \le x_2 < x_1 + (1 - a_{12})\ ,$$
$$x_3 = x_1 + (1 - a_{13})\ , \tag{6.34}$$

if $a_{11} < 1, a_{22} < 1$, *then*

$$a_{12} - a_{13} + a_{23} \ge 1,\ 0 < x_1 \le a_{12} + a_{23} - 1,\ x_2 = x_1 + (1 - a_{12})\ ,$$
$$x_3 = x_1 + 2 - (a_{12} + a_{23}),\ \textit{or}$$
$$a_{12} - a_{13} + a_{23} \le 1,\ 0 < x_1 \le a_{13},\ x_2 = x_1 + (a_{23} - a_{13})\ ,$$
$$x_3 = x_1 + (1 - a_{13}) \tag{6.35}$$

are satisfied.

Proof. Let $A \in \mathcal{I}(3,3)$ satisfy conditions of Theorem 6.19. For convenience, we shall use notation $\mathcal{F}_0^<(A) = \{x \in \mathcal{F}^<(A) | \, x_1 = 0\}$, $\mathcal{F}_1^<(A) = \{x \in \mathcal{F}^<(A) | \, x_1 > 0\}$. Thus, we have $\mathcal{F}^<(A) = \mathcal{F}_0^<(A) \cup \mathcal{F}_1^<(A)$. The assertions of the theorem for $x \in \mathcal{F}_0^<(A)$ and for $x \in \mathcal{F}_1^<(A)$ will be considered separately.

CASE 1 ($x_1 = 0$)

Let $x = (0, x_2, x_3) \in \mathcal{I}^<(3)$ satisfy conditions (6.30) and (6.31). Using the same arguments as in the converse part of Theorem 6.19 we can easily show that $x \in \mathcal{F}_0^<(A)$.

For the proof of the converse implication, suppose that $x \in \mathscr{F}_0^<(A)$. This implies that conditions of Theorem 6.13 are fulfilled. We consider two subcases $a_{22} < 1$ and $a_{22} = 1$.

Subcase 1a. If $a_{22} < 1$, condition (i) of Theorem 6.13 gives $a_{12} \le 1 + x_1 - x_2$, $a_{13} \le 1 + x_1 - x_3$ and $a_{23} \le 1 + x_2 - x_3$. Since $x_1 = 0$ by our assumption, therefore we have $x_2 \le 1 - a_{12}$, $x_3 \le 1 - a_{13}$ and $x_3 \le 1 + x_2 - a_{23}$. Condition (iii) of Theorem 6.13 gives $a_{23} = 1 + x_2 - x_3$, which implies that $x_3 = x_2 + 1 - a_{23}$. By $x_3 = x_2 + 1 - a_{23}$ and $x_3 \le 1 - a_{13}$, we have $x_2 \le a_{23} - a_{13}$. Also, $x_2 < x_3 \le 1 - a_{13}$ implies that $x_2 < 1 - a_{13}$. Therefore we have $0 < x_2 \le \min(1 - a_{12}, a_{23} - a_{13})$, $x_2 < 1 - a_{13}$, $x_3 = x_2 + 1 - a_{23}$.

Subcase 1b. If $a_{22} = 1$, condition (i) of Theorem 6.13 gives $x_2 \le 1 - a_{12}$, $x_3 \le 1 - a_{13}$ and $x_3 \le 1 + x_2 - a_{23}$. Since $0 < x_2 < 1$, then we must have $x_2 < 1 - a_{12}$. Also $x_2 < x_3 \le 1 - a_{13}$ implies that $x_2 < 1 - a_{13}$. Therefore we have $0 < x_2 < \min(1 - a_{12}, 1 - a_{13})$, $x_2 < x_3 \le \min(1 - a_{13}, x_2 + 1 - a_{23})$.

CASE 2 ($x_1 > 0$)

Let us assume that vector $x = (x_1, x_2, x_3) \in \mathscr{I}^<(3)$ with $x_1 > 0$ satisfies conditions (6.32)–(6.35). There are four possible subcases and we show that the conditions (i), (ii) and (iii) of Theorem 6.13 in each subcase are satisfied. In other words, we show that $x \in \mathscr{F}_1^<(A)$.

Subcase 2a. Let $a_{11} = 1$, $a_{22} = 1$. Then from $x_2 \le x_1 + 1 - a_{12}$ we have $a_{12} \le 1 + x_1 - x_2$. Also, $x_3 \le \min(x_1 + 1 - a_{13}, x_2 + 1 - a_{23}, 1) \le x_1 + 1 - a_{13}$ implies $a_{13} \le 1 + x_1 - x_3$, and $x_3 \le \min(x_1 + 1 - a_{13}, x_2 + 1 - a_{23}, 1) \le x_2 + 1 - a_{23}$ implies $a_{23} \le 1 + x_2 - x_3$. That is, (i) of Theorem 6.13 is satisfied. Conditions (ii) and (iii) of Theorem 6.13 are satisfied by the assumption $a_{11} = 1$ and $a_{22} = 1$, respectively.

Subcase 2b. Let $a_{11} = 1$, $a_{22} < 1$. Then from $x_3 = x_2 + 1 - a_{23}$ we have $a_{23} = 1 + x_2 - x_3$. Also, $x_2 \le \min(a_{23}, x_1 + 1 - a_{12}, x_1 + (a_{23} - a_{13}))$ implies $x_2 \le x_1 + 1 - a_{12}$ and $x_2 \le x_1 + (a_{23} - a_{13})$. The first inequality gives $a_{12} \le 1 + x_1 - x_2$. By the second inequality $x_2 \le x_1 + (a_{23} - a_{13})$ and $x_3 = x_2 + 1 - a_{23}$, we have $a_{13} \le 1 + x_1 - x_3$. That is, (i) of Theorem 6.13 is satisfied. Conditions (ii) and (iii) of Theorem 6.13 are satisfied by the assumption $a_{11} = 1$ and by $x_3 = x_2 + 1 - a_{23}$, respectively.

Subcase 2c. Let $a_{11} < 1$, $a_{22} = 1$. If the first part of the disjunction in (6.34) is satisfied, then $x_2 = x_1 + 1 - a_{12}$ implies $a_{12} = 1 + x_1 - x_2$. Also, $x_3 \le \min(x_1 + 2 - (a_{12} + a_{23}), x_1 + 1 - a_{13}, 1)$ implies that $x_3 \le x_1 + 1 - a_{13}$ and $x_3 \le x_1 + 2 - (a_{12} + a_{23})$. The first inequality implies $a_{13} \le 1 + x_1 - x_3$. By $x_2 = x_1 + 1 - a_{12}$ and $x_3 \le x_1 + 2 - (a_{12} + a_{23})$, we have $a_{23} \le 1 + x_2 - x_3$. That is, (i) of Theorem 6.13 is satisfied. Conditions (ii) and (iii) of Theorem 6.13 are satisfied by $x_2 = x_1 + 1 - a_{12}$ and by the assumption $a_{22} = 1$, respectively.

If the second part of the disjunction in (6.34) is satisfied, then $x_3 = x_1 + 1 - a_{13}$ implies $a_{13} = 1 + x_1 - x_3$. Also, $x_1 + (a_{23} - a_{13}) \le x_2 < x_1 + 1 - a_{12}$ implies the inequalities $x_1 + (a_{23} - a_{13}) \le x_2$ and $x_2 < x_1 + 1 - a_{12}$. Using $x_3 = x_1 + 1 - a_{13}$ in the first inequality we have $a_{23} \le 1 + x_2 - x_3$, the second inequality gives $a_{12} \le 1 + x_1 - x_2$. That is, (i) of Theorem 6.13 is satisfied. Conditions (ii) and (iii) of Theorem 6.13 are satisfied by $x_3 = x_1 + 1 - a_{13}$ and by the assumption $a_{22} = 1$, respectively.

Subcase 2d. Let $a_{11} < 1$, $a_{22} < 1$. If the first part of the disjunction in (6.35) is satisfied, then $a_{12} - a_{13} + a_{23} \geq 1$, and $x_2 = x_1 + 1 - a_{12}$ implies $a_{12} = 1 + x_1 - x_2$. By substituting in $x_3 = x_1 + 2 - (a_{12} + a_{23})$, we get $a_{23} = 1 + x_2 - x_3$. Also, $x_3 = x_1 + 2 - (a_{12} + a_{23}) = x_1 + 1 - (a_{12} + a_{23} - 1) \leq x_1 + 1 - a_{13}$. This implies that $x_3 \geq x_1 + 1 - a_{13}$, or $a_{13} \leq 1 + x_1 - x_3$. That is, (i) of Theorem 6.13 is satisfied. Conditions (ii) and (iii) of Theorem 6.13 are satisfied by $x_2 = x_1 + 1 - a_{12}$ and $a_{23} = 1 + x_2 - x_3$, respectively.

If the second part of the disjunction in (6.35) is satisfied, then $a_{12} - a_{13} + a_{23} \leq 1$, and $x_3 = x_1 + 1 - a_{13}$ gives $a_{13} = 1 + x_1 - x_3$. We have, $x_2 = x_1 + (a_{23} - a_{13}) \leq x_1 + 1 - a_{12}$, that is $a_{12} \leq 1 + x_1 - x_2$. Also, $x_2 = x_1 + (a_{23} - a_{13}) = (x_1 + 1 - a_{13}) - (1 - a_{23}) = x_3 - (1 - a_{23})$ implies that $a_{23} = 1 + x_2 - x_3$. That is, (i) of Theorem 6.13 is satisfied. Conditions (ii) and (iii) of Theorem 6.13 are satisfied by $x_3 = x_1 + 1 - a_{13}$ and $a_{23} = 1 + x_2 - x_3$, respectively.

For the proof of the converse implication, suppose that $x \in \mathcal{F}_1^<(A)$. Then conditions (i), (ii) and (iii) of Theorem 6.13 hold true and we show that conditions (6.32)–(6.35) are satisfied. We start with an easy observation condition (i) of Theorem 6.13 implies $x_2 \leq x_1 + 1 - a_{12}$, $x_3 \leq x_1 + 1 - a_{13}$ and $x_3 \leq x_2 + 1 - a_{23}$.

Subcase 2a'. When $a_{11} = 1$, $a_{22} = 1$, then the inequalities in condition (6.32) follow directly from the assumption $x \in \mathcal{I}^<(3)$ and from the above observation.

Subcase 2b'. When $a_{11} = 1$, $a_{22} < 1$, then condition (iii) of Theorem 6.13 gives $x_3 = x_2 + 1 - a_{23}$. Since $x \in \mathcal{I}^<(3)$ then $x_3 = x_2 + 1 - a_{23} \leq 1$ implies $x_1 < x_2 \leq a_{23}$. Using $x_3 = x_2 + 1 - a_{23}$ in $x_3 \leq x_1 + 1 - a_{13}$ we get $x_2 \leq x_1 + (a_{23} - a_{13})$. Thus, we have verified all relations in condition (6.33).

Subcase 2c'. When $a_{11} < 1$, $a_{22} = 1$, then by condition (ii) of Theorem 6.13 either $x_2 = x_1 + 1 - a_{12}$, $x_3 \leq x_1 + 1 - a_{13}$ and $x_3 \leq x_2 + 1 - a_{23}$, or $x_2 < x_1 + 1 - a_{12}$, $x_3 = x_1 + 1 - a_{13}$ and $x_3 \leq x_2 + 1 - a_{23}$ hold true.

Let us consider the first possibility. Since $x \in \mathcal{I}^<(3)$ then $x_2 = x_1 + 1 - a_{12} < 1$, which implies $x_1 < a_{12}$. By $x_2 = x_1 + 1 - a_{12}$ and $x_3 \leq x_2 + 1 - a_{23}$ we have $x_3 \leq x_1 + 2 - (a_{12} + a_{23})$. The remaining relations in the first part of the disjunction in (6.34) follow from the starting observation.

In the second possibility, we have $x_3 = x_1 + 1 - a_{13} \leq 1$ that is, $x_1 \leq a_{13}$. By $x_3 = x_1 + 1 - a_{13}$ and $x_3 \leq x_2 + 1 - a_{23}$, we have $x_2 \geq x_1 + (a_{23} - a_{13})$. We have verified all relations in the second part of the disjunction in (6.34) .

Subcase 2d'. When $a_{11} < 1$, $a_{22} < 1$, then condition (ii) of Theorem 6.13 implies $x_2 = x_1 + 1 - a_{12}$, or $x_3 = x_1 + 1 - a_{13}$. Moreover, condition (iii) of Theorem 6.13 gives $x_3 = x_2 + 1 - a_{23}$. We shall consider two possibilities.

In the first possibility we assume $x_2 = x_1 + 1 - a_{12}$, $x_3 \leq x_1 + 1 - a_{13}$ and $x_3 = x_2 + 1 - a_{23}$. Combining the two equalities we get $x_3 = x_1 + 2 - (a_{12} + a_{23})$, and using this value in the inequality assumption we get $1 \leq a_{12} - a_{13} + a_{23}$. Thus, all relations in the first part of the disjunction in condition (6.35) must hold true.

In the second possibility we assume $x_2 \leq x_1 + 1 - a_{12}$, $x_3 = x_1 + 1 - a_{13}$ and $x_3 = x_2 + 1 - a_{23}$. The two equalities imply $x_2 = x_1 + (a_{23} - a_{13})$. Inserting this value into the inequality assumption we get $a_{12} - a_{13} + a_{23} \le 1$. Thus, all relations in the second part of the disjunction in condition (6.35) have been verified. □

6.4.1 Non-decreasing Eigenvectors

In this subsection the three-dimensional eigenproblem is investigated for non-decreasing eigenvectors. Analogously to the notation introduced in [8] and [16] we denote

$$\mathscr{I}^<(D_{12}, 3) = \left\{x \in \mathscr{I}(3) |\, x_1 = x_2 < x_3\right\} , \tag{6.36}$$

$$\mathscr{I}^<(D_{23}, 3) = \left\{x \in \mathscr{I}(3) |\, x_1 < x_2 = x_3\right\} , \tag{6.37}$$

and

$$\mathscr{F}^<(D_{12}, A) = \left\{x \in \mathscr{I}^<(D_{12}, 3) |\, A \otimes_l x = x\right\} , \tag{6.38}$$

$$\mathscr{F}^<(D_{23}, A) = \left\{x \in \mathscr{I}^<(D_{23}, 3) |\, A \otimes_l x = x\right\} , \tag{6.39}$$

for a given matrix $A \in \mathscr{I}(3, 3)$.

It is easy to see that any non-decreasing vector $x \in \mathscr{I}(3)$ is either increasing or constant, or belongs to one of the sets (6.36), (6.37). Hence,

$$\mathscr{I}^{\le}(3) = \mathscr{I}^<(3) \cup \mathscr{I}^<(D_{12}, 3) \cup \mathscr{I}^<(D_{21}, 3) \cup \mathscr{I}^=(3) , \tag{6.40}$$

$$\mathscr{F}^{\le}(A) = \mathscr{F}^<(A) \cup \mathscr{F}^<(D_{12}, A) \cup \mathscr{F}^<(D_{21}, A) \cup \mathscr{F}^=(A) . \tag{6.41}$$

Proposition 6.10. *Let $A \in \mathscr{I}(3, 3)$, $x \in \mathscr{I}^<(D_{12}, 3)$. Then $x \in \mathscr{F}^<(D_{12}, A)$ if and only if the following hold*

$$1 - a_{13} \ge x_3 - x_1 , \tag{6.42}$$

$$1 - a_{23} \ge x_3 - x_1 , \tag{6.43}$$

$$x_1 = 0 \quad or \quad \max(a_{11}, a_{12}) = 1 \quad or \quad 1 - a_{13} = x_3 - x_1 , \tag{6.44}$$

$$x_1 = 0 \quad or \quad \max(a_{21}, a_{22}) = 1 \quad or \quad 1 - a_{23} = x_3 - x_1 , \tag{6.45}$$

$$a_{33} = 1 . \tag{6.46}$$

Proof. Let $x \in \mathscr{F}^<(D_{12}, A)$, then $0 \le x_1 = x_2 < x_3 \le x_3$. By (6.26) in Proposition 6.8, $a_{13} \otimes_l x_3 \le x_1$. Then by definition $0 \le a_{13} + x_3 - 1 \le x_1$ or $a_{13} + x_3 - 1 \le 0 \le x_1$, implies that $1 - a_{13} \ge x_3 - x_1$ holds in either case. Similarly, $1 - a_{23} \ge x_3 - x_2 = x_3 - x_1$. By (6.27) in Proposition 6.8 we have, $a_{11} \otimes_l x_1 = x_1$ or $a_{12} \otimes_l x_2 = x_1$ or $a_{13} \otimes_l x_3 = x_1$. The first two of the equalities are trivially fulfilled if $x_1 = x_2 = 0$. On the other hand, if $x_1 = x_2 > 0$, then we get $a_{11} = 1$ or

$a_{12} = 1$ or $a_{13} + x_3 - 1 = x_1$. That is, $\max(a_{11}, a_{12}) = 1$ or $1 - a_{13} = x_3 - x_1$. By the same argument we get $x_1 = 0$ or $\max(a_{21}, a_{22}) = 1$ or $1 - a_{23} = x_3 - x_1$. In view of the assumption $x_1 = x_2 < x_3$, equation (6.46) follows from (6.27) in Proposition 6.8.

Conversely, conditions (6.42) and (6.43), in view of Proposition 6.7 are equivalent to $a_{13} \otimes_l x_3 \leq x_1$ and $a_{23} \otimes_l x_3 \leq x_1$ respectively. Three conditions (6.44), (6.45) and (6.46) together imply condition (6.27) in Proposition 6.8. Hence $x \in \mathscr{F}^<(D_{12}, A)$. □

Proposition 6.11. *Let $A \in \mathscr{I}(3,3)$, $x \in \mathscr{I}^<(D_{23}, 3)$. Then $x \in \mathscr{F}^<(D_{23}, A)$ if and only if the following hold*

$$1 - \max(a_{12}, a_{13}) \geq x_2 - x_1 \ , \tag{6.47}$$

$$x_1 = 0 \quad or \quad a_{11} = 1 \quad or \quad 1 - \max(a_{12}, a_{13}) = x_2 - x_1 \ , \tag{6.48}$$

$$\max(a_{22}, a_{23}) = 1 \ , \tag{6.49}$$

$$\max(a_{32}, a_{33}) = 1 \ . \tag{6.50}$$

Proof. Let $x \in \mathscr{I}^<(D_{23}, 3)$ then, $0 \leq x_1 < x_2 = x_3 \leq 1$. Similarly as in previous proof, it is easy to show that conditions (6.47), (6.48), (6.49), (6.50) are equivalent to conditions in Proposition 6.8. □

Theorem 6.21. *Let $A \in \mathscr{I}(3,3)$. Then $\mathscr{F}^<(D_{12}, A) \neq \emptyset$ if and only if the following conditions are satisfied*

(i) $a_{13} < 1$, $a_{23} < 1$,
(ii) $a_{33} = 1$.

Proof. Let there exist $x \in \mathscr{F}^<(D_{12}, A)$. In view of (6.42) of Proposition 6.10, $a_{13} - 1 \leq x_1 - x_3 < 0$ implies that $a_{13} < 1$. Similarly we have $a_{23} < 1$. Condition (ii) is the same as condition (6.46).

Conversely, suppose that conditions (i), (ii) hold true. We show that there exists $x = (x_1, x_2, x_3) \in \mathscr{F}^<(D_{12}, A)$. The choice of $x_1 = x_2 = 0$ and $x_3 = \min(1 - a_{13}, 1 - a_{23})$ satisfies conditions (6.42)–(6.46) of Proposition 6.10. Hence $\mathscr{F}^<(D_{12}, A) \neq \emptyset$. □

Theorem 6.22. *Let $A \in \mathscr{I}(3,3)$. Then $\mathscr{F}^<(D_{23}, A) \neq \emptyset$ if and only if the following conditions are satisfied*

(i) $a_{12} < 1$, $a_{13} < 1$,
(ii) $\max(a_{22}, a_{23}) = 1$,
(iii) $\max(a_{32}, a_{33}) = 1$.

Proof. The proof is analogous to the proof of Theorem 6.21. In the converse part we put $x_1 = 0$, and take an arbitrary value $x_2 = x_3$ in the interval $\langle 1 - \max(a_{12}, a_{13}), 1 \rangle$. □

Theorem 6.23. *Let $A \in \mathscr{I}(3,3)$ and let conditions (i)–(ii) of Theorem 6.21 be satisfied. Denoting $\mathscr{F}_0^<(D_{12}, A) = \{x \in \mathscr{F}^<(D_{12}, A) | \, x_1 = 0\}$, $\mathscr{F}_1^<(D_{12}, A) = \{x \in \mathscr{F}^<(D_{12}, A) | \, x_1 > 0\}$ we have $\mathscr{F}^<(D_{12}, A) = \mathscr{F}_0^<(D_{12}, A) \cup \mathscr{F}_1^<(D_{12}, A)$, where $\mathscr{F}_0^<(D_{12}, A)$ consists exactly of all vectors $x = (x_1, x_2, x_3) \in \mathscr{I}(3)$ satisfying*

$$0 = x_1 = x_2 < x_3 \le 1 - \max(a_{13}, a_{23}) \ , \tag{6.51}$$

and $\mathscr{F}_1^<(D_{12}, A)$ consists exactly of all vectors $x = (x_1, x_2, x_3) \in \mathscr{I}(3)$ satisfying $0 < x_1 = x_2 < x_3$ and conditions

$$\text{if } \max(a_{11}, a_{12}) = 1, \ \max(a_{21}, a_{22}) = 1, \ \text{ then } 1 - \max(a_{13}, a_{23}) \ge x_3 - x_1 \ , \tag{6.52}$$

$$\text{if } \max(a_{11}, a_{12}) < 1, \ \max(a_{21}, a_{22}) = 1, \ \text{ then } 1 - a_{23} \ge 1 - a_{13} = x_3 - x_1 \ , \tag{6.53}$$

$$\text{if } \max(a_{11}, a_{12}) = 1, \ \max(a_{21}, a_{22}) < 1, \ \text{ then } 1 - a_{13} \ge 1 - a_{23} = x_3 - x_1 \ , \tag{6.54}$$

$$\text{if } \max(a_{11}, a_{12}) < 1, \ \max(a_{21}, a_{22}) < 1, \ \text{ then } 1 - a_{13} = 1 - a_{23} = x_3 - x_1 \ . \tag{6.55}$$

Proof. Let $A \in \mathscr{I}(3,3)$ satisfy conditions (i)–(ii) of Theorem 6.21 and let $x = (0, 0, x_3) \in \mathscr{I}^<(D_{12}, 3)$ satisfy condition (6.51). It is easy to see that conditions in Proposition 6.10 are fulfilled, i.e. $x \in \mathscr{F}^<(D_{12}, A)$. Conversely, if $x \in \mathscr{F}_0^<(D_{12}, A)$, then condition (6.51) follows from conditions (6.42) and (6.43) .

Let us assume now that a vector $x = (x_1, x_2, x_3) \in \mathscr{I}(3)$ satisfies $0 < x_1 = x_2 < x_3$ and conditions (6.52)–(6.55). Then $x \in \mathscr{I}^<(D_{12}, 3)$ and conditions (6.52)–(6.55) imply conditions (6.42)–(6.45) in Proposition 6.10. The condition (6.46) of Proposition 6.10 is identical with (ii) in Theorem 6.21. Hence $x = (x_1, x_2, x_3) \in \mathscr{F}^<(D_{12}, A)$, in view of Proposition 6.10.

Conversely, let $x \in \mathscr{F}_1^<(D_{12}, A)$. Then $x_1 > 0$ and conditions (6.42)–(6.46) of Proposition 6.10 are satisfied. Conditions (6.52)–(6.55) then follow immediately.

□

Theorem 6.24. *Let $A \in \mathscr{I}(3,3)$ and let conditions (i)–(ii) of Theorem 6.22 be satisfied. Denoting $\mathscr{F}_0^<(D_{23}, A) = \{x \in \mathscr{F}^<(D_{23}, A) | \, x_1 = 0\}$, $\mathscr{F}_1^<(D_{23}, A) = \{x \in \mathscr{F}^<(D_{23}, A) | \, x_1 > 0\}$ we have $\mathscr{F}^<(D_{23}, A) = \mathscr{F}_0^<(D_{23}, A) \cup \mathscr{F}_1^<(D_{23}, A)$, where $\mathscr{F}_0^<(D_{23}, A)$ consists exactly of all vectors $x = (x_1, x_2, x_3) \in \mathscr{I}(3)$ satisfying*

$$0 = x_1 < x_2 = x_3 \le 1 - \max(a_{12}, a_{13}) \ , \tag{6.56}$$

and $\mathscr{F}_1^<(D_{23}, A)$ consists exactly of all vectors $x = (x_1, x_2, x_3) \in \mathscr{I}(3)$ satisfying $0 < x_1 < x_2 = x_3$ and conditions

$$\text{if } a_{11} = 1, \text{ then } 1 - \max(a_{12}, a_{13}) \geq x_2 - x_1 \ , \tag{6.57}$$

$$\text{if } a_{11} < 1, \text{ then } 1 - \max(a_{12}, a_{13}) = x_2 - x_1 \ . \tag{6.58}$$

Proof. The proof is analogous to the proof of Theorem 6.23. □

6.4.2 General Eigenvectors in Max-Łukasiewicz Algebra

Similarly as it was done in Sect. 6.3.2 for the max-prod algebra, the results derived in Sect. 6.4.1 for the non-decreasing eigenvectors in max-L algebra will be extended for the general eigenproblem in this subsection. To keep the text easily understandable, the explanation is restricted to three-dimensional matrices.

Using the idea from Sect. 6.3.2, we get

$$\mathscr{F}_L(A) = \mathscr{F}(A) = \bigcup_{\varphi \in P_n} \mathscr{F}_\varphi^{\leq}(A) = \bigcup_{\varphi \in P_n} \bigcup_{D \in \mathscr{D}_n} \mathscr{F}_\varphi^{<}(D, A) \ , \tag{6.59}$$

which is formally almost identical with (6.16). However, in the former equation the eigenspaces $\mathscr{F}_\varphi^{<}(D, A)$ are computed in (max, prod) algebra, while here we work in (max-L) algebra, using Theorem 6.23 and Theorem 6.24. The computation of $\mathscr{F}_L(A) = \mathscr{F}(A)$ is shown below in Example 6.7 and Example 6.8.

6.4.3 Examples

In this section the above considerations will be illustrated by computing the complete eigenspace in max-L fuzzy algebra of two given three-dimensional matrices.

Example 6.7. Let us consider the matrix

$$A = \begin{pmatrix} 0.6 & 0.8 & 0.3 \\ 0.5 & 0.9 & 0.4 \\ 0.3 & 0.7 & 1 \end{pmatrix} .$$

Matrix A satisfies the conditions (i), (ii) and (iii) of Theorem 6.19, hence $\mathscr{F}^{<}(A) \neq \emptyset$. By Theorem 6.20, the increasing eigenspace of A is $\mathscr{F}^{<}(A) = \mathscr{F}_0^{<}(A) \cup \mathscr{F}_1^{<}(A)$. Since $a_{22} = 0.9 < 1$, then by condition (6.30) of Theorem 6.20, $\mathscr{F}_0^{<}(A)$ consists exactly of the vectors $(0, x_2, x_3) \in \mathscr{I}(3)$ fulfilling the conditions

$$0 < x_2 \leq \min(1 - a_{12}, a_{23} - a_{13}), \ x_2 < 1 - a_{13}, \ x_3 = x_2 + (1 - a_{23}) \ ,$$

i.e.

$$0 < x_2 \leq \min(1 - 0.8, 0.4 - 0.3),\ \ x_2 < 1 - 0.3,\ \ x_3 = x_2 + (1 - 0.4)\ .$$

Hence,

$$\mathscr{F}_0^<(A) = \{(0, x_2, x_3) \in \mathscr{I}(3) | 0 < x_2 \leq 0.1,\ x_3 = x_2 + 0.6\}\ .$$

Since $a_{11} = 0.6 < 1$, $a_{22} = 0.9 < 1$, and $a_{12} - a_{13} + a_{23} = 0.8 - 0.3 + 0.4 = 0.9 \leq 1$, then we get in a similar way by the condition (6.35) of Theorem 6.20

$$\mathscr{F}_1^<(A) = \{(x_1, x_2, x_3) \in \mathscr{I}(3) | 0 < x_1 \leq 0.3, x_2 = x_1 + 0.1, x_3 = x_1 + 0.7\}\ .$$

Further, Theorem 6.23 implies that

$$\mathscr{F}^<(D_{12}, A) = \mathscr{F}_0^<(D_{12}, A) = \{(0, 0, x_3) \in \mathscr{I}(3) : 0 < x_3 \leq 0.6\}\ .$$

By Theorem 6.22 we get $\mathscr{F}^<(D_{23}, A) = \emptyset$ and according to Theorem 6.17, A has exactly one constant eigenvector $(0, 0, 0)$. By analogous considerations of all matrices $A_{\varphi\varphi}$ for permutations $\varphi \in P_3$, we get that, in view of Theorem 5.1, A has no other eigenvectors. Summarizing we get

$$\begin{aligned}
\mathscr{F}(A) = \mathscr{F}^{\leq}(A) &= \mathscr{F}^{=}(A) \cup \mathscr{F}^<(D_{12}, A) \cup \mathscr{F}_0^<(A) \cup \mathscr{F}_1^<(A) \\
&= \{(0, 0, 0)\} \cup \{(0, 0, x_3) \in \mathscr{I}(3) | 0 < x_3 \leq 0.6\} \\
&\quad \cup \{(0, x_2, x_3) \in \mathscr{I}(3) | 0 < x_2 \leq 0.1,\ x_3 = x_2 + 0.6\} \\
&\quad \cup \{(x_1, x_2, x_3) \in \mathscr{I}(3) | 0 < x_1 \leq 0.3, x_2 = x_1 + 0.1, x_3 = x_1 + 0.7\}\ .
\end{aligned}$$

Example 6.8. In this example we change the entry $a_{22} = 0.9$ to $b_{22} = 1$, and leave the remaining entries unchanged. We start with computing the increasing eigenspace of matrix B

$$B = \begin{pmatrix} 0.6 & 0.8 & 0.3 \\ 0.5 & 1 & 0.4 \\ 0.3 & 0.7 & 1 \end{pmatrix}\ .$$

Matrix B satisfies conditions (i), (ii), (iii) in Theorem 6.19, hence $\mathscr{F}^<(B) \neq \emptyset$. By Theorem 6.20, the increasing eigenspace of B is $\mathscr{F}^<(B) = \mathscr{F}_0^<(B) \cup \mathscr{F}_1^<(B)$. Since $b_{22} = 1$, then by condition (6.31) in Theorem 6.20, $\mathscr{F}_0^<(B)$ consists exactly of the vectors $(0, x_2, x_3) \in \mathscr{I}(3)$ fulfilling the conditions

$$0 < x_2 < \min(1 - b_{12}, 1 - b_{13}),\ \ x_2 < x_3 \leq \min\big(1 - b_{13}, x_2 + (1 - b_{23})\big)\ ,$$

i.e.

$$0 < x_2 \leq \min(1 - 0.8, 1 - 0.3),\ \ x_2 < x_3 \leq \min\big(1 - 0.3, x_2 + (1 - 0.4)\big)\ .$$

Hence,

$$\mathscr{F}_0^<(B) = \{(0, x_2, x_3) \in \mathscr{I}(3) : 0 < x_2 \le 0.2,\ x_2 < x_3 \le \min(0.7, x_2 + 0.6)\} .$$

Since $b_{11} = 0.6 < 1$, $b_{22} = 1$, then we get in a similar way by the condition (6.34) of Theorem 6.20 that $\mathscr{F}_1^<(B)$ consists exactly of the vectors $(0, x_2, x_3) \in \mathscr{I}^<(3)$ fulfilling the conditions

$$0 < x_1 < b_{12},\ x_2 = x_1 + (1 - b_{12}) ,$$
$$x_3 \le \min(x_1 + 2 - (b_{12} + b_{23}), x_1 + (1 - b_{13}), 1), \text{ or}$$
$$0 < x_1 \le b_{13},\ x_1 + (b_{23} - b_{13}) \le x_2 < x_1 + (1 - b_{12}),\ x_3 = x_1 + (1 - b_{13}) .$$

Thus, $\mathscr{F}_1^<(B) =$

$$\{(x_1, x_2, x_3) \in \mathscr{I}(3) | 0 < x_1 < 0.8, x_2 = x_1 + 0.2, x_2 < x_3 \le \min(x_1 + 0.7, 1)\}$$
$$\cup \{(x_1, x_2, x_3) \in \mathscr{I}(3) | 0 < x_1 \le 0.3, x_1 + 0.1 \le x_2 < x_1 + 0.2, x_3 = x_1 + 0.7\} .$$

By Theorem 6.17, B has exactly one constant eigenvector $(0, 0, 0)$, by Theorem 6.23 and Theorem 6.24 we get $\mathscr{F}^<(D_{12}, B) = \mathscr{F}_0^<(D_{12}, B)$, $\mathscr{F}^<(D_{23}, B) = \mathscr{F}_0^<(D_{12}, B) \cup \mathscr{F}_1^<(D_{23}, B)$, where

$$\mathscr{F}_0^<(D_{12}, B) = \{(0, 0, x_3) \in \mathscr{I}(3) | 0 < x_3 \le 0.6\} ,$$
$$\mathscr{F}_0^<(D_{23}, B) = \{(0, x_2, x_3) \in \mathscr{I}(3) | 0 < x_2 = x_3 \le 0.2\} ,$$
$$\mathscr{F}_1^<(D_{23}, B) = \{(x_1, x_2, x_3) \in \mathscr{I}(3) | 0 < x_1 \le 0.8, x_2 = x_3 = x_1 + 0.2\} .$$

Summarizing we get the increasing eigenspace of B in the following form

$$\begin{aligned}
\mathscr{F}^{\le}(B) &= \mathscr{F}^{=}(B) \cup \mathscr{F}_0^<(D_{12}, B) \cup \mathscr{F}_0^<(D_{23}, B) \cup \mathscr{F}_1^<(D_{23}, B) \cup \mathscr{F}_0^<(B) \cup \mathscr{F}_1^<(B) \\
&= \{(0, 0, 0)\} \cup \{(0, 0, x_3) \in \mathscr{I}(3) | 0 < x_3 \le 0.6\} \\
&\quad \cup \{(0, x_2, x_3) \in \mathscr{I}(3) | 0 < x_2 = x_3 \le 0.2\} \\
&\quad \cup \{(x_1, x_2, x_3) \in \mathscr{I}(3) | 0 < x_1 \le 0.8, x_2 = x_3 = x_1 + 0.2\} \\
&\quad \cup \{(0, x_2, x_3) \in \mathscr{I}(3) | 0 < x_2 \le 0.2,\ x_2 < x_3 \le \min(0.7, x_2 + 0.6)\} \\
&\quad \cup \{(x_1, x_2, x_3) \in \mathscr{I}(3) | 0 < x_1 < 0.8, x_2 = x_1 + 0.2, x_2 < x_3 \le \min(x_1 + 0.7, 1)\} \\
&\quad \cup \{(x_1, x_2, x_3) \in \mathscr{I}(3) : 0 < x_1 \le 0.3, x_1 + 0.1 \le x_2 < x_1 + 0.2, x_3 = x_1 + 0.7\} .
\end{aligned}$$

Further eigenvectors of matrix B will be found using Theorem 5.1 with all possible permutations in P_3. E.g., applying permutation $\varphi = \begin{pmatrix} 1\,2\,3 \\ 1\,3\,2 \end{pmatrix}$ to rows and columns of B we get matrix $B_{\varphi\varphi}$ which satisfies conditions (i), (ii) and (iii) of Theorem 6.19

$$B_{\varphi\varphi} = \begin{pmatrix} 0.6 & 0.3 & 0.8 \\ 0.3 & 1 & 0.7 \\ 0.5 & 0.4 & 1 \end{pmatrix} .$$

For simplicity, we use the notation in which the permuted vector x_φ is denoted as $y = (y_1, y_2, y_3) = (x_{\varphi(1)}, x_{\varphi(2)}, x_{\varphi(3)}) = (x_1, x_3, x_2)$. Analogously as above we compute

$$\mathscr{F}_0^<(B_{\varphi\varphi}) = \{(0, y_2, y_3) \in \mathscr{I}(3) | 0 < y_2 < \min(1 - a_{12}, 1 - a_{13}),$$
$$y_2 < y_3 \le \min(1 - a_{13}, y_2 + (1 - a_{23}))\} ,$$
$$\mathscr{F}_1^<(B_{\varphi\varphi}) = \{(y_1, y_2, y_3) \in \mathscr{I}(3) | 0 < y_1 < a_{12}, y_2 = y_1 + (1 - a_{12}),$$
$$y_3 \le \min(y_1 + 2 - (a_{12} + a_{23}), y_1 + (1 - a_{13}), 1)\}$$
$$\cup \{(y_1, y_2, y_3) \in \mathscr{I}(3) | 0 < y_1 \le a_{13}, y_1 + (a_{23} - a_{13}) \le y_2 < y_1 + (1 - a_{12}),$$
$$y_3 = y_1 + (1 - a_{13})\} ,$$
$$\mathscr{F}_0^<(D_{12}, B_{\varphi\varphi}) = \{(0, 0, y_3) \in \mathscr{I}(3) | 0 = y_1 = y_2 < y_3 \le 1 - \max(a_{13}, a_{23})\} ,$$
$$\mathscr{F}_1^<(D_{12}, B_{\varphi\varphi}) = \{(y_1, y_2, y_3) \in \mathscr{I}(3) | 0 < y_1 = y_2 < y_3, 1 - a_{23} \ge 1 - a_{13} = y_3 - y_1\} ,$$
$$\mathscr{F}_0^<(D_{23}, B_{\varphi\varphi}) = \{(0, y_2, y_3) \in \mathscr{I}(3) | 0 = y_1 < y_2 = y_3 \le 1 - \max(a_{12}, a_{13})\} ,$$
$$\mathscr{F}_1^<(D_{23}, B_{\varphi\varphi}) = \{(y_1, y_2, y_3) \in \mathscr{I}(3) | 0 < y_1 < y_2 = y_3, 1 - \max(a_{12}, a_{13}) = y_2 - y_1\} ,$$
$$\mathscr{F}^=(B_{\varphi\varphi}) = \{(0, 0, 0)\} .$$

Putting the entries of matrix $B_{\varphi\varphi}$ into the above formulas we get

$$\mathscr{F}_0^<(B_{\varphi\varphi}) = \{(0, y_2, y_3) \in \mathscr{I}(3) | 0 < y_2 < \min(1 - 0.3, 1 - 0.8),$$
$$y_2 < y_3 \le \min(1 - 0.8, y_2 + (1 - 0.7))\} ,$$
$$\mathscr{F}_1^<(B_{\varphi\varphi}) = \{(y_1, y_2, y_3) \in \mathscr{I}(3) | 0 < y_1 < 0.3, y_2 = y_1 + (1 - 0.3),$$
$$y_3 \le \min(y_1 + 2 - (0.3 + 0.7), y_1 + (1 - 0.8), 1)\}$$
$$\cup \{(y_1, y_2, y_3) \in \mathscr{I}(3) | 0 < y_1 \le 0.8, y_1 + (0.7 - 0.8) \le y_2 < y_1 + (1 - 0.3),$$
$$y_3 = y_1 + (1 - 0.8)\} ,$$
$$\mathscr{F}_0^<(D_{12}, B_{\varphi\varphi}) = \{(0, 0, y_3) \in \mathscr{I}(3) | 0 = y_1 = y_2 < y_3 \le 1 - \max(0.8, 0.7)\} ,$$
$$\mathscr{F}_1^<(D_{12}, B_{\varphi\varphi}) = \{(y_1, y_2, y_3) \in \mathscr{I}(3) | 0 < y_1 = y_2 < y_3, 1 - 0.7 \ge 1 - 0.8 = y_3 - y_1\} ,$$
$$\mathscr{F}_0^<(D_{23}, B_{\varphi\varphi}) = \{(0, y_2, y_3) \in \mathscr{I}(3) | 0 = y_1 < y_2 = y_3 \le 1 - \max(0.3, 0.8)\} ,$$
$$\mathscr{F}_1^<(D_{23}, B_{\varphi\varphi}) = \{(y_1, y_2, y_3) \in \mathscr{I}(3) | 0 < y_1 < y_2 = y_3, 1 - \max(0.3, 0.8) = y_2 - y_1\} ,$$
$$\mathscr{F}^=(B_{\varphi\varphi}) = \{(0, 0, 0)\} .$$

The formulas can be simplified by easy commutation as follows

$$\mathcal{F}_0^<(B_{\varphi\varphi}) = \{(0, y_2, y_3) \in \mathcal{I}(3) | 0 < y_2 < 0.2,\ y_2 < y_3 \leq \min(0.2, y_2 + 0.3))\} ,$$

$$\begin{aligned}\mathcal{F}_1^<(B_{\varphi\varphi}) = \{&(y_1, y_2, y_3) \in \mathcal{I}(3) | 0 < y_1 < 0.3,\ y_2 = y_1 + 0.7, \\ &y_3 \leq \min(y_1 + 1, y_1 + 0.2, 1)\} \\ \cup\ \{&(y_1, y_2, y_3) \in \mathcal{I}(3) | 0 < y_1 \leq 0.8,\ y_1 + (-0.1) \leq y_2 < y_1 + 0.7, \\ &y_3 = y_1 + 0.2)\} ,\end{aligned}$$

$$\mathcal{F}_0^<(D_{12}, B_{\varphi\varphi}) = \{(0, 0, y_3) \in \mathcal{I}(3) | 0 = y_1 = y_2 < y_3 \leq 0.2\} ,$$

$$\mathcal{F}_1^<(D_{12}, B_{\varphi\varphi}) = \{(y_1, y_2, y_3) \in \mathcal{I}(3) | 0 < y_1 = y_2 < y_3,\ 0.3 \geq 0.2 = y_3 - y_1\} ,$$

$$\mathcal{F}_0^<(D_{23}, B_{\varphi\varphi}) = \{(0, y_2, y_3) \in \mathcal{I}(3) | 0 = y_1 < y_2 = y_3 \leq 0.2\} ,$$

$$\mathcal{F}_1^<(D_{23}, B_{\varphi\varphi}) = \{(y_1, y_2, y_3) \in \mathcal{I}(3) | 0 < y_1 < y_2 = y_3,\ 0.2 = y_2 - y_1\} ,$$

$$\mathcal{F}^=(B_{\varphi\varphi}) = \{(0, 0, 0)\} .$$

The results in the final form are

$$\mathcal{F}_0^<(B_{\varphi\varphi}) = \{(0, y_2, y_3) \in \mathcal{I}(3) | 0 < y_2 < y_3 \leq 0.2)\} ,$$

$$\mathcal{F}_1^<(B_{\varphi\varphi}) = \{(y_1, y_2, y_3) \in \mathcal{I}(3) | 0 < y_1 \leq 0.8,\ y_1 < y_2 < y_3 = y_1 + 0.2)\} ,$$

$$\mathcal{F}_0^<(D_{12}, B_{\varphi\varphi}) = \{(0, 0, y_3) \in \mathcal{I}(3) | 0 < y_3 \leq 0.2\} ,$$

$$\mathcal{F}_1^<(D_{12}, B_{\varphi\varphi}) = \{(y_1, y_2, y_3) \in \mathcal{I}(3) | 0 < y_1 = y_2 \leq 0.8,\ y_3 = y_1 + 0.2\} ,$$

$$\mathcal{F}_0^<(D_{23}, B_{\varphi\varphi}) = \{(0, y_2, y_3) \in \mathcal{I}(3) | 0 < y_2 = y_3 \leq 0.2\} ,$$

$$\mathcal{F}_1^<(D_{23}, B_{\varphi\varphi}) = \{(y_1, y_2, y_3) \in \mathcal{I}(3) | 0 < y_1 \leq 0.8,\ y_2 = y_3 = y_1 + 0.2\} ,$$

$$\mathcal{F}^=(B_{\varphi\varphi}) = \{(0, 0, 0)\} .$$

Coming back to the original notation $x_\varphi = (x_1, x_3, x_2) = (y_1, y_2, y_3)$ we get the permuted increasing eigenspace $\mathcal{F}_\varphi^\leq(B)$ of matrix B with $x_1 \leq x_3 \leq x_2$, in the form

$$\begin{aligned}\mathcal{F}_\varphi^\leq(B) = &\mathcal{F}_{0\varphi}^<(B_{\varphi\varphi}) \cup \mathcal{F}_{1\varphi}^<(B_{\varphi\varphi}) \cup \mathcal{F}_{0\varphi}^<(D_{12}, B_{\varphi\varphi}) \cup \mathcal{F}_{1\varphi}^<(D_{12}, B_{\varphi\varphi}) \\ &\cup \mathcal{F}_{0\varphi}^<(D_{23}, B_{\varphi\varphi}) \cup \mathcal{F}_{1\varphi}^<(D_{23}, B_{\varphi\varphi}) \cup \mathcal{F}_\varphi^=(B_{\varphi\varphi}) ,\end{aligned}$$

where

$$\mathcal{F}_{0\varphi}^<(B_{\varphi\varphi}) = \{(0, x_2, x_3) \in \mathcal{I}(3) | 0 < x_3 < x_2 \leq 0.2)\} ,$$

$$\mathcal{F}_{1\varphi}^<(B_{\varphi\varphi}) = \{(x_1, x_2, x_3) \in \mathcal{I}(3) | 0 < x_1 \leq 0.8,\ x_1 < x_3 < x_2 = x_1 + 0.2)\} ,$$

$$\mathscr{F}_{0\varphi}^{<}(D_{12}, B_{\varphi\varphi}) = \{(0, x_2, 0) \in \mathscr{I}(3) | 0 < x_2 \le 0.2\} ,$$

$$\mathscr{F}_{1\varphi}^{<}(D_{12}, B_{\varphi\varphi}) = \{(x_1, x_2, x_3) \in \mathscr{I}(3) | 0 < x_1 = x_3 \le 0.8,\ x_2 = x_1 + 0.2\} ,$$

$$\mathscr{F}_{0\varphi}^{<}(D_{23}, B_{\varphi\varphi}) = \{(0, x_2, x_3) \in \mathscr{I}(3) | 0 < x_3 = x_2 \le 0.2\} ,$$

$$\mathscr{F}_{1\varphi}^{<}(D_{23}, B_{\varphi\varphi}) = \{(x_1, x_2, x_3) \in \mathscr{I}(3) | 0 < x_1 \le 0.8,\ x_3 = x_2 = x_1 + 0.2\} ,$$

$$\mathscr{F}_{\varphi}^{=}(B_{\varphi\varphi}) = \{(0, 0, 0)\} .$$

If another permutation $\psi = \begin{pmatrix} 1\ 2\ 3 \\ 3\ 1\ 2 \end{pmatrix}$ is applied to rows and columns of B, then conditions (i), (ii) and (iii) of Theorem 6.19 are satisfied by the obtained matrix

$$B_{\psi\psi} = \begin{pmatrix} 1 & 0.3 & 0.7 \\ 0.3 & 0.6 & 0.8 \\ 0.4 & 0.5 & 1 \end{pmatrix} .$$

Analogously we compute in notation $z = (z_1, z_2, z_3) = (x_{\psi(1)}, x_{\psi(2)}, x_{\psi(3)}) = (x_3, x_1, x_2)$

$$\mathscr{F}_{0}^{<}(B_{\psi\psi}) = \{(0, z_2, z_3) \in \mathscr{I}(3) | 0 < z_2 \le \min(1 - a_{12}, a_{23} - a_{13}),\ z_2 < 1 - a_{13},$$
$$z_3 = z_2 + (1 - a_{23})\} ,$$

$$\mathscr{F}_{1}^{<}(B_{\psi\psi}) = \{(z_1, z_2, z_3) \in \mathscr{I}(3) | z_1 < a_{23} ,$$
$$z_2 \le \min\big(a_{23}, z_1 + (1 - a_{12}), z_1 + (a_{23} - a_{13})\big),$$
$$z_3 = z_2 + (1 - a_{23})\} ,$$

$$\mathscr{F}_{0}^{<}(D_{12}, B_{\psi\psi}) = \{(0, 0, z_3) \in \mathscr{I}(3) | 0 = z_1 = z_2 < z_3 \le 1 - \max(a_{13}, a_{23})\} ,$$

$$\mathscr{F}_{1}^{<}(D_{12}, B_{\psi\psi}) = \{(z_1, z_2, z_3) \in \mathscr{I}(3) | 0 < z_1 = z_2 < z_3,$$
$$1 - a_{13} \ge 1 - a_{23} = z_3 - z_1\} ,$$

$$\mathscr{F}_{0}^{<}(D_{23}, B_{\psi\psi}) = \{(0, z_2, z_3) \in \mathscr{I}(3) | 0 = z_1 < z_2 = z_3 \le 1 - \max(a_{12}, a_{13})\} ,$$

$$\mathscr{F}_{1}^{<}(D_{23}, B_{\psi\psi}) = \{(z_1, z_2, z_3) \in \mathscr{I}(3) | 0 < z_1 < z_2 = z_3,$$
$$1 - \max(a_{12}, a_{13}) \ge z_2 - z_1\} ,$$

$$\mathscr{F}^{=}(B_{\psi\psi}) = \{(0, 0, 0)\} .$$

Putting the entries of matrix $B_{\psi\psi}$ we get

$$\mathscr{F}_{0}^{<}(B_{\psi\psi}) = \{(0, z_2, z_3) \in \mathscr{I}(3) | 0 < z_2 \le \min(1 - 0.3, 0.8 - 0.7),\ z_2 < 1 - 0.7,$$
$$z_3 = z_2 + (1 - 0.8)\} ,$$

$$\begin{aligned}\mathscr{F}_1^<(B_{\psi\psi}) = \big\{&(z_1, z_2, z_3) \in \mathscr{I}(3) | z_1 < 0.8 \ , \\ &z_2 \le \min\big(0.8, z_1 + (1 - 0.3), z_1 + (0.8 - 0.7)\big), \\ &z_3 = z_2 + (1 - 0.8)\big\} \ ,\end{aligned}$$

$$\mathscr{F}_0^<(D_{12}, B_{\psi\psi}) = \big\{(0, 0, z_3) \in \mathscr{I}(3) | 0 = z_1 = z_2 < z_3 \le 1 - \max(0.7, 0.8)\big\} \ ,$$

$$\begin{aligned}\mathscr{F}_1^<(D_{12}, B_{\psi\psi}) = \big\{&(z_1, z_2, z_3) \in \mathscr{I}(3) | 0 < z_1 = z_2 < z_3, \\ &1 - 0.7 \ge 1 - 0.8 = z_3 - z_1\big\} \ ,\end{aligned}$$

$$\begin{aligned}\mathscr{F}_0^<(D_{23}, B_{\psi\psi}) = \big\{&(0, z_2, z_3) \in \mathscr{I}(3) | 0 = z_1 < z_2 = z_3 \le \\ &1 - \max(0.3, 0.7)\big\} \ ,\end{aligned}$$

$$\begin{aligned}\mathscr{F}_1^<(D_{23}, B_{\psi\psi}) = \big\{&(z_1, z_2, z_3) \in \mathscr{I}(3) | 0 < z_1 < z_2 = z_3, \\ &1 - \max(0.3, 0.7) \ge z_2 - z_1\big\} \ ,\end{aligned}$$

$$\mathscr{F}^=(B_{\psi\psi}) = \big\{(0, 0, 0)\big\} \ .$$

Simple computation gives

$$\mathscr{F}_0^<(B_{\psi\psi}) = \big\{(0, z_2, z_3) \in \mathscr{I}(3) | 0 < z_2 \le 0.1, \ z_2 < 0.3, \ z_3 = z_2 + 0.2)\big\} \ ,$$

$$\begin{aligned}\mathscr{F}_1^<(B_{\psi\psi}) = \big\{&(z_1, z_2, z_3) \in \mathscr{I}(3) | z_1 < 0.8, \ z_2 \le \min\big(0.8, z_1 + 0.7, z_1 + 0.1\big), \\ &z_3 = z_2 + 0.2)\big\} \ ,\end{aligned}$$

$$\mathscr{F}_0^<(D_{12}, B_{\psi\psi}) = \big\{(0, 0, z_3) \in \mathscr{I}(3) | 0 = z_1 = z_2 < z_3 \le 0.2\big\} \ ,$$

$$\mathscr{F}_1^<(D_{12}, B_{\psi\psi}) = \big\{(z_1, z_2, z_3) \in \mathscr{I}(3) | 0 < z_1 = z_2 < z_3, \ 0.3 \ge 0.2 = z_3 - z_1\big\} \ ,$$

$$\mathscr{F}_0^<(D_{23}, B_{\psi\psi}) = \big\{(0, z_2, z_3) \in \mathscr{I}(3) | 0 = z_1 < z_2 = z_3 \le 0.3\big\} \ ,$$

$$\mathscr{F}_1^<(D_{23}, B_{\psi\psi}) = \big\{(z_1, z_2, z_3) \in \mathscr{I}(3) | 0 < z_1 < z_2 = z_3, \ 0.3 \ge z_2 - z_1\big\} \ ,$$

$$\mathscr{F}^=(B_{\psi\psi}) = \big\{(0, 0, 0)\big\} \ .$$

The final form is

$$\mathscr{F}_0^<(B_{\psi\psi}) = \big\{(0, z_2, z_3) \in \mathscr{I}(3) | 0 < z_2 \le 0.1, \ z_3 = z_2 + 0.2)\big\} \ ,$$

$$\begin{aligned}\mathscr{F}_1^<(B_{\psi\psi}) = \big\{&(z_1, z_2, z_3) \in \mathscr{I}(3) | 0 < z_1 < z_2 \le \min\big(0.8, z_1 + 0.1\big), \\ &z_3 = z_2 + 0.2)\big\} \ ,\end{aligned}$$

$$\mathscr{F}_0^<(D_{12}, B_{\psi\psi}) = \big\{(0, 0, z_3) \in \mathscr{I}(3) | 0 < z_3 \le 0.2\big\} \ ,$$

$$\mathscr{F}_1^<(D_{12}, B_{\psi\psi}) = \big\{(z_1, z_2, z_3) \in \mathscr{I}(3) | 0 < z_1 = z_2 < z_3 = z_1 + 0.2\big\} \ ,$$

$$\mathscr{F}_0^<(D_{23}, B_{\psi\psi}) = \big\{(0, z_2, z_3) \in \mathscr{I}(3) | 0 < z_2 = z_3 \le 0.3\big\} \ ,$$

$$\mathscr{F}_1^{<}(D_{23}, B_{\psi\psi}) = \{(z_1, z_2, z_3) \in \mathscr{I}(3) | 0 < z_1 < z_2 = z_3 \le z_1 + 0.3\} ,$$
$$\mathscr{F}^{=}(B_{\psi\psi}) = \{(0, 0, 0)\} .$$

Coming back to the original notation $x_\psi = (x_3, x_1, x_2) = (z_1, z_2, z_3)$ we get the permuted increasing eigenspace $\mathscr{F}_{\psi}^{\lneqq}(B)$ of matrix B with $x_2 \le x_3 \le x_1$, in the form

$$\mathscr{F}_{\psi}^{\lneqq}(B) = \mathscr{F}_{0\psi}^{<}(B_{\psi\psi}) \cup \mathscr{F}_{1\psi}^{<}(B_{\psi\psi}) \cup \mathscr{F}_{0\psi}^{<}(D_{12}, B_{\psi\psi}) \cup \mathscr{F}_{1\psi}^{<}(D_{12}, B_{\psi\psi})$$
$$\cup\, \mathscr{F}_{0\psi}^{<}(D_{23}, B_{\psi\psi}) \cup \mathscr{F}_{1\psi}^{<}(D_{23}, B_{\psi\psi}) \cup \mathscr{F}_{\psi}^{=}(B_{\psi\psi}) ,$$

where

$$\mathscr{F}_0^{<}(B_{\psi\psi}) = \{(x_1, x_2, 0) \in \mathscr{I}(3) | 0 < x_1 \le 0.1,\ x_2 = x_1 + 0.2)\} ,$$
$$\mathscr{F}_1^{<}(B_{\psi\psi}) = \{(x_3, x_1, x_2) \in \mathscr{I}(3) | 0 < x_3 < x_1 \le \min(0.8, x_3 + 0.1),$$
$$x_2 = x_1 + 0.2)\} ,$$
$$\mathscr{F}_0^{<}(D_{12}, B_{\psi\psi}) = \{(0, x_2, 0) \in \mathscr{I}(3) | 0 < x_2 \le 0.2\} ,$$
$$\mathscr{F}_1^{<}(D_{12}, B_{\psi\psi}) = \{(x_1, x_2, x_3) \in \mathscr{I}(3) | 0 < x_3 = x_1 < x_2 = x_3 + 0.2\} ,$$
$$\mathscr{F}_0^{<}(D_{23}, B_{\psi\psi}) = \{(x_1, x_2, 0) \in \mathscr{I}(3) | 0 < x_1 = x_2 \le 0.3\} ,$$
$$\mathscr{F}_1^{<}(D_{23}, B_{\psi\psi}) = \{(x_1, x_2, x_3) \in \mathscr{I}(3) | 0 < x_3 < x_1 = x_2 \le x_3 + 0.3\} ,$$
$$\mathscr{F}^{=}(B_{\psi\psi}) = \{(0, 0, 0)\} .$$

It is easy to verify that, of all six possible permutations in P_3, only the identical permutation and permutations φ, ψ have the property that the permuted matrix satisfies conditions (i), (ii) and (iii) of Theorem 6.19, hence there are no further permuted increasing eigenvectors of B. By Theorem 6.21 and Theorem 6.22 we can also verify that no further permuted non-decreasing eigenvectors exist. As a consequence, the eigenspace of matrix B in max-L fuzzy algebra is equal to

$$\mathscr{F}(B) = \mathscr{F}^{\lneqq}(B) \cup \mathscr{F}_{\varphi}^{\lneqq}(B) \cup \mathscr{F}_{\psi}^{\lneqq}(B) .$$

6.5 Conclusions

Presented results are part of the research aimed on investigation of steady states of systems with fuzzy transition matrix in max-T fuzzy algebras with various triangular norm T. Steady states correspond to eigenvectors of the transition fuzzy matrix. For matrices in max-min algebra the eigenvectors were described in [8]. The investigation in this chapter has been oriented to drastic t-norm and to Łukasiewicz t-norm, i.e. to matrices in max-drast algebra and in max-Łukasiewicz algebra.

The structure of the eigenspace of a given fuzzy matrix in max-drast algebra has been completely described. The importance of the result follows from the fact that the eigenvectors in max-drast algebra correspond to steady states of such discrete-events systems in which extreme reliability is required. The eigenspaces of fuzzy matrices in max-Łukasiewicz algebra have been described for the dimensions ≤ 3.

By the fact that every eigenvector can be permuted to a non-decreasing eigenvector of some permuted fuzzy matrix, the investigation of the complete eigenspace structure has been reduced to sets of all non-decreasing eigenvectors, called monotone eigenspaces. Necessary and sufficient conditions were formulated under which the monotone eigenspace of a given matrix is non-empty. Using these conditions, the structure of the monotone eigenspace has been described in detail. Examples of how the knowledge of the monotone eigenspace structure is used to describe the complete eigenspace of a given matrix are presented and illustrated by figures and examples.

References

1. Casasnovas, J., Mayor, G.: Discrete t-norms and operations on extended multisets. Fuzzy Sets Syst. **159**, 1165–1177 (2008)
2. Cechlárová, K: Eigenvectors in bottleneck algebra. Lin. Algebra Appl. **175**, 63–73 (1992)
3. Cechlárová, K: Efficient computation of the greatest eigenvector in fuzzy algebra. Tatra Mt. Math. Publ. **12**, 73–79 (1997)
4. Cohen, G., Dubois, D., Quadrat, J.P., Viot, M.: A linear-system-theoretic view of discrete event processes and its use for performance evaluation in manufacturing. IEEE Trans. Automat. Contr. **AC-30**, 210–220 (1985)
5. Cuninghame-Green, R.A.: Describing industrial processes with interference and approximating their steady-state behavior. Oper. Res. Quart. **13**, 95–100 (1962)
6. Cuninghame-Green, R.A.: Minimax Algebra, Lecture Notes in Economics and Mathematical Systems, vol. 166. Springer, Berlin (1979)
7. Cuninghame-Green, R.A.: Minimax algebra and application. In: Hawkes, P.W. (ed.), Advances in Imaging and Electron Physics, vol. 90. Academic Press, New York (1995)
8. Gavalec, M.: Monotone eigenspace structure in max–min algebra. Lin. Algebra Appl. **345**, 149–167 (2002)
9. Gavalec, M., Rashid, I.: Monotone eigenspace structure of a max-drast fuzzy matrix. In: Proc. of the 28th Int. Conf. Mathematical Methods in Economics, University of South Bohemia, České Budějovice 2010, pp. 162–167 (2010)
10. Gavalec, M., Rashid, I., Cimler, R.: Eigenspace structure of a max-drast fuzzy matrix. Fuzzy Sets Syst. (2013). Doi:10.1016/j.fss.2013.10.008
11. Gondran, M.: Valeurs propres et vecteurs propres en classification hiérarchique. R. A. I. R. O. Informatique Théorique **10**, 39–46 (1976)
12. Gondran, M., Minoux, M.: Eigenvalues and eigenvectors in semimodules and their interpretation in graph theory. In: Proc. 9th Prog. Symp., pp. 133–148 (1976)
13. Gondran, M., Minoux, M.: Valeurs propres et vecteurs propres en théorie des graphes. Colloques Internationaux, pp. 181–183. CNRS, Paris (1978)

14. Gondran, M., Minoux, M.: Dioïds and semirings: Links to fuzzy sets and other applications. Fuzzy Sets Syst. **158**, 1273–1294 (2007)
15. Olsder, G.: Eigenvalues of dynamic max–min systems. In: Discrete Events Dynamic Systems, vol. 1, pp. 177–201. Kluwer, Dordrecht (1991)
16. Rashid, I.: Stability of discrete-time fuzzy systems, 99 pp. Doctoral thesis, University of Hradec Kralove, Czech Republic (2011)
17. Rashid, I., Gavalec, M., Sergeev, S.: Eigenspace of a three-dimensional max-Łukasiewicz fuzzy matrix. Kybernetika **48**, 309–328 (2012)
18. Sanchez, E.: Resolution of eigen fuzzy sets equations. Fuzzy Sets Syst. **1**, 69–74 (1978)
19. Tan, Y.-J.: Eigenvalues and eigenvectors for matrices over distributive lattices. Lin. Algebra Appl. **283**, 257–272 (1998)
20. Tan, Y.-J.: On the powers of matrices over a distributive lattice. Lin. Algebra Appl. **336**, 1–14 (2001)
21. Zimmermann, U.: Linear and Combinatorial Optimization in Ordered Algebraic Structure. In: Ann. Discrete Math., vol. 10. North Holland, Amsterdam (1981)

Index

M. Gavalec et al., *Decision Making and Optimization*, Lecture Notes in Economics and Mathematical Systems 677, DOI 10.1007/978-3-319-08323-0

The manufacturer's authorised representative in the EU is Springer Nature Customer Service Centre GmbH, Europaplatz 3, 69115 Heidelberg, Germany. If you have any concerns regarding our products, please contact ProductSafety@springernature.com

Printed and bound by CPI Group (UK) Ltd, Croydon, CR0 4YY
23/03/2026
02076668-0013